D0500041

RADIATION BIOLOGY

N875

PRENTICE-HALL INTERNATIONAL, INC., *London*
PRENTICE-HALL OF AUSTRALIA, PTY., LTD., *Sydney*
PRENTICE-HALL OF CANADA, LTD., *Toronto*
PRENTICE-HALL OF INDIA PRIVATE LTD., *New Delhi*
PRENTICE-HALL OF JAPAN, INC., *Tokyo*

RADIATION BIOLOGY

Alison P. Casarett

Department of Physical Biology
New York State Veterinary College
Cornell University

Prepared under the direction of the
American Institute of Biological Science
For the Division of Technical Information
United States Atomic Energy Commission

Prentice-Hall, Inc.
Englewood Cliffs, New Jersey

© 1968 by
United States Atomic Energy Commission
Washington, D. C.

All rights reserved. No part of this book
may be reproduced in any form or by
any means without permission in writing
from the publisher. Copyright assigned
to the General Manager of the United
States Atomic Energy Commission. All
royalties from the sale of this book
accrue to the United States Govern-
ment.

Current printing (last digit):

10 9 8 7 6

Library of Congress Catalog Card Number 68-22702
Printed in the United States of America

FOREWORD

As an Advisory Committee responsible to three masters — the author, the American Institute of Biological Sciences, and the Atomic Energy Commission — we were pleased with the ease with which all three worked together in the development of this text. We believe that this publication admirably meets the needs that we felt were not yet met in existing material. We hope that the reader will capture some of the excitement that permeates this discipline, along with a fundamental understanding of the principles that underlie Radiation Biology.

Carl P. Swanson, Chairman
The Johns Hopkins University

Walter D. Claus
U. S. Atomic Energy Commission

Charles W. Shilling
George Washington University

J. Newell Stannard
University of Rochester

John R. Totter
U. S. Atomic Energy Commission

PREFACE

It has been known since the early 1900's that ionizing radiations can interact with biological systems to produce change. The accelerated use of radioactive sources in industry, medicine, and research during the past two decades has promoted a special interest in this subject. As a result there has been a huge outpouring of data and ideas aimed at quantitating and explaining these changes. During this period of rapid growth much of the information has been accumulated in sources readily available to the trained radiobiologist. There are few textbooks, however, which provide a broad, unified coverage in moderate depth of the basic elements of many different aspects of the biological effects of radiation. This book was specifically conceived to fulfill the needs of people with little or no knowledge of the field. It is intended for use as an introductory textbook on the graduate or undergraduate level, as supplementary reading at the high school or early college level, and as a source of general information for those with a major interest in other fields of science.

The book begins with a brief historical introduction to the field with an account of the contributions of early radiobiologists. For an understanding of the later material, the next chapters describe the principles of radiation physics—the major characteristics of various types of radiation and their detection and measurement. Typical experimental facilities for exposure of different biological materials and problems of dosimetry and dosage calculation are also discussed in Chapter 3. This gives the uninitiated reader some idea of how the experimental information is obtained and stresses the importance of understanding the pattern of energy deposition in the irradiated material. Chapter 4 contains a general description of radiation chemistry, providing a bridge between the relatively well understood features of radiation physics and the more nebulous aspects of radiation biology. An understanding of the effects of radiation on chemical systems of varying complexity can then be utilized for understanding the biological responses. The remainder of the book is devoted to the effects of radiation on biological material, starting with simple systems and progressing in order of increasing biological complexity. Emphasis is on the effects on mammalian systems and mammals, but other biological forms are considered whenever pertinent. The final chapter discusses radiation in our environment—sources, uses, exposure levels, and possible risks to the human population. It stresses the relationship of the experimental effects described in preceding chapters to the effects that may occur in humans as a result of the expanding use of radioactivity.

To aid the reader in understanding the changes which are described in the various systems, basic biological background information is given. It is hoped that this will make it unnecessary to refer extensively to basic biology texts. Furthermore, since it has obviously been impossible to discuss all of the important and interesting radiation topics in depth, specific and general references are included at the end of each chapter and at the end of the book. These are for the convenience of the reader who wishes to obtain a more detailed knowledge of the subject area.

The experimental nature of the field is stressed by the inclusion of many illustrative studies. These are generally investigations with which the author is personally familiar. For example, exposure facilities and a number of experiments from the University of Rochester and Cornell University are described. Much of the illustrative material for radiobotany was provided by Dr. Arnold Sparrow from his work at the Brookhaven National Laboratory. The studies described have necessarily been selective.

Many individuals have contributed generously to the preparation of this book with critical comments and encouragement. I would particularly like to thank the members of my advisory committee, Drs. J. N. Stannard, W. Claus, C. Shilling, C. Swanson, and J. Totter. I am especially grateful to Dr. J. Newell Stannard at whose suggestion the book was written, and who encouraged and helped me throughout its preparation. I am most grateful to Dr. Arnold Sparrow for his major assistance on Chapter 13 and Dr. E. L. Powers for his help on sections of Chapter 7. Among my other friends and colleagues who have read and criticized portions of the book are Drs. C. L. Comar, G. W. Casarett, T. R. Noonan, E. B. Ehrle, J. C. Thompson, R. K. Coomes, J. M. Hanson, R. Z. Korman, and B. Wallace. Nevertheless the responsibility for any shortcomings is mine. Grateful acknowledgment is also made to the United States Atomic Energy Commission and to the American Institute of Biological Sciences for sponsorship of the book and for invaluable aid in its preparation. The illustrations were prepared by Marion Newson and R. K. Coomes.

It is my hope that this text will introduce the reader to the many complex and important aspects of radiation biology, that it will assist him in obtaining information on the subject, and that it will stimulate him to inquire further into areas of his specific interest. I trust it will be of help to the beginning radiobiologist as well as to the specialist in another scientific discipline who seeks an introduction to the rapidly developing field of radiation biology.

ALISON P. CASARETT

Ithaca, New York

CONTENTS

CELL SENSITIVITY   159

        Criteria of Sensitivity   160
        Factors Influencing Sensitivity   161
        Categories of Mammalian Cell Sensitivity   167
        Specific Classifications of Mammalian Cell Sensitivity 168

9 RADIATION EFFECTS ON
MAJOR ORGAN SYSTEMS
OF MAMMALS   171

        Interdependence of Cells and Tissues   172
        Generalized Tissue Changes   173
        Effects on Specific Organ Systems   174
        Effects on Stress Reaction   211
        Effects on Immune Mechanisms   212

10 ACUTE RADIATION EFFECTS
IN WHOLE ANIMALS   217

        Lethality   217
        Acute Radiation Syndrome in Mammals   220
        Radiation Effects on Prenatal Development   226
        Radiation Effects on Regeneration   232

11 MODIFICATION OF
RADIATION INJURY   236

        Physical Modification of Radiation Exposure   236
        Biological Factors Which Modify the Radiation
            Response   249
        Chemical Protection   256
        Treatment of Radiation Injury   262

12 LATE EFFECTS
OF RADIATION   266

        Radiologic Aging   266
        Life-shortening   268

INTRODUCTION

1

Life has evolved in a world in which a major source of the energy essential for biological processes is in the form of radiant energy, or radiation. Radiation impinges on living material in a variety of ways. For example, sunlight provides heat, light, and the energy for photosynthesis; radiowaves give a means of communication. Most of these radiations are not only harmless in ordinary quantities but are actually necessary for life. Certain types of high energy or "ionizing" radiations, however, are not so harmless. They are known to produce deleterious effects in all forms of life from the relatively simple unicellular plants and animals to the complex higher organisms.

Changes produced by these radiations may be grossly apparent and may be visible soon after exposure of the living organism. Or, the radiation may not appear on casual examination to have affected the organism at all. Even then, however, there may be small changes which can be detected by careful chemical or microscopic study, or which are only apparent after many years, perhaps in the offspring of the irradiated organism. The subject matter of *Radiation Biology* is concerned with the description and explanation of these many changes which radiation produces in biological material — gross or microscopic, lethal or nonlethal, immediate or delayed. From these changes it may be possible to deduce explanations for a broad spectrum of biological mechanisms and reactions — explanations that provide leads for the formation of hypotheses and further observations and for the development of instruments and methods for radiation protection and useful applications of radiation to research, medicine, and agriculture. All of these subject areas make up the science of radiation biology.

DISCOVERY OF RADIATION AND
RADIOACTIVE MATERIALS

Radiation biology had its start in the year 1895, with the discovery of "x-rays" by the German professor Wilhelm Conrad Röntgen.[1] While studying the electrical discharge through a simple cathode-ray tube, he noted a fluorescent glow in some barium platinocyanide nearby. Investigation showed that this was caused by rays originating in a fluorescent area of his cathode tube. Röntgen found that these rays traveled in straight lines and, curiously, that they penetrated some materials but not others. Familiar objects cast very strange shadows when placed between the tube and a platinocyanide screen.

When he discovered that these rays exposed photographic plates, Röntgen photographed the shadows from his hand and thus produced the first x-ray pictures. The bones of his hand were boldly revealed on the photographic plate.

Henri Becquerel, a French physicist, was intrigued by Röntgen's x-ray pictures and by the observation that they arose in the fluorescent spots on the wall of the cathode tube. He thought that x-rays might be produced in other types of fluorescences and proceeded to examine a variety of materials. Fortunately, one of the materials chosen was potassium uranyl sulfate. He placed a well-wrapped photographic plate in the sunlight, with the uranium crystals on top. As expected, the plate was exposed where the crystal had been, and Becquerel hypothesized that the sun had produced a fluorescence in the crystals and that the x-rays were a normal part of this fluorescence.

Sometime later, a series of cloudy days made it impossible for Becquerel to expose his plates to sunshine, and he left his preparation in a drawer. Upon developing the plates, he found that they had been exposed! He tried other fluorescent materials and observed that plates were exposed whenever uranium was present, whether or not the compound was fluorescent. Uranium was the important factor, not fluorescence! However, the source of the energy required to expose the plates was still a mystery.

Marie Curie became interested in the reports of Becquerel and set out to systematically examine other materials for the presence of this "emanation." She used a simple electrometer as a detector, utilizing Becquerel's observation that the rays discharged electrified bodies. To her surprise, pitchblende, an ore of uranium, gave a discharge more powerful than uranium crystals, although it contained less uranium per volume than did the crystals. Recognizing the possibility that impurities might be responsible for this radiation, she subjected the pitchblende to a series of chemical separations and finally

---

[1]See Romer, A. (1960) for a survey of the discovery of radiation.

isolated a new radioactive element, polonium. Sometime later, Marie Curie isolated another new radioactive element, radium.

At about the same time, Ernest Rutherford and his co-workers at Mc-Gill University became interested in thorium, another naturally occurring radioactive element. Between 1898 and 1903, they performed a series of experiments which demonstrated that thorium, in the process of emitting its radiation, was changed to a new element, called thorium X. Moreover, thorium X was also radioactive. Thus, the existence of radioactive series (see Chapter 2) was recognized.

During the first three decades of this century, several types of radiations were recognized and characterized; structures were proposed for the atom and for the atomic nucleus; and the various radioactive elements and their products were located in the periodic table of elements. Physicists and chemists had uncovered a new and exciting area of science.

BIOLOGICAL EFFECTS
OF RADIATION

Physicians in both Europe and America were immediately interested in Röntgen's report and especially in his x-ray pictures of the bones of living individuals. X-ray photography became a common practice. Physicians photographed their patients, families, and colleagues to demonstrate normal structures, to look for abnormalities, or just to satisfy curiosity seekers. In April 1896, Dr. J. Daniels of Vanderbilt University reported that irradiation of the skull of one of his colleagues had resulted in the loss of hair.[2] Thus, the ability of radiation to damage cells was detected within 4 months of Röntgen's initial report!

Dr. Leopold Freund, a young Viennese physician, read Daniels' paper and attempted to use this epilatory effect for therapeutic purposes. In 1897, he was able to report the successful removal, by radiation, of an extensive hairy birth mark. As a result of the publicity that followed, many attempts were made to cure pathologic growths of the skin. In 1899, two Swedish physicians, Drs. J. T. Stënbeck and T. A. V. Sjøgren, claimed the first radiation cure, the removal of a skin tumor from the tip of a patient's nose. An early effect obtained on tissue within the body was the decrease in spleen size of a leukemic patient, reported in 1903 by Dr. Nicholas Senn of Chicago.

The use of x-irradiation for therapy was producing some unfortunate results. Skin erythema (reddening) appeared in many of the irradiated patients. Physicians developed erythema and skin ulcerations of the hands as

[2]Ellinger, F. (1943) contains a survey of early human exposures to radiation.

a result of exposure during treatments. They commonly checked cathode-ray tubes by holding the left hand close to the tube and examining the image on a fluorescent screen held in the right hand. The practice resulted in ulceration and malignant tumors in the left hand of many of these individuals. Indeed, a number of scientists (including the French biologist Bergonié) eventually lost their lives as a result of radiation exposure. After a succession of amputations, beginning with fingers and extending through the hands and arms, they usually died from a spreading of the original skin cancer to the internal organs.

Despite the biological damage which was obviously resulting from ir-radiation, the accidental and deliberate exposure of humans continued throughout the first few decades of this century. Miners in several of the Bohemian mines were exposed to uranium and its degradation products; careless radiologists received considerable amounts of radiation while treat-ing patients; patients drank "radium water" prescribed by physicians for the treatment of various conditions such as gout; and girls ingested radium in the process of painting the luminous dials on watches and clocks. These expo-sures were unfortunate. They occurred partly through disregard of hazards due to the exigencies of World War I and partly through a lack of apprecia-tion of the long-term, delayed effects of irradiation. Nevertheless, they have provided much of our information on the effects of low doses of irradiation on humans.

At the same time that physicians were using x-rays to examine human structures and to treat pathologic conditions, biologists were irradiating other living organisms. Both x-rays and radiation from radium were used for experimentation on plants and animals. Abnormal forms resulted from irradiation of plant seeds and frog or bird eggs, and the lethal effect of ir-radiation on bacteria was noted. A systematic study of pathologic changes in mice, guinea pigs, rabbits, and dogs was reported by H. Heineke in 1905. The reproductive organs of irradiated rodents were examined in detail by H. E. Albers-Schönberg (1903), L. A. Halberstaedter (1905), and J. L. Bergonié and L. Tribondeau (1906). Many other comprehensive pathologic studies followed.

Nearly 20 years later, two geneticists, H. J. Muller working with *Drosophila* and L. J. Stadler experimenting with corn, demonstrated the genetic effects of radiation. One of the results of these studies was the scheme proposed by Muller in 1927 for lethal gene mutations in entire chromosomes. This made the quantitative study of mutation rates feasible.

Many diverse and apparently contradictory reports appeared in the early literature. For example, radiation was reported to shorten the life of some types of protozoa, but to lengthen the life of others; radiation increased the number of blood cells in some animals, but markedly decreased the blood count in others. The species, exposure conditions, and even the individual investigator seemed to have a marked effect on the response observed.

One major reason for apparent contradictions was the difficulty in quantitating the radiation received by the system. Many different methods were tried for measuring the radiation exposure. Chemically coated discs were used which gradually changed color when irradiated (Holzknecht pastilles), but the dose range was wide for a small color change and it was difficult to compare results from different laboratories. Numerous attempts were made to standardize radiation exposures by the response of "biological dosimeters" such as plants, unicellular organisms, parasites, seeds, eggs, and fruit flies. Again, the results were inconsistent because of the variability of the organisms from one laboratory to another and the difficulties encountered in quantitating the responses.

The "skin erythema dose" (SED) was another attempt at biological standardization. This was the amount of radiation needed to produce visible reddening of human skin in 1 to 3 weeks. The obvious disadvantages of this method were many. A delay of 3 weeks was necessary before the radiation dose was known; exposure of humans was necessary; and the variability in human responses was considerable. For example, thin-skinned blondes are more sensitive to erythema production than brunettes; the forearms are more sensitive than the palms of the hands. While using the SED did permit some quantitation of exposure, a more precise dosimeter was needed which did not depend on reproducibility in inherently variable biological systems. Finally, in 1928, the roentgen (R) unit, a measure of ionization in air, was established as an International Standard of Radiation Exposure.

Another major advance in radiation biology was made in 1914 with the development of the Coolidge hot filament tube. It provided a constant source of radiation which could be duplicated and fairly well controlled and which yielded amounts of radiation which were unavailable with the gas tube. The Coolidge tube has served as the model for x-ray tubes up to the present day.

Thus, by 1940 many sources of radiation had been identified, and the amount of radiation delivered to an organism could be measured. An understanding of the biological effect of radiations was growing both in terms of the changes produced and of the dosage associated with a given effect.

The real impetus to research in radiation biology, however, came with the development of nuclear fission and the need to understand the effects of the resulting radiation on our populations. On December 2, 1942, in a squash court under the stadium at the University of Chicago, the first man-made, self-sustaining nuclear chain reaction occurred. The atomic bombs subsequently dropped on Nagasaki and Hiroshima in 1945 and the bomb testing that has occurred since have introduced a potential source of high-intensity radiation. Nuclear technology is becoming increasingly important in the development of atomic energy as a source of power for production of electricity and for propulsion and in the development of radioactive materials as tools for medicine, industry, research, and agriculture. It appears certain that man-made sources of radiation have become an inevitable part of our civilization.

Concomitant with these developments in the production and use of radiations has been the expanded study of their biological effects. Radiation biology may now be subdivided into areas such as radiation chemistry and biochemistry, radiation microbiology, radiation physiology, radiation genetics, radiation botany, radiation pathology. Much has been learned about the effects of radiation within these different disciplines; much still has to be learned. The following chapters will present an introduction to the information which has been gathered in some of the fields which make up radiation biology.

GENERAL REFERENCES

Colwell, H. A., and Russ, S. *Radium, X-Rays and the Living Cell*. With Physical Introduction. G. Bell and Sons, Ltd., London (1915).

Duggar, B. M. *Biological Effects of Radiation*. Mechanism and Measurement of Radiation, Applications in Biology, Photochemical Reactions, Effects of Radiant Energy on Organisms and Organic Products. Vols. 1 & 2. McGraw-Hill Book Co., Inc., New York (1936).

Ellinger, F. Lethal Dose Studies with X-rays I. *Medical Times*, May (1943).

Romer, A. *The Restless Atom*. Anchor Books, Doubleday & Company, Inc., Garden City, New York (1960).

Warren, S. L. The Physiological Effects of Radiation Upon Normal Body Tissues. *Physiol. Rev.*, **8:** 92–129 (1928).

ADDITIONAL REFERENCES CITED

Albers-Schönberg, H. E. Ueber eine bisher unbekannte Wirkung der Röntgenstrahlen auf den Organismus der Tiere. *Münch. Med. Wochschr.*, **50:** 1859–1860 (1903).

Bergonié, J., and Tribondeau, L. Action des rayons X sur le testicule. *Arch. d'electr. medicalé*, **14:** 779–791, 823–846, 874–883, 911–927 (1906).

Halberstaedter, L. Die Einwirkung der Röntgenstrahlen auf Ovarien, Berlin. *Klin. Wochschr.*, **42:** 64–66 (1905).

Heineke, H. Experimentelle Untersuchungen über die Einwirkung der Röntgenstrahlen auf innere Organe. *Mitt. Grenzg. Med. Chir.*, **14:** 21–94 (1905).

Muller, H. J. Artificial Transmutation of the Gene. *Science*, **66:** 84–87 (1927).

Stadler, L. J. Mutations in Barley Induced by X-rays and Radium. *Science*, **68:** 186–187 (1928).

RADIATION
PHYSICS

2

THE ATOM

The concept of the atom as a small unit of mass is not new. As long ago as the 5th Century B.C., in Ancient Greece, Democritus proposed that all matter is composed of very small particles, which he called *atoms*. His approach was philosophical, however, not scientific, and was challenged by the Greek philosopher Aristotle. He supported Empedocles' theory that all matter was composed of four "principles" or "elements" — earth, water, air, and fire. Aristotle believed that different substances consisted of different relative amounts of each of these elements. His view persisted through the Middle Ages and Renaissance period until the development of quantitative chemistry in the last half of the 18th Century. In 1789, the French chemist Lavoisier substituted 33 of what he believed were elemental substances for Aristotle's four "elements."

An English chemistry teacher, John Dalton, started the modern atomic theory in 1803 with a publication in which he described his determinations of the relative weights of the "ultimate particles," or atoms. His ideas were based on the contention that all matter is composed of a number of elemental substances. Each of these is composed of identical atoms which differ only from element to element. During the next 50 years, many substances were positively identified as truly elemental, and their chemical properties and atomic weights were determined. To the Russian Mendeléeff goes the main credit for arranging the known elements according to a periodic function which correlates their chemical properties with their atomic weights. To keep this arrangement internally self-consistent, it was necessary to leave many spaces in the table. Nearly a century was to elapse before all the spaces were filled.

It is now known that elements are indeed atomic and that the atoms themselves have a structure. Only a brief discussion of this extensive subject will be presented in this chapter.

ATOMIC STRUCTURE

The atom may be viewed as containing a positively-charged nucleus and one or more negative electrons (e) which are arranged in circular or elliptical orbits around the nucleus. The nucleus may be simply[1] considered as made up of singly-charged positive protons (p), each with a mass of 1.0076 atomic mass units (amu),[2] and uncharged neutrons (n), each with a mass of 1.0089 amu. The protons and neutrons are held together in the nucleus by a force which is referred to as the *binding energy*, and may be demonstrated as follows:

The nucleus of the helium atom is made up of two protons and two neutrons. The total mass of its individual components is thus:

$$\left. \begin{array}{l} \text{Mass of two protons } = 2 \times 1.0076 \\ \text{Mass of two neutrons} = 2 \times 1.0089 \end{array} \right\} = 4.0330 \text{ amu}$$

However, the actual mass of the helium nucleus has been determined as 4.0028 amu. The difference in mass (4.0330 − 4.0028 = 0.0302 amu) can be considered equivalent to a certain amount of energy (28.2 Mev),[3] based on the Einstein mass-energy relationship of $E = mc^2$.[4] In the combination of two protons and two neutrons to form a helium nucleus, 28.2 Mev of energy would be released. Likewise, the same amount of energy must be supplied to separate the nucleus into its four components. This is the "binding energy."

Energy levels are associated with nuclear structure; transitions between the energy levels occur. If, for example, a nucleus goes from one energy

[1]There are more than 30 known elementary particles and antiparticles. Their involvement in the constitution of nuclei (except for protons, neutons, and possibly pi-mesons) is obscure (see Frisch and Thorndike, 1963) and their importance to radiation biology is unknown.

[2]An atomic mass unit (amu) is defined as $1/_{12}$ of the mass of carbon-12 — 1.66 × 10⁻²⁴ gram.

[3]Energy can be expressed in terms of electron volts. An electron volt is the energy acquired by an electron when it passes through a potential difference of 1 volt; 1 ev is equal to 1.602 × 10⁻¹² erg. The kiloelectron volt (kev), million electron volt (Mev), and billion electron volt (Bev) are commonly used extensions of this unit.

[4]E is energy in ergs; m is mass in grams; c is the velocity of light, approximately 3 × 10¹⁰ cm/sec.

state to a lower state, energy may be given off in the form of electromagnetic or particulate radiation in an amount equal to the energy difference.

The number of protons in an atomic nucleus is called the atomic number (Z) and is usually represented by the appropriate numeral as a subscript to the left of the element symbol. The number of protons plus neutrons is called the mass number (A) and is approximately, but not quite, equal to the atomic weight. The mass number is represented by the appropriate numeral as a superscript to the left.[5] Thus, one would represent hydrogen, which has one proton and no neutrons in the nucleus (that is, $Z = 1$, $A = 1$), as $_1^1H$ and helium with two protons and two neutrons as $_2^4He$. Some simple atomic nuclei may be pictured as shown in Figure 2.1.

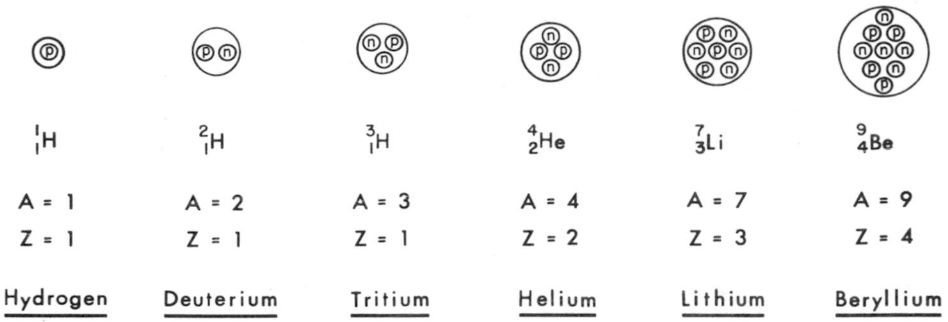

$_1^1H$	$_1^2H$	$_1^3H$	$_2^4He$	$_3^7Li$	$_4^9Be$
A = 1	A = 2	A = 3	A = 4	A = 7	A = 9
Z = 1	Z = 1	Z = 1	Z = 2	Z = 3	Z = 4
Hydrogen	Deuterium	Tritium	Helium	Lithium	Beryllium

Fig. 2.1. Diagrammatic representation of some simple atomic nuclei.

In an uncharged atom, the number of orbital electrons is equal to the number of protons in the nucleus. Electrons are particles which carry a unit charge of negative electricity and have a mass of about 0.00055 amu. They orbit the atomic nucleus in precisely defined paths, each path being characterized by its own unique level of energy. Electrons may move from one level to another unoccupied level. If to a lower level, the jump may be accompanied by the release of an amount of energy, in the form of electromagnetic radiation, equal to the difference in energy between the levels. If the jump is to a level of higher potential energy, the electron must absorb energy from some source. The application of wave-mechanical theory to the orbital electrons has introduced a factor of uncertainty into the position of the electrons, but has not otherwise changed the concept of the electron orbits.

[5]In the notation used heretofore, A is shown at the upper right. A recent international recommendation which is being generally adopted now calls for it to be noted in the upper left (Joint Subcommittee on Symbols, Units and Notations, of International Unions of Pure and Applied Physics, and of Pure and Applied Chemistry).

NUCLIDES[6]

In small atoms with low Z numbers, the number of neutrons in the nucleus is approximately the same as the number of protons; in atoms with higher Z numbers, the number of neutrons is greater than (up to about 1.6 times) the number of protons. For example, ^4_2He has 2 protons and 2 neutrons, and $^{238}_{92}U$ has 92 protons and 146 neutrons.

The chemical identity of an atom or nuclide is determined by the number of protons in the nucleus. There may be several nuclides of the same element with different numbers of neutrons in their nuclei. For example, two naturally occurring hydrogen nuclides have one proton each in their nuclei, but one has no neutron while the other has one neutron (see Figure 2.1). The latter (the relatively rare deuterium or "heavy hydrogen") is about twice as heavy as the more commonly occurring form of hydrogen. As another example, all atoms with 82 protons are lead, and all atoms of the element lead have 82 protons. Different atoms of lead, however, have 122, 124, 125, or 126 neutrons, that is, $^{204}_{82}Pb$, $^{206}_{82}Pb$, $^{207}_{82}Pb$, and $^{208}_{82}Pb$. Nuclides which differ only in the number of nuclear neutrons are called *isotopes* of the element. Stated differently, an isotope is one of a group of two or more nuclides having the same atomic number, that is, the same number of protons.

The proportion of the various isotopic constituents of an element is called the *abundance ratio*, and expresses the relative amount or abundance of each isotope as a percentage. The relative abundances of each of the hydrogen isotopes listed above are 99.985% (^1H) and 0.015% (^2H); the relative abundances of each of the lead isotopes are 1.5% (^{204}Pb), 23.6% (^{206}Pb), 22.6% (^{207}Pb), and 52.3% (^{208}Pb).

RADIONUCLIDES

Many nuclides have stable nuclei; others are unstable, or radioactive. All elements above bismuth ($Z = 83$) in the periodic table are naturally unstable and a few of the lighter elements have one or more unstable isotopes. In addition, recent developments in atomic science have made possible the artificial production of unstable nuclides which are isotopes of virtually all of the known elements. These range from hydrogen-3 (tritium) to heavy nuclides such as uranium-240. They are termed *artificial radionuclides*.

Unstable nuclei stabilize their structure by a change in nuclear energy

---

[6]Nuclide is the general term for any atomic configuration which has a measureable lifetime, usually considered to be greater than 10^{-9} sec. It is a general term for a particular type of atom which is characterized by the constitution of its nucleus, that is, by the numbers of protons and neutrons which it contains.

levels accompanied by the release of particulate and/or electromagnetic radiation. Some of these emit beta particles (electrons) to attain a stability. Others emit alpha particles which are identical to helium nuclei (2 protons + 2 neutrons). Gamma rays (short wavelength electromagnetic radiation) are frequently given off with the particulate alpha and beta emissions. These gamma rays are similar to the x-rays discovered by Röntgen and are often highly energetic (of the order of million electron volts).

ALPHA EMISSION

The loss of an alpha particle from a nucleus represents a release of 2 protons (2 positive charges) and 2 neutrons, or a total of 4 mass units. Therefore, the atomic weight A is reduced by 4 and the atomic number Z is reduced by 2. For example, an atom of $^{210}_{84}Po$ (polonium) becomes $^{206}_{82}Pb$. The atom has changed or "decayed" into a different element. This change of an atom from one element to another by the release of particulate radiation is called a *radioactive transformation*.

The energy of the alpha particle released in a radioactive transition is characteristic of the particular transition involved. This energy probably represents the difference between the energy of the emitting, or parent, atomic nucleus and the energy of an excited state of the resulting, or daughter, nucleus. The alpha particle which is emitted in the transition of polonium-210 to lead-206 has an energy of 5.3 Mev.

The gamma radiation which accompanies the alpha emission in some radioactive transformations is probably due to the transition of the daughter nucleus from a higher energy level (excited state) to the ground state. This can be represented diagrammatically, as shown in Figure 2.2. The horizontal lines represent the energy level of the parent element (At. No. = Z) and the higher, or excited, energy level and ground energy level of the daughter element (At. No. = $Z - 2$). The alpha particle emission is depicted as a line slanted to the left to indicate a *decrease* in Z, and the gamma ray emission, as a vertical line to indicate no change in Z.

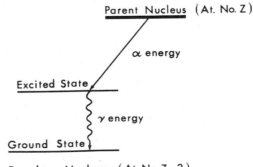

Parent Nucleus (At. No. Z)

α energy

Excited State

γ energy

Fig. 2.2 Diagrammatic representation of the energy change associated with alpha particle emission.

Ground State

Daughter Nucleus (At. No. Z-2)

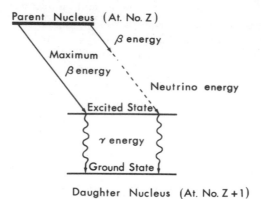

Parent Nucleus (At. No. Z)

β energy

Maximum

β energy

Neutrino energy

Excited State

γ energy

Ground State

Daughter Nucleus (At. No. Z + 1)

Fig. 2.3. Diagrammatic representation of the energy change associated with beta particle emission.

BETA EMISSION

A beta particle has a single negative charge and the very small mass of about 9.108×10^{-28} gram (5.5×10^{-4} amu). The emission of a beta particle from a nucleus may be viewed as equivalent to changing a neutron into a proton, that is, there is a negligible change in atomic weight, but an *increase* of one in the atomic number. Thus, $^{234}_{90}$Th (thorium) decays to $^{234}_{91}$Pa (protactinium) with the emission of a beta particle.

The energies of the beta particles for a given transformation are not all the same. They represent a continuous spectrum with a definite maximum. The upper limit is considered to represent the difference between the initial and resulting energy levels; the intermediate energies represent the same transition, but with only part of the energy retained by the beta particle. The remainder is carried off by a neutrino, a particle of very small mass and zero charge. The average energy of the beta particles from a particular radionuclide is approximately one-third of the maximum energy for that transformation. Gamma radiation may accompany beta emission, to bring the daughter nucleus to a ground state. Figure 2.3 represents the process of beta particle emission. As in Figure 2.2, the horizontal lines represent the energy levels of the parent and daughter nuclei. Beta particle emission is shown as a line to the right to represent an *increase* in atomic number.

PHYSICAL HALF-LIFE

When a radioactive atom emits a beta or an alpha particle, it becomes a different element. Thus, the number of atoms of the original element is decreased by one. This nuclear decay is a random process; that is, there is a probability (called the *decay constant*, λ) that an atom will undergo decay in a certain interval of time. In a population of many atoms, the number disintegrating in a short time interval will be approximately the total number of atoms present at that time (N) multiplied by the probability of decay for each atom, or λN. It can then be shown that the number of the original atoms (N_0) which are present at any time t is:

$$N = N_0 e^{-\lambda t}$$ (2-1)

The decay constant, λ, is characteristic of the specific element being measured, and does not depend on environmental factors such as pressure, temperature, or chemical state of the element.

This process of radioactive decay can also be considered from a different viewpoint. In a certain interval of time, half of the atoms of a radioactive element will give off their emissions and be changed to the daughter nucleus. In the next equal interval of time, half of the remaining atoms of the isotope will decay, and so on. This interval is known as the *physical*, or *radioactive*, *half-life* ($T_{1/2}$) of the radioactive element. It may be represented graphically as indicated in Figure 2.4, and can be shown to be related to the decay constant as follows:

$$T_{1/2} = \frac{0.693}{\lambda}$$ (2-2)

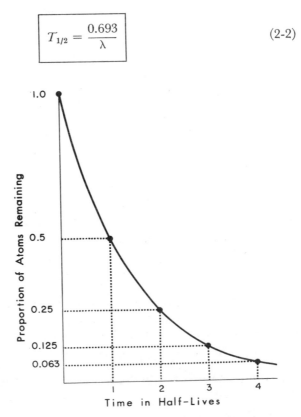

Fig. 2.4. Radioactive decay, illustrating the exponential nature of the process.

The half-lives of various radioactive isotopes vary greatly. For example, the decay of $^{226}_{88}$Ra to $^{222}_{86}$Rn has a half-life of 1620 years while the transition of $^{214}_{84}$Po to $^{210}_{82}$Pb has a half-life of 1.64×10^{-4} second.

ARTIFICIAL RADIONUCLIDES

Stable nuclei can be made unstable, or artificially radioactive, by the introduction of energetic particles such as electrons, deuterons, neutrons, protons, or alpha particles. High-energy gamma rays are also capable of producing excited atoms which disintegrate by radioactive decay. The alpha radiations from natural radioactive elements were the first source of such bombarding particles. The recent development of cyclotrons, linear accelerators, and betatrons, plus reactors with their high neutron fluxes has provided additional means for producing and utilizing high-energy particles.

An incident particle impinges on a nucleus, and is absorbed into it, resulting in the formation of a "compound nucleus" (nucleus plus particle). This structure is usually unstable, and emits either a gamma ray or a particle to become a different element. The new nucleus may then be radioactive and will give off a particle with a characteristic energy and half-life. Almost any nucleus can be made radioactive in this way.

Most of the radioactive nuclides commonly used in research are produced artificially. For example, phosphorous-32 which decays with the emission of a beta particle to sulfur-32 is produced by exposing sulfur-32 to the neutrons from a reactor. A neutron may be absorbed to produce a compound nucleus from which a proton is ejected to form the phosphorous isotope. The process, abbreviated ^{32}S(n, p)^{32}P, can be written as follows:

$$\boxed{^{32}_{16}\text{S} + ^{1}_{0}\text{n} \rightarrow ^{32}_{15}\text{P} + ^{1}_{1}\text{p}} \qquad (2\text{-}3)$$

Several radioactive isotopes of some elements can be produced. For example, sodium-22 is produced by the bombardment of magnesium by cyclotron-accelerated deuterons in the reaction ^{24}Mg(d, α)^{22}Na, and sodium-24 is produced by impinging reactor neutrons on the stable sodium-23 in the reaction ^{23}Na(n,γ)^{24}Na.

Different nuclei which eject particles of differing types and energies can be produced from bombardment of a single type of nucleus, depending on the energy and characteristics of the incident particle and of the initially ejected particle. For example, bombardment of stable cobalt, cobalt-59, by slow neutrons will result in cobalt-60, according to the process ^{59}Co(n, γ) ^{60}Co. If fast neutrons are used as bombarding particles, however, some iron-59 will result by ^{59}Co(n, p)^{59} Fe.

Nuclear fission is another example of the production of radioactive materials by high-energy bombardment. When neutrons impinge on and are absorbed by uranium-235 molecules, the resulting nuclei are most unstable. Instead of emitting a single particle, as in the case of the production of phosphorous-32, these nuclei each emit several neutrons plus gamma radiation and usually break into two distinct, smaller nuclei, both of which may be radioactive. The unstable uranium nuclei do not always break the same way and a variety of fission product nuclei will be formed. Certain breaks are more probable than others so that some fission products are formed more often than others. A representative reaction is as follows:

$$^{235}_{92}U + ^1_0n \longrightarrow ^{236}_{92}U \rightarrow ^{140}_{54}Xe + ^{94}_{38}Sr + 2\,^1_0n + \gamma \qquad \text{(2-4)}$$

If sufficient uranium-235 molecules are present, the neutrons produced in one fission will be absorbed by other uranium nuclei and a chain or sustained process will result.

Some artificially produced radioactive isotopes undergo a process called *K capture*, in which the nucleus of an atom absorbs an orbital electron. The effect can be viewed as the conversion of a nuclear proton to a neutron; a neutrino is emitted:

$$_{+1}^{1}\text{proton} + _{-1}^{0}\text{electron} \rightarrow _0^1\text{neutron} + _0^0\text{neutrino} \qquad \text{(2-5)}$$

Most often the electron is from the inner, or K, shell; consequently, the name K electron capture, or simply K capture, is used for this process. Gamma radiation may be emitted.

RADIOACTIVE SERIES

When a radionuclide decays to its daughter nuclide, the daughter may be stable or may itself be radioactive, and may decay to another element with a radiation of characteristic energy and half-life. The daughter of this decay may, in turn, be radioactive and undergo further decay by giving off another, sometimes different, particle with different energy and half-life characteristics. Such a series of radioactive disintegrations will continue until a stable element is reached. Three such radioactive series are known to occur in nature: the actinium series starting with uranium-235 and decaying finally to stable lead-207, the thorium series starting with thorium-232 and decaying to lead-208, and the uranium series starting with uranium-238 and decaying to stable lead-206. There is also a similar series, the neptunium series, which

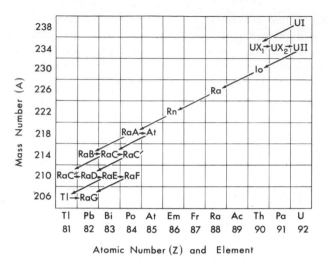

Fig. 2.5. The uranium series. The arrows downward to the left indicate the emission of alpha particles (A decreased by 4, Z decreased by 2). Arrows to the right indicate beta emission (no change in A, Z increased by 1). The general chemical names given to atoms of the various atomic numbers are shown at the bottom; the common names for the particular isotopes are shown on the chart. Note that the three polonium ($Z = 84$) isotopes are often called Radium A, Radium C', and Radium F.

begins with artificially produced plutonium-241 and decays finally to bismuth-209. Most of the naturally occurring radioisotopes are found in one or another of these decay schemes. Figure 2.5 indicates the one known as the uranium series which starts with uranium-238 (frequently referred to as U-I) and ends with stable lead-206 (called RaG). Both alpha and beta particles are given off at different stages in this series. Gamma radiation accompanies most of the alpha and beta emissions. The half-lives of the various steps range from more than a billion years for U-I(uranium-238)to less than a millisecond for RaC' (polonium-214).

INTERACTION OF RADIATION WITH MATTER

The radiations which will be discussed here are those which have a great deal of energy and which are capable of ionization, that is, producing ions by ejection of orbital electrons from the atoms of the material through which they travel. These radiations may be divided into two categories: (1) particulate radiation such as alpha particles, electrons, protons, neutrons, and certain fission fragments, and (2) electromagnetic radiation of specific wavelengths.

ELECTRONS

There are several sources of ionizing electrons: (1) electrons which have been ejected from the nuclei of unstable atoms are known as *beta-minus*[7] or, simply, *beta particles;* (2) beams of high-energy electrons may be produced by particle accelerators; and (3) when other types of radiation, such as gamma rays, pass through matter, energetic electrons are released. These are referred to as *secondary electrons.* Regardless of their source, high-energy electrons undergo the same interactions with matter.

Electrons in the energy range of beta particles or secondary electrons interact primarily with orbital electrons. When a moving electron passes near an orbital electron, a powerful repulsive force (coulomb force) will exist between the two negatively charged particles. If this force is sufficient, the orbital electron may be repelled with enough momentum to move it to a higher energy orbit of the same atom (excitation) or to dislodge or "kick" it away from the atom (ionization). Some of the energy of the original moving electron has thus been transferred to the orbital electron. If the electron has been ejected from the atom, it is called a *secondary electron.* It may have been given sufficient energy to produce excitation or ionization in other molecules. This process is identical to the ionization produced by the initial electron, and results in short secondary paths of ionization projecting from the path of the original electron. These paths are called *delta rays.*

It has been determined that one ionization occurs, on the average, for every 32.5 electron volts (ev) of energy transferred from the electron to the medium.[8] When the electron has expended all of its energy by repeated ionizations and excitations, it becomes attached to a molecule as an orbital electron. It may then be said to have been "absorbed" by the medium.

An electron will travel a certain distance in an absorber, depending on the energy of the electron and the composition of the absorber. The path of the electron will not be straight, however, since the incident electron will be deflected or scattered in inelastic collisions with orbital electrons and may undergo many changes in direction (indicated in Figure 2.6). Accordingly, the majority of the electrons are scattered from the beam, and the relative

---

[7]Positrons, or beta-plus particles, are similar to beta-minus particles in mass but have a single positive charge. They are emitted in certain radioactive decay processes and are formed in pair production (see p. 25). They give up energy to matter in a manner similar to electrons. Upon coming to rest, a positron combines with an electron, both are annihilated, and the mass energy is converted into two photons.

[8]This value is approximate and is given as 34 ev or 35 ev by some authors.

Beta Particle Alpha Particle

Fig. 2.6. Diagrammatic representation of tracks of beta and alpha particles. The secondary electrons ejected by the beta and alpha particles are represented by e–.

number of electrons decreases with distance. As shown in Figure 2.7, the relative intensity will decrease almost linearly in a beam of monoenergetic electrons. The maximum distance traveled, or range (R), can be determined by extrapolation and represents the total linear path length traveled. This value is characteristic of a given energy particle and is usually expressed in terms of the thickness (mg/cm² or gm/cm²) of a specific absorber.

Beta particles emitted from a specific nuclide have a continuous spectrum of energies (see previous section, p. 12). The relative intensity of a beam of such particles decreases more rapidly than with monoenergetic electrons and has been shown to approximate an exponential curve (Figure 2.7).[9] The

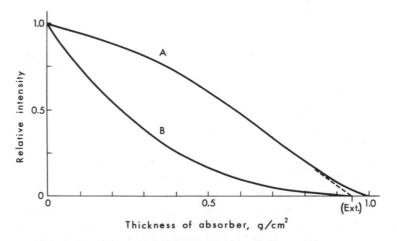

Fig. 2.7. Absorption curves for electrons. (A) Monoenergetic (1.9 Mev) electrons. (B) Beta particles with continuous energy spectrum (E_{max} = 1.9 Mev). The extrapolation of the linear portion (Ext) gives the maximum range of the monoenergetic electrons. It is not possible to determine the maximum range of beta particles by this type of plot.

---

[9]The fact that the beta absorption curve follows an exponential shape is apparently fortuitous. The distribution of energy of the beta particles combined with the absorption characteristic for each energy lead to the exponential function.

maximum range of such a beam can be determined and is characteristic of the particular nuclear transformation.

The total path length or maximum range of an electron depends on the density of the molecules in the absorber, although approximately the same number of ion pairs are formed by a particle, and approximately the same amount of energy is used to form an ion pair when the particle is traveling through air, tissue, or a dense metal. For example, the total length of the track of most electrons is several meters in air, several centimeters in tissue, and less than a centimeter in aluminum.

A second mechanism of energy transfer from electrons involves inter-action with atomic nuclei. When a fast-moving electron approaches the nucleus of an atom, the electrical interaction causes the electron to be ac-celerated and deflected from its original path. This process must be accom-panied by the emission of energy as electromagnetic radiation in accordance with Maxwell's classical electromagnetic theory. The loss of energy results in the slowing down of the electron. The radiation is known as *bremsstrahlung*, or *braking radiation*, and is quite analogous to x-radiation. Bremsstrahlung constitutes a small proportion of the energy transfer for beta particles and secondary electrons. Since the bremsstrahlung radiation produces secondary ionization, there will be ionization at a distance beyond the maximum range of the electron. This is sometimes "counted" as an indirect method of quan-titating the beta radiation. For very high-energy electrons, the proportion of bremsstrahlung to ionization may be high. Except for the effect from bremsstrahlung radiation, ionization from electrons will only be produced within a short distance of the electron source.

ALPHA PARTICLES

Alpha particles (helium nuclei) lose energy by excitation and ionization of the atoms in the material through which they pass. A strong attractive (coulomb) force will exist between a positively charged alpha particle and a negatively charged orbital electron. The electron may be "pulled" either into a higher energy orbit (excitation) or completely away from the atom (ionization) by this attractive force. Because the alpha particle is heavy, it will not be scattered by this interaction, and the path traveled will be rela-tively straight (see Figure 2.6). Delta rays will project from the path of an alpha particle as secondary electrons produce additional ionizations. When most of the energy has been dissipated by ionization and excitation, an alpha particle will capture 2 electrons in its vacant orbit and become a helium atom.

If one measures the ionization at various distances from a source of alpha particles, a characteristic curve called a *Bragg curve* is obtained

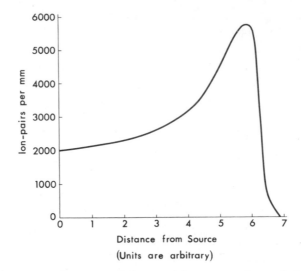

Fig. 2.8. "Bragg curve" showing the rate of formation of ion pairs at various distances from a source of alpha particles.

(Figure 2.8). As an alpha particle traverses its path, it loses energy and slows down. As it spends more time in the vicinity of each atom, it has a greater probability of producing an ionization. Thus, there is an increase in the number of ion pairs formed. Finally, the alpha particle captures 2 electrons, becomes neutral, and can no longer ionize. The rate of ionization falls abruptly.

All alpha particles from a particular nuclear transformation have the same energy and, therefore, the same range and relative distribution of ionization.

The overall path length is short for alpha particles since they are heavy and have a double charge. Therefore, they travel slowly and the rate of energy loss is high. An ordinary piece of paper will stop most alpha particles. The range in tissue is less than 100 microns. A source of alpha particles must be in almost direct contact with a material in order to produce ionization within it.

NEUTRONS

Neutrons are uncharged particles which are emitted by artificial radionuclides (see previous section, p. 14). The most common source of large quantities of neutrons is the uranium fission in nuclear reactors.

Since neutrons have no charge, there is no electromagnetic field associated with their movement and they do not ionize molecules directly. As a result of their lack of charge, however, they are not repelled by positively charged atomic nuclei and are thus able to penetrate close to the nuclei of the material through which they are traveling. They may be captured to form a new compound nucleus or may be scattered or deflected away from the target nuclei in a process referred to as *elastic scattering*. This is similar to the way that a moving billiard ball will be deflected by contact with another. In elastic scattering, some of the energy of the neutron is transferred to the "target" nucleus and appears as kinetic energy — that is, energy of motion of the target nucleus. A larger proportion of the neutron energy will be transmitted to target nuclei of low atomic weight than to nuclei of high atomic weight. Accordingly, materials containing many hydrogen atoms, such as water or paraffin, are very effective in removing some of the energy from the neutrons, thus slowing or "moderating" them.

In a series of these elastic collisions, the neutron will lose more and more of its energy until it moves very slowly, and has approximately the same kinetic energy as the molecules of the medium through which it is moving. At this point it is called a *thermal* neutron.[10] Thermal and slow neutrons can be readily captured by most atomic nuclei. An unstable structure is produced, and gamma radiation is usually given off. The atom which results is frequently radioactive itself, as described earlier in this chapter.

Thus, by impact, energy from a neutron is frequently transferred to a hydrogen nucleus in the form of kinetic energy. The result is a moving *charged* particle (a proton) which is capable of ionizing the molecules of the medium through which it is passing. This ionization by protons is probably the most important mechanism by which biological effects are produced by neutrons.

FISSION FRAGMENTS

When a uranium nucleus breaks up or fissions (see previous section, p. 15), the resulting fragments have a high initial velocity (of the order of 10^9 cm/sec) and much energy (between 65 and 100 Mev/fragment). They are also highly charged, since about 40 of the 92 orbital electrons of the uranium atom are stripped off in the process of fission. This leaves each fragment with about 20 electrons less than the normal atom. Consequently, the fragments are capable of ionizing molecules. The path length of a fission fragment is only 2 to 3 cm in air, and the ionizations are produced very close together. Only material in very close contact with the fission process will be ionized.

ELECTROMAGNETIC RADIATION

An entire spectrum of electromagnetic radiations[11] is present in our environment in the form of radiowaves, infrared, visible, ultraviolet, x-rays, and

---

[10]The kinetic energy of ordinary molecules is temperature dependent and is, thus, often called the *thermal* energy. For this reason, a neutron (about 0.025 ev) with a kinetic energy approximately that of the surrounding molecules is called a *thermal* neutron.

[11]Electromagnetic radiations are so named because they are conceived to consist of oscillating electric and magnetic fields.

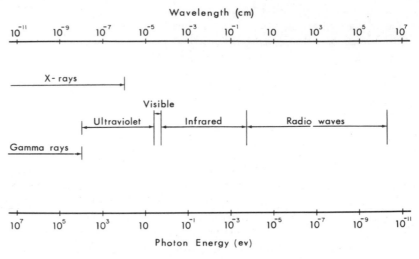

Fig. 2.9. Approximate wavelengths and photon energies of major types of electromagnetic radiation.

gamma rays. The wavelengths and photon energies associated with each of these types of radiation are indicated in Figure 2.9.

Electromagnetic radiation of wavelengths in the visible region from about 4×10^{-5} to 8×10^{-5} cm [4000 to 8000 A (Angstroms)] produces photoreactions which are of importance to life. Photosynthesis, growth of plants, and vision all depend on visible light. Ultraviolet radiation (10^{-8} to 4×10^{-5} cm or 1 to 4000 A) is important for a few biological processes, and has also been shown to have a detrimental effect on certain biological systems.[12] Ultraviolet light does not transmit its energy to atoms or molecules by ionization, but rather almost entirely by excitation. Some of the effects produced by it may resemble the changes resulting from ionizing radiation, and some of the mechanisms may be similar when excitation contributes to the effect produced by the higher energy radiation.

The electromagnetic radiations with shorter wavelengths — x-rays and gamma rays — produce both ionization and excitation in the media through which they travel. The biological effects of this radiation apparently result largely from the ionization which is produced.

X-rays and gamma rays. These are electromagnetic radiations with wavelengths in the range 10^{-11} to 10^{-7} cm (10^{-3} to 10 Angstrom units). Those radiations which originate from atomic nuclei are termed gamma rays; those which originate outside the atomic nucleus are termed x-rays. In all

[12]See Hollaender (1955), Burton *et al.* (1960), and many other books or articles for a discussion of ultraviolet radiation effects. The subject will not be discussed in this book.

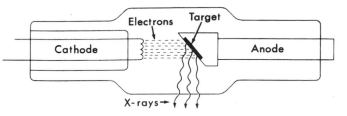

Fig. 2.10. Diagrammatic representation of simple x-ray tube. Electrons from the cathode impinge on the target; x-rays are emitted.

respects other than origin, these two radiations are identical and will be considered together in the following discussions.

Gamma rays are produced when an unstable atomic nucleus releases energy to gain stability. X-rays are produced when high-energy charged particles impinge on a suitable target such as tungsten (see Figure 2.10). When fast-moving electrons approach the coulomb fields around the nuclei of the atoms of the target material, these electrons are deflected from their path. Energy in the form of electromagnetic radiation is emitted. These x-rays may have any energy from zero to a maximum which is determined by the kinetic energy of the impinging electrons. Thus, a whole continuous spectrum of x-rays of various energies is produced when large numbers of impinging electrons are involved. In addition, certain characteristic x-rays are produced that have an energy dependent on the specific material used as a target. These are produced when a high-speed electron impinging on the target material knocks out an orbital electron (ionization) from a target atom. When this electron is from an inner shell, its place is immediately taken by the electron from an outer shell. During this latter transition, an x-ray is given off. The energy of that x-ray represents the difference in energies of the inner and outer orbital electron levels. Since each element has specific energy levels, the energies of these "characteristic" x-rays are specific for the element bombarded. Figure 2.11 indicates a typical x-ray spectrum with continuous and characteristic energies shown.

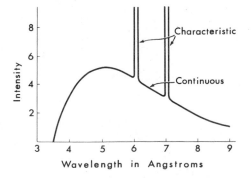

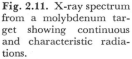

Fig. 2.11. X-ray spectrum from a molybdenum target showing continuous and characteristic radiations.

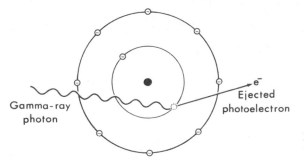

Fig. 2.12. The photoelectric effect

The interaction of x- and gamma rays with matter is primarily through three mechanisms — namely, the photoelectric effect, Compton scattering, and pair production. When a low-energy gamma photon interacts with an atom, all the energy of the gamma ray may be transferred to an electron. Figure 2.12 shows a schematic representation of this photoelectric effect. Some of the energy is used to remove the electron from the orbit, and the remainder is transferred to the electron in the form of kinetic energy. Thus, an ionization is produced by the gamma ray. These photoelectrons may carry considerable kinetic energy and are, therefore, capable of ionizing molecules. There may be a whole series of secondary ionizations along the path of an ejected photoelectron. Because there may be many of these secondary ionizations, they are of more importance in producing biological effects than is the single primary ionization which is produced by the gamma ray.

The Compton effect may occur when a gamma photon of somewhat higher energy approaches a molecule. The photon gives up only part of its energy in ejecting an orbital electron and is deflected with a longer wavelength, as depicted in Figure 2.13. The less energetic secondary gamma photon may then interact with another molecule ejecting another electron. This process continues until all the energy has been transferred to electrons. The electrons, in turn, will each ionize other molecules and will produce paths of secondary ionizations.

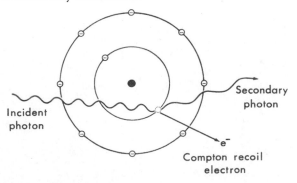

Fig. 2.13. The Compton process

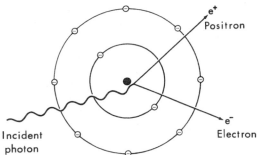

Fig. 2.14. Pair production

High-energy gamma photons transfer their energy primarily by "pair production" (Figure 2.14). The mechanism is particularly important when the medium contains elements with a high mass number. A high-energy gamma ray passing close to a nucleus suddenly disappears, and an electron and a positron (the latter has the same characteristics as an electron, except the positron has a single *positive* charge, see footnote, p. 17) appear in its place. This process requires a minimum of 1.02 Mev. This amount of energy is equivalent to the rest mass of the positron and electron and is used up in the production of these particles (see previous section on mass-energy conversion, p. 8). Additional energy of the gamma ray will appear as kinetic energy of the two particles and will be transferred to the medium by excitation and ionization.

The absorption of x- and gamma rays can be demonstrated by measurement of the beam intensity which remains after passage through various thicknesses of absorber. The relationship which one obtains for a narrow beam of monoenergetic gamma rays is

$$\frac{I}{I_0} = e^{-\mu d}$$

(2-6)

I/I_0 is the fraction of photons remaining in the beam after passage through an absorber of thickness d. Figure 2.15 illustrates this relationship graphically, with intensity plotted on a linear scale and also on a logarithmic scale to further demonstrate the exponential nature of the absorption of monoenergetic radiation.[13]

If a broad beam of electromagnetic radiation impinges on an absorber, some of the radiation will be scattered out of the direct path to the detector, but even more of the peripheral radiation will be scattered into the measuring device from interaction with the absorber. This radiation would not be

---

[13]If the radiation is not monoenergetic, as is the case with x-rays, the absorption curve will not be exponential. Rather, it will represent a combination of curves from all the energies involved.

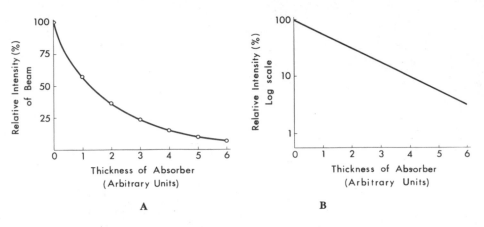

Fig. 2.15. Absorption of gamma rays. (A) Linear plot. (B) Logarithmic plot.

counted if the scatter had not occurred. To account for the scatter radiation, a "build-up" factor must be added to the right-hand side of equation 2-6 if a broad beam is considered.

The quantity μ is the total linear absorption coefficient and represents the absorption of photons by all of the three processes described above. The relative contributions of the three mechanisms depend primarily on the energy of the incident photons and on the nature of the absorber. Figure 2.16 shows the relationship between the absorption coefficient and photon energy for aluminum, including partial absorption coefficients for the several mechanisms.

In the absorption of electromagnetic radiation, it is not possible to specify the range of a photon, but only the probability that it will travel no further than a given distance. The term half-value layer is used as a measure of penetrability. This is the absorber thickness which is required to absorb half of the photons and is given by the equation

$$\text{HVL} = \frac{0.693}{\mu} \qquad (2\text{-}7)$$

For example, the half-value thickness for 1 Mev gamma rays in lead is 0.90 cm. Thus, 0.90 cm of lead will decrease the intensity of a beam of 1 Mev gamma rays to half its initial value. The HVL in air is about 25 meters for most gamma rays which have an energy greater than 1 Mev. It is obvious from this value that electromagnetic radiation can travel a great distance in air and, therefore, can interact with matter at a considerable distance from its source.

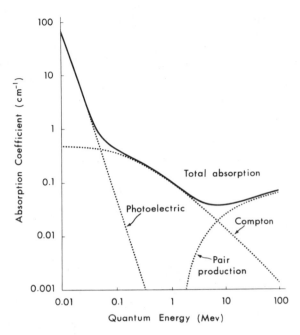

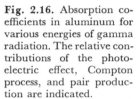

Fig. 2.16. Absorption co-efficients in aluminum for various energies of gamma radiation. The relative contributions of the photoelectric effect, Compton process, and pair production are indicated.

COMPARISON OF
IONIZING RADIATIONS

The nature of the radiation makes a considerable difference, quantitatively, in the biological response which is produced in a system, even when the same amount of energy has been released, or the same number of ion pairs have been formed in the system. In order to understand the reasons for this and, indeed, to understand the mechanism of radiation action on biological systems, it is necessary to have a means of comparing the different radiations.

It will be recalled that ionizing radiations transfer their energy to the media through which they travel primarily by the formation of ion pairs and excitation. Ionization seems to be the most important of the two for producing biological changes and will, therefore, be given major consideration. Heavy particulate radiations such as alpha particles and fission fragments have a short path length, with ion pairs produced close together. Particles of small mass such as beta particles have a longer path length and produce ion pairs farther apart, even when the same total amount of energy is transferred. Electromagnetic radiations have an even longer primary path although the secondary paths from the Compton or photoelectrons may be similar in length and density to those of primary beta radiations. The radiations which have the greatest charge — alpha particles and fission fragments — will produce the paths of densest ionization.[14]

Radiations can be compared on the basis of the average number of ion pairs formed per unit track length. That is called *linear ion density* or *specific ionization*. It is customary to express the specific ionization as the *average number of ions* formed over the entire track, although the actual rate of ionization varies greatly over the track. As energy is given up to the medium, a particle moves more slowly. It then has greater probability of interacting with the orbital electrons in its path, so that the number of ionizations per unit path length increases. Such an effect with alpha particles was illustrated in Figure 2.8. As an example of this variation, the average specific ionization of the secondary electrons from certain x-rays has been given as 50 ion pairs/micron. The initial and terminal values, however, are approximately 5 and 1000 ion pairs/micron, respectively.

Certain disadvantages exist in the use of specific ionization to compare radiations. The actual number of ion pairs formed in the medium is not known. Moreover, specific ionization emphasizes the process of ionization to the exclusion of the process of excitation. Excitation may play only a small part in the biological effects produced by the radiations, but as one process of energy transfer, it should not be ignored until the full extent of its effect is understood.

The average energy released per unit track length, called the *linear energy transfer* or LET, has become a more common basis for comparison of radiations. The LET is still an average value, but it does include both energy of excitation and energy of ionization and does not imply a knowledge of the number of ionizations produced. Linear energy transfer is usually expressed in units of kev/μ (thousands of electron volts per micron of path length). Some values for linear energy transfer in water of a number of types of radiation are given in Table 2.1.

Table 2.1. LET VALUES FOR SEVERAL TYPES OF RADIATIONS

Radiation	LET (kev/μ)
3 Mev x-ray	0.3
^{60}Co γ radiation (1.2–1.3 Mev)	0.3
250 kev x-ray	3.0
0.6 kev β from ^{3}H	5.5
Recoil p from fission n	45.0
5.3 Mev α from Po	110.0
Fission fragments	4000–9000

[14]Both galactic and solar cosmic rays contain particles which produce very dense ionization. While the spectrum of secondary radiations is very complex, it may be possible in the future to analyze the energy transfer patterns from these particles.

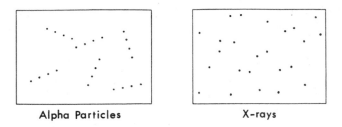

Alpha Particles   X-rays

Fig. 2.17. Distribution of ionization from alpha radiation and x-rays. The total number of ionizations is the same. Ionizations from alpha particles are clustered; those from x-rays are diffusely distributed.

The pattern of ionization from the different radiations must be viewed not only in terms of the distribution along the path of an individual particle but also in terms of the overall distribution of ionization in the medium irradiated. If a block of material is irradiated, for example, by impregnating it with a source of monoenergetic alpha particles, a number of short tracks will be produced, each of which contains a high concentration of ion pairs. A great portion of the medium may have no ionizations occurring within it (see Figure 2.17).

If, instead, a similar block of material is exposed to x-rays and the same number of x-ray photons is absorbed (each of which has the same energy as did each of the alpha particles), the same number of tracks are produced and the same amount of total energy is transferred to the system. Each x-ray-produced track will be longer, however, with a lower concentration of ionizations. Thus, the ionization will be distributed more diffusely over a larger volume.

Almost any biological system can be viewed as made up of blocks or targets in which a certain amount of energy must be absorbed in order to produce the effect being studied. Sometimes these units are very, very small, the size of a single molecule; sometimes they are as large as a cell or even an entire tissue. Sometimes a single ionization will produce the effect that is being studied in the block; sometimes many ionizations are required. If the blocks are large, and if only one ionization is needed to produce the effect, the most "efficient" type of radiation is obviously one of low LET, where the energy is distributed widely and randomly and "hits" several blocks. If, however, the blocks are small and many ionizations are required in each to produce the measured effect, the most effective radiations are those with a high LET and high concentrations of ionization.

Conversely, if a high LET radiation is most efficient in producing a given effect, it may be assumed that a number of ionizations are required within a certain small volume. Information on the relative efficiencies of various radiations can, in this way, contribute toward an understanding of the mechanism of action of ionizing radiation.

GENERAL REFERENCES

Burton, M., Kirby-Smith, J. S., and Magee, J. L. (eds.). *Comparative Effects of Radiation*, Chapters 1, 4, 5. John Wiley & Sons, Inc., New York (1960).

Frisch, D. H., and Thorndike, A. M. *Elementary Particles*. D. Van Nostrand Co., Inc., Princeton, New Jersey (1963).

Glasstone, S. *Sourcebook on Atomic Energy*, 2d ed. D. Van Nostrand Co., Inc., Princeton New Jersey (1958).

Hollaender, A. *Radiation Biology*. Vol. II, Ultraviolet and Related Radiations. McGraw-Hill Book Co., Inc., New York (1955).

Hollaender, A. *Radiation Biology*. Vol. III, Visible and Near-Visible Light. McGraw-Hill Book Co., Inc., New York (1956).

Kinsman, S. (ed.). *Radiological Health Handbook*. U.S. Dept. of Commerce, Office of Technical Services, Washington, D. C. (1960).

Lapp, R. E., and Andrews, H. L. *Nuclear Radiation Physics*, 2d ed. Prentice-Hall, Inc., Englewood Cliffs, New Jersey (1954).

Lea, D. E. *Actions of Radiations on Living Cells*, 2d ed. Cambridge University Press, Cambridge (1955).

Price, W. J. *Nuclear Radiation Detection*. McGraw-Hill Book Co., Inc., New York (1958).

RADIATION DETECTION
AND DOSIMETRY

3

In order to study and understand the effects of ionizing radiation on biological systems, it is necessary to know the total amount of radiation to which a system is exposed and the pattern of dissipation of energy within the system. This chapter will deal with radiation detection, measurement, and dosimetry and with general techniques of irradiation.

RADIATION UNITS

Many different units have been proposed and used for the various radiation quantities. In 1962, the International Commission on Radiological Units and Measurements (ICRU) redefined many of the concepts and quantities of radiation measurement, proposed some new special units, and defined other units more precisely. Some of the units which are in use now are briefly described in Table 3.1.

The roentgen (R) relates to the quantity of ionization produced by x- or gamma radiation. This quantity is called *exposure*. The roentgen is based on ionization produced in air and is defined now as 2.58×10^{-4} coulomb per kilogram of dry air. This value is identical with the old specification of 1 esu of charge per 0.001293 gm, or 1 cc, of dry air. The unit is not.applicable to particulate radiation, such as alphas, betas, or neutrons. Multiples of the roentgen commonly used are the kiloroentgen (kR), representing a thousand R (10^3 R), and the milliroentgen (mR), a thousandth of an R (10^{-3} R). Exposure rate is usually expressed as roentgens per hour (R/hr), roentgens per minute (R/min), roentgens per second (R/sec), etc.

Table 3.1. RADIATION UNITS

R (Roentgen) is a unit of *exposure* of x- or gamma radiation based on the ionization that these radiations produce in air. An exposure of one roentgen results in 2.584×10^{-4} coulomb per kilogram of air, or 1 esu per cc of air at standard temperature and pressure.

Rad is a unit of *absorbed dose* for any ionizing radiation. One rad is 100 ergs absorbed per gram of any substance.

Rem is a unit of *dose equivalent* which is numerically equal to the dose in rads multiplied by appropriate modifying factors such as RBE (or QF) or DF.

RBE (Relative Biological Effectiveness) is a factor expressing the relative effectiveness of radiations with differing linear-energy-transfer (LET) values, in producing a given biological effect. This unit is now limited to use in radiobiology.

QF (Quality Factor) is another name for a linear-energy-transfer dependent factor by which absorbed doses are to be multiplied to account for the varying effectiveness of different radiations. This unit is used for purposes of radiation protection and is similar to the RBE unit used in radiobiology.

DF (Dose Distribution Factor) is a factor expressing the modification of biological effect due to nonuniform distribution of internally deposited radionuclides.

Ci (Curie) is a unit of *activity* of a radioactive nuclide which is equal to 3.7×10^{10} radioactive disintegrations per second.

The rad is a unit of absorbed dose and expresses the energy absorbed per gram of material from any ionizing radiation. One rad is 100 ergs absorbed per gram of any substance or 0.01 joule per kilogram. When water and soft tissues absorb x- or gamma radiation of an energy between 100 kev and 3 Mev, the absorbed dose per roentgen is between 0.93 and 0.98 rad. Therefore, the number of rads is approximately equal to the number of roentgens.[1,2] The rad is used for particulate as well as electromagnetic radiation. Absorbed dose in rads is usually determined indirectly, often from ionization measurements.

[1]The energy absorbed per gram of bone is considerably higher than per gram of fat, muscle, or other soft tissue for low energy radiation (below about 100 kev). An appropriate conversion chart can be found in many sources, such as the *Radiological Health Handbook*.
[2]The rad is a relatively recent term, so that dosage has been expressed in R units in much of the radiobiology literature. In the following chapters, dosages will usually be as given in the original papers, either in rad or R units, without an attempt to convert. In most cases, the terms are almost equivalent, numerically.

The rem is a unit of dose equivalent. It was proposed to make allowances for the fact that the same dose in rads from different types of radiation does not necessarily produce the same degree of biological effect. In consideration of permissible exposure levels where a mixture of radiations has to be considered, allowance for the difference in biological effectiveness is made by the quality factor which relates the effect of other radiations to that of gamma rays from cobalt-60. Some quality factors are listed in Table 3.2. The values

Table 3.2. QUALITY FACTORS

Radiation	QF
X-rays, gamma rays, beta rays and electrons of all energies	1.0
Fast neutrons and protons up to 10 Mev	10
Alpha particles	10

listed are not experimentally established figures but, rather, are estimates of the probable effectiveness of the radiations in producing several types of changes in man. For purposes of radiation protection, the dose equivalent in rems is the absorbed dose in rads multiplied by the quality factor.

In experimental radiobiology, the term relative biological effectiveness (RBE) is used instead of quality factor to express the relative effectiveness of various radiations. The RBE values are usually experimentally derived values which pertain only to the system under study (see Chapter 11).

The dose distribution factor (DF) is used in the calculation of the rem when internally deposited radionuclides are considered. Nonuniform body distribution of a given radionuclide may result in a high concentration in certain organs. The dose distribution factor takes this into account.

The curie (Ci) is a unit of radioactivity and indicates the rate at which the atoms of a radioactive substance are decaying. Originally defined as the activity of one gram of radium, it has been redefined arbitrarily so that one curie is 3.7×10^{10} disintegrations per second. Smaller units have been defined such as the millicurie (mCi) which is one-thousandth (10^{-3}) part of a curie, the microcurie (μCi) which is a millionth (10^{-6}) of a curie, the nanocurie (nCi) which is a thousandth of a millionth (10^{-9}) of a curie, and the picocurie (pCi) which is a millionth of a millionth (10^{-12}) of a curie. These correspond to activities of 3.7×10^7, 3.7×10^4, 37, and 3.7×10^{-2} disintegrations per second, respectively. A conversion which is often used is the fact that 1 μCi $= 2.2 \times 10^6$ dpm (disintegrations per minute).

The specific activity of a radioactive substance can be defined as the number of curies per gram of the element present, where the weight includes both stable and radioactive isotopes of the element.

RELATIONSHIP BETWEEN
ACTIVITY AND ABSORBED DOSE

EXTERNAL SOURCE

An external source of radiation is one that is not contained within the material being exposed. Alpha or beta rays are not ordinarily used for external radiation because of their relatively short path lengths in air. Gamma rays have a long path length in air and are usually suitable. X-rays are also a common source of external radiation. It is possible to estimate the radiation exposure from gamma rays with energies between about 0.3 and 3.0 Mev by the following equation:

$$R_f = 6 \text{ Ci } E \tag{3-1}$$

where R_f is the exposure rate, R/hr at 1 ft.
    Ci is the number of curies of activity.
    E is the average gamma energy per disintegration, Mev.

If more than one gamma ray is emitted per disintegration, this must be taken into consideration as follows. If the radiations are emitted in cascade, the two energy values are added to give the value of E; for cobalt-60, $E = 1.2 + 1.3 = 2.5$ Mev. If there are several modes of disintegration, then the weighted fractions of the energy values are used.

As an example of this method for estimating the radiation exposure, one can approximate the exposure rate 1 foot from a 10 curie source of cobalt-60 as follows:

$$R_f = 6 \text{ Ci } E = 6 \times 10 \times 2.5 = 150 \text{ R/hr} \tag{3-2}$$

Another relationship that can also be used with gamma sources is the inverse square law, which states that the radiation intensity varies inversely as the square of the distance from the source. Thus, an exposure rate of 150 R/hr at 1 foot would be reduced to $150 \times 1^2/3^2$ which is approximately 17 R/hr at 3 feet.

INTERNAL SOURCE

The absorbed dose can be estimated for the radiation from a source that is contained within a system. Radionuclides which emit alpha or beta rays are frequently used as internal emitters. Because of the short path lengths of these radiations in tissue (usually 1 cm or less), it can usually be assumed that most of the energy has been absorbed within the specimen. The general formula for the dose rate is:

$$
\begin{aligned}
&D \text{ (rads/day)} \\
&= \frac{A\ (\mu Ci) \times 2.2 \times 10^6 (dpm/\mu Ci) \times 1440 (min/day) \times \bar{E}(Mev)}{W\ (gm) \times 6.24 \times 10^5 (Mev/erg) \times 100 (ergs/gm/rad)}
\end{aligned}
\tag{3-3}
$$

where A is the total activity in the system or portion of the system (such as a specific tissue).

W is the weight of the system or portion in which the activity A is distributed.

D is the average dose rate in the system (or part of system) of weight W, containing activity A. This factor does not take into account any nonuniform distribution.

$\bar{E}$ is the average energy per disintegration. This is equal to the total energy for an alpha emitter or one-third the maximum energy for a beta emitter.

The total dose which is absorbed over a period of time can then be calculated from the initial dose rate if the effective half-life of the radionuclide is known. The dose accumulated up to a given time (t) after the introduction of the nuclide is:

$$
\begin{aligned}
\text{Dose (rads)} &= 1.44 \times T_E \text{ (days)} \times D \text{ (initial rads/day)} \\
&\times (1 - e^{-0.693 t/TE})
\end{aligned}
\tag{3-4}
$$

If time, t, is very great relative to the effective half-life, the total absorbed dose is:

$$
\text{Dose (rads)} = 1.44 \times T_E \text{ (days)} \times D \text{ (initial rads/day)}
\tag{3-5}
$$

Effective half-life includes both physical half-life and biological half-life (time to reduce activity to half due to loss by biological processes, such as excretion). Effective half-life is given by the formula:

$$
T_E = \frac{T_p \times T_b}{T_p + T_b}
\tag{3-6}
$$

where T_p is physical half-life and T_b is biological half-life.

If either the time considered (t) or the effective half-life (T_E) is short, appropriate factors can be introduced into formulas 3-3, 3-4, and 3-5 to give dose rates per hour or per minute and the total dose over a relatively short period of time.

RADIATION DETECTION
AND MEASUREMENT

There are many different methods for detecting and measuring radiation. Some detectors, called *ionization chambers*, operate as a result of the ionization which is produced in them by the passage of charged particles. Excitation of atoms by ionizing radiation, and the resulting luminescence, form the basis for instruments such as scintillation detectors and luminescent dosimeters. Certain chemical reactions can be quantitatively related to the amount of radiation received by the system, and heat production from radiation can be measured by calorimetric means. Latent images in photographic film and visible tracks in cloud or bubble chambers mark the passage of an ionizing particle or photon. Each of these detection methods will be considered briefly. Emphasis will be primarily on those detectors which are used in radiobiology for measuring radiation distribution and exposure.

IONIZATION CHAMBERS

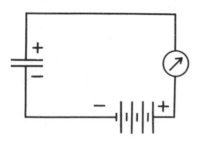

Fig. 3.1. Simplified circuit of an ionization detector (see text).

The most common method of measuring radiation exposure is with an ionization chamber. For this reason, this method will be considered in greatest detail. All ionization chambers operate by collecting the ionization which is produced in them by radiation. They can be designed in many sizes and shapes and can be used to measure all types of radiation that produce ionization.

The principle of an ionization chamber can be illustrated by reference to the simplified circuit shown in Figure 3.1. The circuit consists of two parallel plates or electrodes across which a potential can be impressed by a battery. When radiation produces ionization in the space between the plates, each component of the ion pair is attracted to the plate of opposite sign; that is, the electrons move to the positive plate (anode); the positive ions move to the negative plate (cathode). A small current results in the system. This can be read on a meter.

The number of ions collected and the resultant current flow (for a given set of electrodes and a specified gas) depend on the amount of ionization produced by the radiation and the potential between the plates. These relationships are illustrated in Figure 3.2. Notice that the ordinate is the number of ions collected when a *single* ionizing particle or photon passes through the chamber.

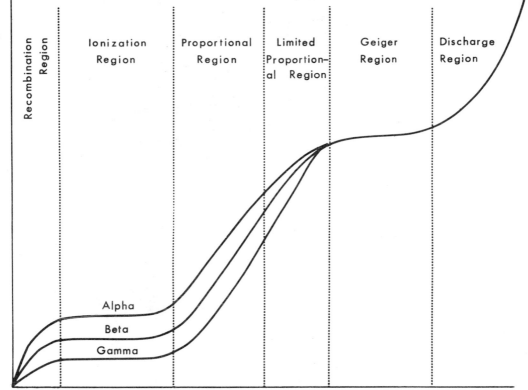

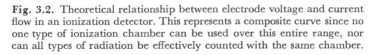

Fig. 3.2. Theoretical relationship between electrode voltage and current flow in an ionization detector. This represents a composite curve since no one type of ionization chamber can be used over this entire range, nor can all types of radiation be effectively counted with the same chamber.

If no potential is applied to the electrodes, recombination of the electrons and positive ions occurs, no ions are collected on the plates, and no current flows. If a small potential is applied, some of the electrons are attracted to the anode (positive) and some of the positive ions are attracted to the cathode (negative). Those electrons and ions which do not reach the electrode recombine to form neutral atoms. As the potential difference of the electrode is increased, the number of electrons and ions reaching the plates increases. At some potential, all of the electrons and ions reach the electrodes; no recombination takes place. Small additions of potential do not cause a further increase in the number of ions collected as all of the electrons and ions are already reaching the plates. This is the ionization region.

Because alpha particles lose their energy at a higher rate (higher LET) than do beta particles (lower LET), a larger portion of the total path of an

Fig. 3.3. Victoreen Model 570 Condenser R-meter and chamber. (Courtesy Victoreen Instrument Co.)

alpha will be between the plates than will that of a beta. The number of ion pairs produced between the plates by an alpha particle is, therefore, greater than the number produced by a beta particle or gamma photon, and the number of ions collected is greatest for an alpha particle.

As the potential across the electrodes is further increased, there is an increase in the number of ions collected as a result of the formation of secondary ion pairs. The primary electrons are accelerated with sufficient energy to ionize other atoms and thus produce additional secondary electrons and positive ions to be collected on the plates. This is called *gas amplification*. The number of secondary ionizations, and, therefore, the number of collected ions, is proportional to the number of primary ion pairs produced by the radiation. For this reason, the region is called the *proportional region*. Alpha particles can still be distinguished from beta particles or gamma photons, since the total number of ions is still proportional to the number of primary ionizations, although the numbers have been increased by many orders of magnitude. The amplification factor is the same for each type of radiation.

Gas amplification reaches a limit with further increase in voltage. This limit is regulated by the total number of ionizations (primary and secondary) which can be produced between the plates. Beyond this limit, all radiations, regardless of the number of primary ionizations, will yield the same maximum number of ions collected. In this voltage range, called the *Geiger region*, it is not possible to distinguish between types of radiation. At extremely high voltages, continuous discharge will occur across the plates. In this region, continuous ion collection is not related to the passage of radiation.

There are several ways to measure the amount of ionization which is produced in a chamber. Ions are collected on the electrodes in bursts or pulses, corresponding to the ionization by successive particles or photons. In an integrating circuit, pulses are smoothed out, and the recording instrument will measure the average voltage change or current flow through the system. In other so-called pulse-type circuits, each pulse is registered separately, and the total *number* of pulses is recorded, thereby indicating the number of ionizing radiations entering the detector. Chambers operating in the ionization range may be operated either as pulse-type or integrating instruments. Counters operating in the proportional or Geiger regions are usually pulse-type.

In most radiobiology laboratories, x- or gamma-ray exposures are measured with chambers operating in the ionization region. The condenser R-meter (Figure 3.3) is extensively used for such measurements. It consists, essentially, of a thimble-type ionization chamber and a separate charger-charge-reader unit containing an electrometer (see Figure 3.4).

The chamber contains a well-insulated, thimble-shaped ionization chamber connected in parallel with a fixed capacitor. The chamber electrodes consist of a central rod and the conducting layer on the inner surface of the bakelite covering of the thimble. In use, the chamber is attached to the charger unit and is charged to a known voltage. It is then disconnected from the charger and exposed to x- or gamma radiation for a measured time. The chamber is partially discharged by the ionization produced by the radiation. It is reinserted into the unit and the loss of charge read directly in

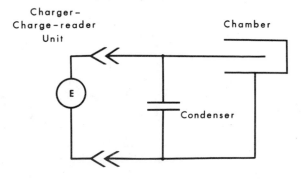

Charger–
Charge-reader
Unit

Chamber

E

Condenser

Fig. 3.4. Schematic diagram of condenser R-meter chamber.

roentgen units. The exposure rate can be determined by dividing the total exposure by the time during which the chamber was exposed to the radiation. By changing the size and shape of the ionization chamber and the composition of the thimble walls, chambers of this type can be calibrated and used to measure different total doses or different energies of x- or gamma rays. Pencil-shaped chambers of this type are extensively used as personnel dosimeters. They can be readily carried in the pocket by workers and visitors in a radiation area.

Similar ionization chambers can be used with devices to measure the *average* rate of ionization rather than the total ionization. Such instruments are widely used for dose-rate meters, giving readings directly in roentgens per hour or milliroentgens per hour.

Pulse-type chambers are used for counting individual ionizing particles or photons. They may be made to operate in any region (see Figure 3.2) but are mostly used in the Geiger region to provide maximum pulse size per ionizing particle and minimum dependence on voltage stability. (Note that small variations in voltage in the Geiger region in Figure 3.2 have very little effect on the numbers of ions collected.) The Geiger-Müller, Geiger, or G-M counter has wide application in tracer experiments and is sometimes utilized in radiobiological studies with internal emitters.

Proportional region pulse-type counters require a very stable voltage and are not usually used in radiobiology for measuring radiation exposures. Special types of proportional counters have been designed to detect thermal or fast neutrons and can, therefore, be used for neutron dosimetry.

LUMINESCENT DOSIMETERS

Several types of luminescent dosimeters have been developed in recent years. They are based on the idea that certain substances, which are not normally luminescent when treated in certain ways, can be made to give off light when so treated by prior exposure to ionizing radiation.

In a typical photoluminescent system, silver-containing metaphosphate glass is exposed to radiation. Some of the loosely bound electrons which are released from the negative ions of the glass are trapped by silver ions (Ag^+) to form metastable silver atoms (Ag°). When the silver atoms are exposed to ultraviolet light, they absorb energy and fluoresce upon returning to the stable ground state. One can measure either the shift in the ultraviolet absorption spectrum as a result of the presence of metastable silver atoms or the emission spectrum when the atoms return to ground state. With proper calibration, the radiation exposure can be determined from either of these measurements.

Thermoluminescent dosimetry is based on a similar process. When a crystalline material composed of a phosphor, lithium fluoride (LiF), is exposed to ionizing radiation, electrons will be excited into higher energy levels. They then fall into "electron traps" and a metastable state results. When energy is applied in the form of heat, the trapped electrons return to the ground state with an emission of light. The radiation exposure can be determined from the intensity of emitted light.

A commonly used glass rod dosimeter (photoluminescent system) consists of a small rod 1 mm in diameter and 6 mm long. The thermoluminescent LiF powder can be incorporated into Teflon and then formed into small rods or discs. The small size of each of these dosimeters makes them very valuable for many applications, particularly in tissue implantation. By incorporation of a number of the rods on or within an irradiated specimen, the exposure to various parts or organs can be estimated.

SCINTILLATION DETECTORS

When an ionizing particle or photon passes through a suitable crystal or liquid phosphor, energy is absorbed, resulting in excitation of electrons in the scintillator. The energy is re-emitted as flashes of light or scintillations which are absorbed by the photocathode of a photomultiplier tube. Photoelectrons are emitted and the number of electrons is amplified by the photomultiplier tube. A current pulse is produced. The size of the pulse is proportional to the number of scintillations produced, and thus to the energy lost by the particle or photon in the scintillator. This detector can, in this way, measure the energy distribution of the incident radiations as well as counting them. It can distinguish, for example, between the beta particles from tritium ($E_{max} = 0.0189$ Mev) and the beta particles from carbon-14 ($E_{max} = 0.155$ Mev) when a mixture containing these two radionuclides is counted. The multichannel analyzers now in use can separately record the number of particles or photons in each of many energy intervals. These counters are not generally used for radiation dosimetry.

CALORIMETERS

A calorimeter is a thermal, or heat measuring, device. It can be used for measurement of the energy absorbed from radiation, since ultimately the absorbed energy is degraded into heat.

Practically, the method is not very sensitive since the rate of heat production from most sources of radiation is very small. It is suitable for measuring radiation sources of high activity, for determining the energy of particles produced by particle accelerators, and for measurement of the complex radiations from reactors.

CHEMICAL DOSIMETERS

Details of many of the radiochemical reactions are not clearly understood. However, a number of reactions can be calibrated and used as a means of detecting and measuring a radiation exposure. The ferrous-ferric system, often called the *Fricke Dosimeter*, is one of these. A well-oxygenated solution of pure ferrous ammonium sulfate (10^{-3} M) in dilute sulfuric acid (0.8 M) is exposed to radiation. Ferrous ions are oxidized to ferric ions in an amount which is dependent on the total amount of energy deposited in the system. The quantity of ferric ions produced is then determined in a UV spectrophotometer. The most suitable range for the dosimeter is between 5000 and 50,000 rads, which makes the Fricke dosimeter useful above the range of doses of interest for most biological effects.

Another interesting type of chemical dosimeter has been developed for the estimation of the distribution of the radiation exposure within a biological specimen. The Potsaid-Irie Chemical Dosimeter is a combination of paraffin wax with chloroform-methyl yellow. The resultant mixture has a density close to that of tissue and can be molded into any desired shape, such as a head or hand. During exposure to radiation, its color changes from yellow to red in proportion to the amount of ionizing radiation absorbed. The color change is believed to result from a linkage between nitrogen atoms in the dye and chloride atoms in the chloroform. Radiation supplies the energy for this reaction. The sensitivity of the Potsaid-Irie Dosimeter is low. One can barely distinguish visibly between the color changes from exposures differing by about a thousand roentgens. Its chief value lies in the three-dimensional color picture of the distribution of the radiation. When high radiation exposures are used to visualize this distribution, it can be assumed that the same distribution is produced by lower radiation exposures.

PHOTOGRAPHIC FILM

Photographic film was one of the first radiation detectors and is still widely used for personnel monitoring and for visualization of gross or cellular localization of certain radionuclides. Two types of film are used primarily: film which can be calibrated for exposure in terms of gross blackening, and nuclear emulsions in which the tracks of ionizing particles can be visualized microscopically.

Small crystals of a silver halide (usually silver bromide) are suspended in a gelatin base. Ionizing radiations passing through the film produce free electrons. These electrons are trapped in the crystal lattice structure of the film and reduce the silver ions. When the film is developed, silver bromide grains which contain reduced silver ions are converted to grains of metallic silver; undeveloped silver salts are removed by the fixer. If there are a

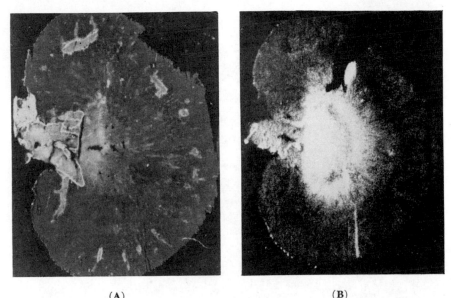

(A)   (B)

Fig. 3.5. (A) Section of kidney from rat which had received sodium-22. (B) Autoradiogram prepared from adjacent kidney section, using non-screen x-ray film. ((A) Courtesy R. H. Wasserman. (B) Courtesy Mercer [1964] and AAAS)

sufficient number of grains developed, the film will appear to be grossly blackened. The degree of blackening can be quantitated and with appropriate calibration will give an estimate of the radiation exposure. With special types of film, it is possible to detect and quantitate fast neutrons, thermal neutrons, electrons, alpha particles, and gamma or x-rays.

Film can also be used to study the localization of radionuclides in tissues, in the process known as *autoradiography* (literally self-radiography). If a tissue from a plant or animal containing a radionuclide is placed next to a film, the radiation from decaying atoms of the radionuclide will expose the film. When the film is developed, the dark (exposed) areas of film will correspond to tissue areas of greatest concentration of the radionuclide. Figure 3.5 illustrates this type of autoradiogram. On the left (A) is a photomicrograph of a tissue section from the kidney of a rat which had been given radioactive sodium-22. The photograph at the right (B) is an autoradiogram of an adjacent kidney section which had been placed next to the film for several weeks. The film was exposed by radiation given off by the radioactive decay of atoms of sodium-22 in the tissue. When the tissue section was removed and the film developed, the pattern of exposed grains corresponded to the distribution of sodium-22 in the kidney. If the two photographs are compared, it is apparent that the sodium-22 has localized deep in the inner

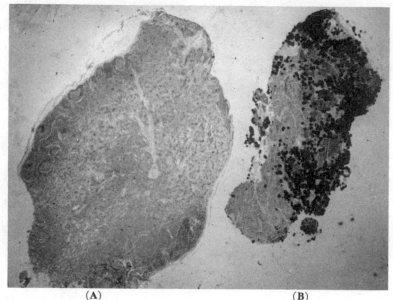

(A)   (B)

Fig. 3.6. Autoradiogram of two pulmonary lymph nodes from a dog exposed to plutonium-239 (an alpha emitter). Both the stained tissue and the superimposed developed emulsion are shown. The black spots are clusters of tracks (developed silver grains) in the emulsion from the alpha particles emitted by the plutonium-239. Notice how much more plutonium is contained in node B. (From Morrow and Casarett [1961], courtesy L. Casarett and Pergamon Press)

medulla and at the junction between the inner and outer medulla of the kidney.

Autoradiograms can also be prepared without separating the tissue and film after exposure. This technique is used for a more precise intraorgan or intracellular localization of the radionuclides, especially if a source of alpha or low-energy beta radiation is used. A special type of fine grain emulsion is placed over a microscope slide which has on it a tissue section containing a radionuclide. After a suitable exposure time, the emulsion is developed and the tissue stained, without separating the emulsion and tissue. By proper focusing with a microscope, the pattern of developed grains in the emulsion can be correlated with tissue structures.

Figure 3.6 shows an autoradiogram of two pulmonary lymph nodes from a dog given plutonium-239 (an alpha emitter). Notice that the right node contains a great deal of plutonium; the left node apparently contains very little. The difference in plutonium content in the two nodes illustrates the nonuniform distribution which can occur with internal emitters, even in tissues of the same type. In such a situation one would expect that the effects of the alpha irradiation would be much greater in the node on the right.

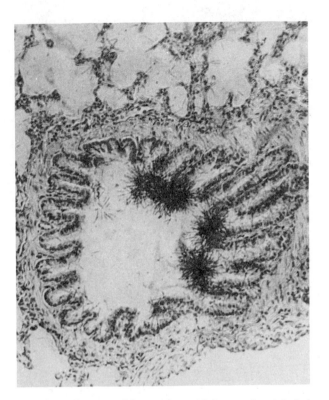

Fig. 3.7. Autoradiogram of lung of rat 12 hours after inhalation of polonium-210, an alpha emitter, showing aggregation of polonium colloids at the surface of the bronchiole. (Courtesy of L. Casarett [1964] and Academic Press, Inc.)

Figure 3.7 is a photomicrograph of an autoradiogram of the lung of a rat given polonium-210. The many tracks are clearly clustered on the surface of the bronchiole, giving a considerable radiation exposure to the bronchiolar surface. Figure 3.8 is a photomicrograph at a higher power of another lung autoradiogram. This shows many alpha particle tracks (in the emulsion) originating from a lung macrophage (in the tissue). Other tracks come from other portions of the tissue.

Figure 3.9 shows the association of exposed grains with the cells and chromosomes of human white blood cells in a preparation containing tritium (^{3}H), a low-energy beta particle emitter. A single grain corresponds to the passage of a beta particle in the emulsion, in contrast to the track of grains produced by the alpha particles from polonium-210. This is largely a result of the lower linear-energy-transfer rate of the beta particle, as well as the energy of the beta radiation.

The autoradiographic technique is valuable in radiobiology primarily to determine the intraorgan and intracellular distribution of internal emit-

Fig. 3.8. Alpha particle tracks originating from a lung macrophage in rabbit given polonium-210. The cells are slightly out of focus, since the emulsion is at a different level than the tissue. (Courtesy L. Casarett [1964] and Academic Press, Inc.)

ters. It often helps the radiobiologist to interpret puzzling tissue changes in light of the nonuniform radiation exposure received by tissues. In addition, the tissue distribution of small quantities of radionuclides ("tracer levels") can be visualized. It can be assumed that this distribution is the same as the distribution of the stable isotopes of the same elements. This technique makes it possible to predict the behavior of stable elements under a variety of experimental treatments. Such information contributes to our knowledge of normal biological processes.

CLOUD CHAMBERS AND BUBBLE CHAMBERS

Cloud and bubble chambers are both used to visualize the passage of an ionizing particle or photon. The operation of a cloud chamber depends on

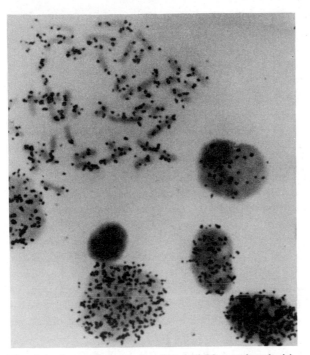

Fig. 3.9. Autoradiogram showing variable uptake of tritiated (^{3}H) thymidine in human white blood cells in tissue
culture. One cell contains almost no tritium. Others have
incorporated tritium during DNA replication. The individual chromosomes are from a cell which was in division when
the smear was prepared. Note the association of grains with
chromosomes. (Courtesy R. Herzog)

the presence of a supersaturated region in a gas which contains a condensable
vapor. In the absence of centers, such as dust particles or ions, on which
condensation can occur, the supersaturated condition can exist with practically no condensation occurring. When centers occur, vapor condenses on
them, producing visible spots. The path of an ionizing radiation shows up in
the supersaturated region as a vapor trail of ions on which condensation has
occurred.

In a bubble chamber, ions formed by the passage of ionizing radiations
through a superheated liquid create condensation nuclei for the formation of
bubbles. If the chamber is filled with a high-density liquid, even a highly
energetic particle has a limited range. For this reason, the bubble chamber
has its chief use with high-energy particle accelerators.

Neither the cloud chamber nor the bubble chamber are particularly
useful for detecting and measuring the radiations which are of primary interest in radiobiology.

DOSIMETRY

Most of the dosimeters which have been described are used to detect the presence of radiation or to measure the radiation intensity either at a particular spot or in some small volume. A series of these measurements can be used to determine the distribution of the radiation exposure within an irradiated system. Consider, for example, the radiation exposure to various tissues of an animal. If an internal emitter (with short path length) is used as a radiation source, various organs can be removed and "counted" individually with a Geiger or scintillation system. Intraorgan distribution can be determined by autoradiography of the tissues. Microdosimeters of the glass rod or LiF types can be implanted in the tissues to measure exposure from either internal or external sources.

The ionization chambers, such as the R-meter discussed in a previous section, p. 39, are commonly used for dosimetry when the radiation source is external. They are accurate but have the disadvantage of being inconvenient to use with biological specimens. Because of their relatively large size, they are not suitable for insertion into small animals. In addition, they should not be placed in direct contact with tissue and thus require a protective cover.

Partially as a result of some of the difficulties with measurements in living tissues, "phantom" measurements are commonly used to measure internal doses. Water-filled or masonite blocks (phantoms) are used which have holes at appropriate spots into which the chambers can be placed. The exposures at various locations are determined. The values so obtained can be used to estimate tissue dosage since water and masonite have about the same density as tissue. The following discussion will draw mainly from data obtained from x-rays. Most of the considerations apply equally to other sources of external radiation.

X-RAY DOSIMETRY

Many factors influence the radiation dose which a tissue receives. These include the depth of the tissue below the skin surface, the thickness and type of material below the tissue, the area of the radiation field, the quality of the radiation (HVL), and the distance from the radiation source to the skin surface. No attempt will be made to discuss these factors in detail, but brief mention will be made of how they influence the final radiation exposure.

A radiation exposure from an external source is usually measured with a dosimeter positioned in air at some specified location. This is known as the *air dose*. When a specimen or phantom is introduced into the beam, so that

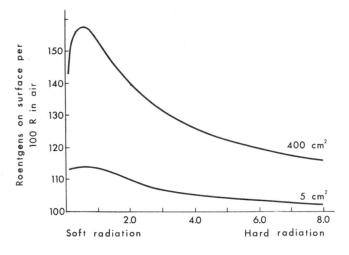

Fig. 3.10. Surface dose relative to air dose for a wide range of qualities of x-rays. Surface areas of 5 cm² (highly columnated beam) and 400 cm² (broad beam) are shown to illustrate the greater scatter with large field size. (Modified from Glasser *et al.* [1961], courtesy Paul B. Hoeber, Inc.)

the surface is at this same position, the exposure at the surface is usually called the *surface dose*, or *skin dose*. The exposure at some location within the specimen is called the *depth dose* at that site. Skin dose can be determined by placing a dosimeter on the surface of the skin, or it can be calculated from the air dose if the characteristics of the radiation are known. Depth dose can be measured with implanted dosimeters, or it can be determined from either the air dose or skin dose with the use of experimentally determined depth-dose tables or charts.(See Fig. 3.11).

The surface dose is greater than the air dose by the amount of radiation which is scattered back from the underlying material. The amount of this backscatter depends on the quality of the radiation, the size of the field, and the depth of the underlying material. Figure 3.10 illustrates the first two of these factors. Consider first the lower curve (5 cm² field), which uses a highly columnated beam for irradiation of a small area. With low-energy (soft) radiation (low half-value layer), there is considerable backscatter, and the skin dose relative to the air dose is maximum. With harder radiation (higher HVL), there is a greater tendency for scatter to be in a generally forward direction, therefore the skin dose decreases. The maximum backscatter is not found with the softest radiation, since the absorption of both the primary and secondary radiation reduces the amount which can get back to the surface. Radiations with half-value layers of about 0.5 mm copper give maximum backscatter.

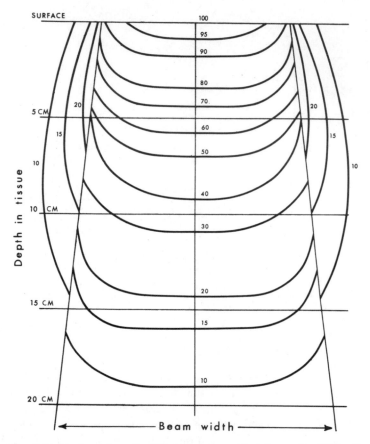

Fig. 3.11. Isodose chart for 200 kvp x-rays with 1.5 mm Cu HVL, 10 cm × 10 cm field, 70 cm TSD. (Modified from Glasser *et al.* [1961], courtesy Paul B. Hoeber, Inc.)

A comparison of the upper and lower curves in Figure 3.10 illustrates the effect of field size on surface dose. A larger field results in a higher surface dose due to "sidescatter." Notice that the maximum surface dose still occurs at a half-value layer of about 0.5 mm copper. Since much of the experimentation in radiobiology involves total-body or large-volume irradiation, the large-field size data may be more applicable to the radiobiologist.

At points within the specimen, the intensity differs from that at the surface because of absorption and scattering within the tissues and because of increased distance from the source. By making measurements at different locations in the specimen or phantom, a chart can be constructed in which points of equal dose can be joined. Such an isodose chart is shown in Figure 3.11. Notice that doses are expressed as a percentage of the surface dose.

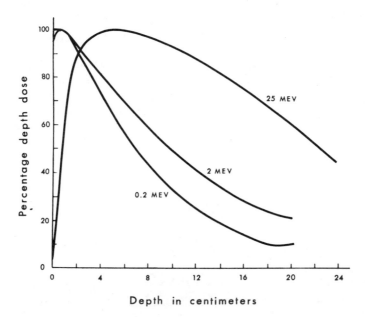

Fig. 3.12. Central axis dose as a function of depth in water for peak x-ray energies of 0.2 Mev ("typical" x-ray machine), 2 Mev (Van de Graaff), and 25 Mev (Betatron). (Modified from Behrens [1959], courtesy Williams & Wilkins Co.)

The dose is greatest along the central axis of the beam and decreases toward the boundaries. Scatter by the tissue within the beam results in a small dose received by tissue outside of the beam. The isodose chart in Figure 3.11 is only applicable for the given characteristics — that is, 200 kvp (peak, or maximum energy in kilovolts) filtered so as to give a HVL (half-value layer) of 1.5 mm copper, with a field size (on the surface of the tissue) of 10 × 10 cm, and with a 70 cm TSD (Target of x-ray machine to Surface of specimen Distance). Isodose charts can be constructed for any desired combination of characteristics and are commonly used in radiation therapy to estimate depth doses.

The depth-dose distributions are markedly influenced by the energy of the incident radiation. Figure 3.12 illustrates the depth doses in water along the central axis of beams having peak x-ray energies of 0.2 Mev, 2 Mev, and 25 Mev. The two lower energy beams have a maximum dose at or very near the surface of the water; the 25 Mev beam has a maximum intensity about 5 cm below the surface. The dose from the 25 Mev beam is still appreciable at a distance of 20 cm below the surface, but the dose at that point is very low from the other two beams. When high-energy beams of this type become more generally available, their greater penetrability will make them useful tools, especially in radiation therapy, for irradiation of internal tissues.

GAMMA RAY DOSIMETRY

A beam of gamma rays differs from a beam of x-rays primarily because the gamma rays are monoenergetic or, in some cases, consist of photons with a limited number of different energies. Without the "soft" radiation component of most x-ray beams, the gamma rays have less surface backscatter and greater penetration. Phantom measurements can be made with gamma radiation, and isodose charts compiled in the same way as was described for x-rays.

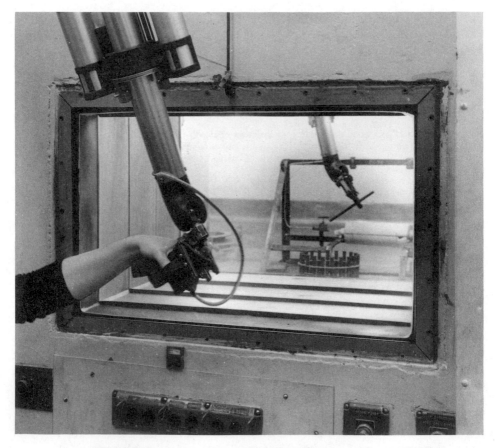

Fig. 3.13. Gamma irradiation cell at Cornell University Reactor Facility. The cell is designed for the use of sealed cobalt-60 sources up to a total of 10 kilocuries in strength. The concrete walls and glass viewing window are both $3\frac{1}{2}$ feet thick. Sources and experimental material inside the cell are manipulated by remote control from outside, as shown. (Cornell University)

Fig. 3.14. Total-body exposure of rats to 250 kvp x-rays. The lower portion of the x-ray machine is shown at the top of the photograph. A Victoreen R-chamber placed in one section of the rotating lucite wheel measures the exposure rate. Rats are placed in the other sections. (University of Rochester)

EXPOSURE OF
BIOLOGICAL SYSTEMS

When a specimen is to be irradiated, many factors must be considered. The choice of radiation source depends on the total dose required, the desired distribution of dose of the radiation within the system, the type of radiation (alpha, beta, x-rays, etc.), and the dose rate required. The possibilities for exposure set-ups are endless. A few typical experimental arrangements are described in this section.

Fig. 3.15. Exposure of legs of rats to x-rays. Anesthetized rats are covered by a lead shield with only the legs exposed. The lower portion of the x-ray machine is at the top of the photograph. (University of Rochester)

The gamma irradiation cell shown in Figure 3.13 is suitable for irradiation of relatively small specimens when a high exposure rate is required. Chemical solutions, microorganisms, or small seeds are positioned in the cell. Using remote control manual manipulators, an operator working from outside the cell removes "pencils" containing cobalt-60 from a shielded "well" which is in one corner of the room. The pencils are placed at a predetermined distance from the specimen and left for the time required to give the desired exposure. The exposure rate in air can be calculated from the activity of the pencils or can be experimentally determined with a dosimeter in the specimen position.

Another experimental arrangement is shown in Figure 3.14. This shows rats receiving a total-body radiation exposure from a 250 kvp x-ray machine. The rats are placed in wedge-shaped sections of a rotating lucite wheel in order to assure equal exposure to each animal. The exposure rate in air is determined with a Victoreen R-chamber placed in one of the sections of the wheel. The same arrangement can be used with other rodents. Larger animals, such as rabbits, can be exposed individually. An x-ray machine is also suitable for partial-body or localized irradiation.

Figure 3.15 shows a method for irradiating only the legs of rats, using the same x-ray machine as in Figure 3.14. A lead shield is designed to cover all tissue except the legs, and to fit between the legs to minimize back and side scatter. Lead shields can be constructed to localize the radiation to almost any area of a specimen.

The illustration in Figure 3.16 is of a gamma field at Cornell University, which is used to irradiate large animals, such as cows or sheep. Cesium-137 and cobalt-60 sources are kept at the bottom of the round, water-filled moat. Any one or combination of the sources can be raised by remote control to a suitable height above the floor for irradiation. Usually eight cobalt-60 sources are used to irradiate a cow. They are so positioned in the circle that a reasonably uniform total-body dose is delivered to the cow when it is restrained in the center of the island. Smaller animals, such as goats, are often placed in a sling between two sources and irradiated simultaneously from both sides. This geometric arrangement achieves a reasonably uniform distribution of dose at a higher dose rate (closer to sources) than is available in the center of the island.

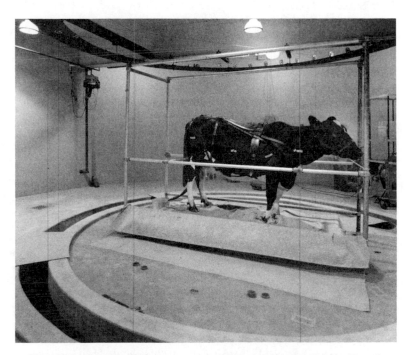

Fig. 3.16. Gamma field for exposure of large animals. Cobalt-60 and cesium-137 sources are stored in the circular water-filled moat. They are raised by remote control during the exposure period. (Cornell University)

Several arrangements for chronic irradiation of plants are shown in Chapter 13. These sources are used at Brookhaven National Laboratory for irradiation of a natural pine-oak forest (Figure 13.16) and of flowering plants (Figure 13.7). Obviously, all of these radiation sources can be used to expose many types of biological material.

Many other equally good set-ups are in use in radiobiology laboratories throughout the world. The examples shown were chosen to illustrate a few of the experimental arrangements that are presently in use.

GENERAL REFERENCES

Crafts, A. S., and Yamaguchi, S. *The Autoradiography of Plant Materials* (Manual 35). University of California Press, Berkeley (1964).

Glasser, O., Quimby, E. H., Taylor, L. S., Weatherwax, J. L., and Morgan, R. H. *Physical Foundations of Radiology*, 3d ed. Paul B. Hoeber, Inc., New York (1961).

Hine, G. J., and Brownell, G. L. (eds.). *Radiation Dosimetry*. Academic Press, Inc., New York (1956).

Johns, H. E. *The Physics of Radiology*, 2d ed. Charles C Thomas, Springfield, Illinois (1961).

Kinsman, S. (ed.). *Radiological Health Handbook*. U.S. Dept. of Commerce, Office of Technical Services, Washington, D. C. (1960).

Price, W. J. *Nuclear Radiation Detection*. McGraw-Hill Book Co., Inc., New York (1958).

ADDITIONAL REFERENCES CITED

Behrens, C. F. (ed.). *Atomic Medicine*, 3d ed. Williams & Wilkins Co., Baltimore (1959).

Casarett, L. J. Distribution & Excretion of Polonium-210. XII. Autoradiographic Observations After Inhalation of Polonium-210 in Rats, pp. 187–204 in Metabolism and Biological Effects of an Alpha Particle Emitter, Polonium-210. Stannard, J. N., and Casarett, G. W. (eds.). *Radiation Res.*, Suppl. 5 (1964).

Casarett, L. J., and Morrow, P. E. Distribution and Excretion of Polonium-210. XI. Autoradiographic Studies after Intratracheal Administration in the Rabbit, pp. 175–186 in Metabolism and Biological Effects of an Alpha Particle Emitter, Polonium-210. Stannard, J. N., and Casarett, G. W. (eds.). *Radiation Res.*, Suppl. 5, (1964).

International Commission on Radiological Units and Measurements (ICRU), Report 10a, *Radiation Quantities and Units*. Handbook 84, National Bureau of Standards, U.S. Dept. of Commerce, Washington, D. C. (1962).

Mercer, P. F., and Wasserman, R. H. Autoradiographic Distribution of Radioactive Sodium in Rat Kidney. *Science*, **143:** 695–696 (1964).

Morrow, P. E., and Casarett, L. J. An Experimental Study of the Deposition and Retention of Plutonium-239 Dioxide Aerosol. In Davies, C. N. (ed.), *Inhaled Particles and Vapours*. Pergamon Press, New York (1961).

RADIATION
CHEMISTRY

GENERAL RADIATION CHEMISTRY

The ionization and excitation of molecules by radiation has been described in the preceding chapters. This chapter will be concerned with the molecules that are ionized, with the changes produced in them as a result of the ionization, and with the changes they produce, secondarily, in other atoms and molecules. The interpretation of chemical effects in terms of the physical events (ionization, excitation) represents the conclusions that have been drawn in the light of our present knowledge. These conclusions may be altered when more is known.

In general, all types of ionizing radiations produce qualitatively similar effects. Throughout this chapter, x- or gamma rays will be implied as the radiation source, unless otherwise stated.

DIRECT VS. INDIRECT EFFECT

It is important, first, to distinguish between a molecule which has received energy directly from the incident radiation — that is, has been ionized or excited by it — and a molecule which has received the energy by transfer from another molecule. These two processes are referred to as the direct effect of radiation and the indirect effect of radiation, respectively. The indirect effect is especially important in aqueous systems, where a water molecule may be ionized and then may transfer its acquired energy to another molecule. The change in the latter is *indirect*.

IONIZATION AND EXCITATION

The two major mechanisms for energy transfer, as has been indicated, are ionization and excitation. In the former, an orbital electron is ejected from the molecule; in the latter, an electron is raised to a higher energy level.

It is commonly stated that an average of one ionization occurs for each 32.5 electron volts (ev) dissipated in a system.[1] This is called the *average energy loss per ionization*. However, the actual amount of energy needed to eject an electron from a molecule (ionization potential) ranges from 10 ev to 25 ev. The extra energy which is expended, between 7.5 and 22.5 ev, is used to form excited molecules. Sometimes excited molecules will dissociate and different chemical forms will result. More often, however, especially when large molecules are involved, the energy is distributed over the entire molecule, with too little energy concentrated at any one bond to cause its rupture. The energy is removed from the system as oscillation energy or the like. For this reason, excitation energy is usually considered to have a low efficiency for the production of changed molecules.[2]

Ionizations are seldom produced singly, but rather as double or triple events. Within the past few years the suggestion has been made that chemically and biologically significant events occur in connection with these clusters of ionizations, or "primary ionizations," rather than in relation to single events. Accordingly, based on the assumption that an average of three ionizations occur per cluster, the figure of 100 ev (3 ionizations/cluster $\times$ 32.5 ev/ionization) per primary ionization is often used when discussing energy transfer. This tentative value of 100 ev may be modified when more is known about the pattern of ionization.

RADIATION CHEMICAL YIELD

Almost any chemical reaction can probably be induced by ionizing radiation. Some types of reactions occur readily. Others, however, while theoretically possible, are seldom if ever observed. Two different terms are extensively used to compare and quantitate the radiation effect on chemical systems.

The G value is usually defined as the number of molecules changed per 100 ev transferred to the system. That is,

$$G = \frac{\text{number of molecules changed}}{100 \text{ ev}} \qquad (4\text{-}1)$$

[1]This number of 32.5 ev/ion pair is subject to some doubt. For the purposes of this book it can be assumed to be approximately correct.

[2]Recent results from experiments with certain enzymes have suggested that the excitation process may be more important than has previously been realized. Absorption of subionizing quantities of energy in the range of 4.5 to 5 ev has produced changes in disulfide bonds with an unexpectedly high efficiency.

If 100 ev is approximately the average energy transferred to irradiated matter per primary ionization, G represents the number of radicals formed for each primary ionization.

The M/N value, or *ionic yield*, is defined as the number of molecules changed per ionization or

$$\frac{M}{N} = \frac{\text{number of molecules changed}}{\text{number of ionizations}} \qquad (4\text{-}2)$$

Since about 32.5 ev are expended each time an ionization occurs, the two terms are approximately related by the expression

$$G = 3\frac{M}{N} \qquad (4\text{-}3)$$

For example, if each ionization results in one changed molecule, M/N equals 1, and G equals 3. A very wide range of G or M/N values has been measured. A number of examples are given in Table 4.1.

Table 4.1. REPRESENTATIVE VALUES FOR REAC-
TION EFFICIENCIES FOR GASES IRRADIATED WITH
BETA RAYS

Reaction	M/N^a	G
NH_3 decomposition	1.20	3.60
NH_3 synthesis	0.2	0.6
C_2H_2 polymerization	26.0	78.0
CO_2 decomposition	0.04	0.12

[a]Lea (1955) contains many additional examples of the ionic yields (M/N values) of various types of reactions.

In determining G values, it is most important to consider the system being studied. For example, suppose radiation will cause the breakdown of molecule A into B or C. Measurement of the disappearance of A will give a true indication of the overall change. If, however, only the formation of B is measured, the extent of the total reaction will not be known. The formation of B and C must be known. Failure to recognize this may suggest a radioresistance of the breakdown of A, which is not correct.

Very large ionic yields can be explained, in part, by the occurrence of chain reactions and by the process of excitation, which may be an important factor in some reactions.

FORMATION OF ION PAIRS

When an electron is ejected from a molecule, the result is a positive ion (A^+) and a negative electron (e^-), both of which contain a great deal of energy.

$$A \rightarrow A^+ + e^- \tag{4-4}$$

The electron cannot exist free for very long and is rapidly captured by another molecule to yield a negative ion (B^-):

$$e^- + B \rightarrow B^- \tag{4-5}$$

The overall result of these events is the formation of two ions, one negative and one positive.

$$A + B \rightarrow A^+ + B^- \tag{4-6}$$

Sometimes the positive ion and the negative ion are referred to as the *ion pair* formed by the radiation, although the products indicated in equation 4-4 are the first ions formed.

The relative electron affinities of molecules near the ejected electron will determine which one will capture the electron and become the negative ion. Since oxygen has a high electron affinity, it is one of the most common electron acceptors. The importance of this in biological systems will be stressed in later chapters.

The electron may also be recaptured by the positive ion from which it was ejected. This process is called *charge neutralization:*

$$A^+ + e^- \rightarrow A^* \tag{4-7}$$

Charge neutralization will occur in the unusual event that no other molecule is readily available which has a high electron affinity. It represents an uncommon situation in biological systems in terms of proportion of events, but because of the large number of total ionizations, a considerable number of recombinations occur. The molecule formed by this recombination (A^*) has more energy than that associated with normal stability, and almost always dissociates immediately. At least one of the resulting dissociation products is usually a very reactive free radical.[3]

$$A^* \rightarrow C^\circ + D^\circ \qquad (4\text{-}8)$$

FORMATION OF FREE RADICALS[4]

Ion pairs last only a very short time (less than 10^{-10} second). They then undergo one of many possible reactions to form free radicals. For example, an ion EF^- can dissociate into another ion E^- which does not contain an excess of energy and a free radical F° which contains much energy:

$$EF^- \rightarrow E^- + F^\circ \qquad (4\text{-}9)$$

Free radicals are almost always intermediaries between ion pairs and final chemical products. They are extremely reactive as a result of an unpaired electron in one of their outer orbits.

Thus, somewhat less than 10^{-10} second following the passage of an ionizing radiation, a few excited molecules, a few ions, and a number of very reactive free radicals will be along the path of the radiation. A number of radical reactions are possible depending on the concentration and reactivity of the surrounding molecules and on the concentration of the radicals themselves.

Two free radicals can join and thus share their unpaired electrons in a chemical bond. Sometimes the original products are reunited (recombination).

$$RH \rightarrow R^\circ + H^\circ \rightarrow RH \qquad (4\text{-}10)$$

Sometimes formerly unassociated radicals join (radical combination).

$$R^\circ + S^\circ \rightarrow RS \qquad (4\text{-}11)$$

[3]A free radical is an electrically neutral molecule which has an unpaired electron in the outer orbit. It is extremely reactive. Free radicals are usually indicated by a small circle as a superscript to the right, to show there is *no* charge.

[4]Studies utilizing the techniques of electron spin resonance have confirmed the presence and identity of free radicals following irradiation of many simple chemical systems. This method is now being applied to complex chemical and biological systems in order to identify the free radicals formed by irradiation. The technique measures the absorption spectra of free radicals by detecting the presence of the unpaired magnetic moment of the unpaired valence electron. See Blois (1961) for additional details of the technique.

In the first situation (4-10) there may be no net effect on the system; in the second situation (4-11) a new molecule is formed whose presence may markedly influence the system studied. For example, a small alteration in a single molecule may result in a genetic change which, in turn, may alter an entire organism.

Radicals can also react with ordinary molecules. Oxygen is extremely reactive, and combines with a free radical to form the peroxyl radical.

$$R° + O_2 \rightarrow RO_2°$$ (4-12)

This is a very reactive form and represents a *new* molecule to the system. Accordingly, the changes resulting from formation of a single peroxyl radical may be widespread.

RADIOCHEMISTRY OF WATER

Most biological systems are about 80% water. Hence, most primary ionizations in biological material occur in the water molecules. For this reason, the radiochemistry of water is chosen to illustrate the generalizations on radiation chemistry which were presented in the preceding section.

The effects of radiation on water were noted as early as 1901 when Curie and Debierne observed the evolution of oxygen and hydrogen gases from a solution of radium salts. Additional investigation by these and other workers indicated that the alpha radiation from radium was causing a dissociation of water and that hydrogen peroxide (H_2O_2) was being formed in addition to gaseous hydrogen and oxygen. More recently, the mass spectrometer has been used to determine the ions formed by the impact of electrons on water vapor. This instrument has demonstrated the presence of HO_2^+, HO^+, and H_2O^+ ions and small amounts of H_2^+ and O^+ ions in the bombarded vapor.

The irradiation breakdown, or radiolysis, of water is understood in considerable detail. The generally accepted reaction for the radiolysis of water is

$$H_2O \rightarrow H° + OH°$$ (4-13)

The steps between the radiation and the formation of the free radicals have been the subject of much discussion. One of the most likely possibilities is as follows:

An electron is ejected from a water molecule as was indicated in the prototype equation 4-4.

$$H_2O \rightarrow H_2O^+ + e^- \tag{4-14}$$

The electron is picked up by another water molecule as indicated in equation 4-5.

$$e^- + H_2O \rightarrow H_2O^- \tag{4-15}$$

In this way a positive ion (H_2O^+) and a negative ion (H_2O^-) are formed. Each of these ions, in the presence of another molecule of water, decompose into an ion and a free radical.

$$H_2O^+ \xrightarrow{\text{with } H_2O} H^+ + OH^\circ \tag{4-16}$$

$$H_2O^- \xrightarrow{\text{with } H_2O} OH^- + H^\circ \tag{4-17}$$

The ions H^+ and OH^- do not contain an excessive amount of energy and will recombine to form water. The free radicals OH° and H° are highly reactive.

The free radical formation may be better visualized with the classical picture of electron configurations that follows.

$$H \; :\!\overset{\cdot\cdot}{\underset{\cdot\cdot}{O}}\!: \; H \rightarrow [H \; \cdot\overset{\cdot\cdot}{\underset{\cdot\cdot}{O}}\!: \; H]^+ + e^- \tag{4-18}$$

$$e^- + H \; :\!\overset{\cdot\cdot}{\underset{\cdot\cdot}{O}}\!: \; H \rightarrow [\; \cdot H \; :\!\overset{\cdot\cdot}{\underset{\cdot\cdot}{O}}\!: \; H]^- \tag{4-19}$$

$$[H \; :\!\overset{\cdot\cdot}{\underset{\cdot\cdot}{O}}\!\cdot \; H]^+ \rightarrow [H]^+ + [\; \cdot\overset{\cdot\cdot}{\underset{\cdot\cdot}{O}}\!: \; H]^\circ \tag{4-20}$$

$$[\; \cdot H \; :\!\overset{\cdot\cdot}{\underset{\cdot\cdot}{O}}\!: \; H]^- \rightarrow [\; \cdot H]^\circ + [\; :\!\overset{\cdot\cdot}{\underset{\cdot\cdot}{O}}\!: \; H]^- \tag{4-21}$$

Note that equations 4-18 and 4-19 are the same as equations 4-14 and 4-15, and that the free radicals indicated in equations 4-20 and 4-21 do, indeed, contain an unpaired electron in the outer orbit.

In a pure water solution, many of the free radicals will react with each other to form H_2, H_2O, and H_2O_2, according to equations 4-22 to 4-24.

$$\boxed{H° + OH° \rightarrow H_2O} \qquad (4\text{-}22)$$

$$\boxed{H° + H° \rightarrow H_2} \qquad (4\text{-}23)$$

$$\boxed{OH° + OH° \rightarrow H_2O_2} \qquad (4\text{-}24)$$

The relative probability of these three reactions depends to a large degree on the spatial distribution of the radicals; that is, a radical is more apt to react with another which is near it than one which is at a greater distance.

An alternate mechanism for formation of free radicals has been suggested. This assumes that the ejected electron is not moving fast enough to escape from the vicinity of the (H_2O^+) ion from which it was ejected. The electron is recaptured by the (H_2O^+) ion to form a *very* energetic molecule (H_2O^*), which immediately dissociates into the free radicals OH° and H°.

The chief difference in these two theories is in the spatial distribution of the free radicals formed. In the first theory, illustrated in equations 4-14 to 4-17, the path of the ionizing particle can be visualized as a track of OH° radicals formed from the heavier H_2O^+ ions (equation 4-16). In a cylinder surrounding these OH° radicals are the H° radicals formed from the reactions of the ejected electrons, according to equations 4-15 and 4-17. This model assumes that the electrons were ejected with considerable energy. Figure 4.1 indicates this model on a two-dimensional level. The reader should visualize a three-dimensional cylinder of hydrogen radicals surrounding the core of hydroxyl radicals. The second model assumes that each electron remains near its positive ion and is recaptured. This may be visualized

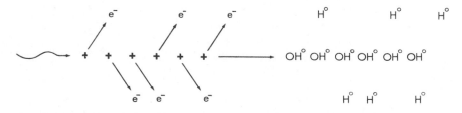

Fig. 4.1. Formation of free radicals along the path of ionizing radiation according to the first proposed mechanism (see text).

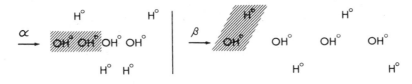

a. Alpha particle path   b. Beta particle path

Fig. 4.2. Relative distribution of radicals along the paths of an alpha particle and a beta particle (in pure water). The most likely radical interactions are indicated.

as a mixture of the positive ions and electrons, rather than the cylindrical arrangement described in the first model.

The mechanism of free radical formation in water may be clarified by a comparison of peroxide formation by radiations with different rates of energy loss (LET). When an alpha particle produces a path of ionization in pure water, the ionizations occur close together, giving a dense track of hydroxyl radicals and a cylinder of hydrogen radicals which are relatively closer to each other than to the hydroxyl radicals (see Figure 4.2a). A considerable number of hydroxyl radicals react with other hydroxyl radicals to form H_2O_2, and hydrogen radicals react with other hydrogen radicals to form H_2. However, with beta irradiation (Figure 4.2b), the radicals are farther apart along the core. With this distribution, the probability is greater for a hydroxyl radical of the core to react with a hydrogen radical in the cylinder to form H_2O than it is with the distribution of ionization which is produced by alpha irradiation. Since alpha radiation does produce a greater yield of hydrogen peroxide than does beta radiation, this is taken as evidence for the validity of the first mechanism of radical formation which was proposed. If the second model (formation of H_2O^* and then dissociation) were correct, there would be no difference in the amounts of H_2O_2 formed from the two types of radiation.

HYDROGEN PEROXIDE FORMATION

The combination of two OH° radicals into H_2O_2 has been described. Since hydrogen peroxide is an active oxidizing agent, its quantitation is important when considering the radiation effect on an aqueous system. In biological systems where peroxide destroying enzymes such as catalase and peroxidases are usually present, formation of H_2O_2 may not be significant. However, the mechanism of its formation can serve as a prototype for the formation of organic peroxides which may be of greater biological importance.

Even in pure water, H_2O_2 formation is not as simple as has been indicated, but is complicated by the chain reactions indicated in equations 4-25 and

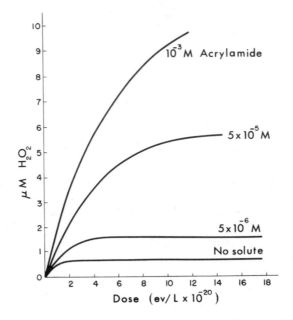

Fig. 4.3. Formation of H_2O_2 in irradiated aqueous solutions of acrylamide. (Modified from Collinson *et al.* [1957], courtesy Faraday Society.)

4-26. Both of the molecular products (H_2O_2 and H_2) are removed by reactions with free radicals.

$$H° + H_2O_2 \rightarrow OH° + H_2O \qquad (4\text{-}25)$$

$$OH° + H_2 \rightarrow H° + H_2O \qquad (4\text{-}26)$$

Thus, H_2O_2 is formed in the system, as shown in equation 4-24, but it is removed by the remaining free radicals. The net result is a low yield of H_2O_2.

If, however, other molecules are present in the water which are capable of reacting with the free radicals H° and OH°, competition will exist for these radicals between the reacting material and the molecular products. Accordingly, the H_2O_2 formed may not be destroyed. An example of such a material is acrylamide.[5] Figure 4.3 indicates the effect on peroxide formation of different concentrations of acrylamide. As more of this material is added, more of the free radicals are used up and less of the peroxide is removed by the reaction indicated in equation 4-25.

[5]Acrylamide is an organic compound with the formula $H-\overset{\overset{\displaystyle H}{|}}{C}=\overset{\overset{\displaystyle H}{|}}{C}-\overset{\overset{\displaystyle O}{\|}}{C}-NH_2$ It probably reacts with free radicals by addition at the C=C double bond.

If, now, molecular oxygen (O_2) is added to the irradiated water, the oxygen rapidly combines with the hydrogen radical ($H°$) to form the peroxyl radical ($HO_2°$).

$$H° + O_2 \rightarrow HO_2° \qquad (4\text{-}27)$$

This is a less powerful oxidizing agent than $OH°$ or H_2O_2, but has a longer life and is, therefore, capable of diffusing farther from its site of formation to undergo further reactions. Two peroxyl radicals can react to form hydrogen peroxide.

$$HO_2° + HO_2° \rightarrow H_2O_2 + O_2 \qquad (4\text{-}28)$$

This adds to the total amount of peroxide formed in the system.

If, in addition to oxygen, some other material is present which is capable of being oxidized (giving up an electron), the peroxyl radical will readily accept an electron:

$$HO_2° + e^- \rightarrow HO_2^- \qquad (4\text{-}29)$$

This HO_2^- is the anion of H_2O_2, so that, unless the solution is strongly alkaline, a hydrogen peroxide molecule will be produced.

$$HO_2^- + H^+ \rightarrow H_2O_2 \qquad (4\text{-}30)$$

There are, then, two additional ways that H_2O_2 can be produced by irradiation of an aqueous system in which oxygen is present. The reaction of equation 4-28 requires two $HO_2°$ radicals, while the reactions in equations 4-29 and 4-30 require only one $HO_2°$ for each molecule of H_2O_2 produced.

The production of H_2O_2 in an aqueous system can be summarized as follows (Figure 4.4):

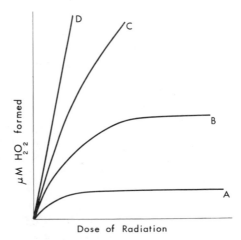

Fig. 4.4 Production of H_2O_2 in aqueous systems. (A) Pure water; (B) water plus "free radical scavenger"; (C) water plus O_2; (D) water plus O_2 plus oxidizable additives.

1. With pure water and no O_2, there is little measurable H_2O_2 since that which is formed is destroyed by the free radicals present (curve A).
2. With material such as acrylamide present that will react with free radicals, but with no O_2 present, some H_2O_2 can be measured. That which was formed in the molecular reaction of equation 4–24 will remain (curve B).
3. In the presence of oxygen, additional H_2O_2 will be formed by the reaction in equation 4-28 (curve C).
4. In the presence of oxygen plus additives which are oxidized by $HO_2°$, the production of H_2O_2 will be high (curve D).

RADIATION EFFECTS ON SIMPLE CHEMICAL SYSTEMS

When ionizing radiation impinges on a system, the molecules ionized will be those which are actually in the path of the radiation, without regard for their chemical composition or "reactivity." Thus, the largest or most numerous molecules will most often be ionized, as they will occupy the greatest proportion of the volume. In a dilute, aqueous solution, the vast majority of the molecules are H_2O. Most of the ionizations will occur in water molecules. The effect on the solute will be primarily "indirect" and will be mediated through the water molecules.

The $H°$, $OH°$, and $HO_2°$ radicals which are formed by the irradiation of water molecules can have a variety of effects on other molecules present in the system. The $H°$ radical is a powerful reducing agent; that is, it readily gives up its unpaired electron. In fact, it is more powerful as a reducing agent than the $OH°$ radical is as an oxidizing agent. The net effect in such a system, however, is oxidation, due to the greater total quantity of oxidizing agents ($OH°$ radical, $HO_2°$ radical from the interaction of O_2 with $H°$ radical, and H_2O_2 from the molecular reaction). The $H°$ radical can also initiate polymerization or remove H atoms from organic molecules.

The primary reaction of the $OH°$ radical is oxidation, that is, removal of an electron from another molecule to pair with its unpaired electron. A typical reaction is indicated in equation 4-31.

$$Fe^{++} + OH° \rightarrow Fe^{+++} + OH^- \qquad (4\text{-}31)$$

The free radical is changed to the hydroxyl ion. Hydroxyl radicals can also initiate polymerization, add at double bonds of organic compounds, reduce powerful oxidizing agents, and remove hydrogen atoms from organic molecules (equation 4-32).

$$RH + OH° \rightarrow R° + H_2O \qquad (4\text{-}32)$$

In the reaction shown in equation 4-32, a hydroxyl radical has removed a hydrogen atom (with electron) from the organic molecule. The organic molecule is then a radical and is capable of a variety of reactions, many of which might result in a *different* molecule. There will be a change in the chemical constituents of the system.

The $HO_2°$ radical, like the $OH°$ radical, is capable of oxidation and also of removing H atoms from organic substances. In addition, it can donate an oxygen atom to a molecule which contains a lone pair of electrons, as illustrated in equation 4-33.

$$RNH_2 + HO_2° \rightarrow \overset{\displaystyle O}{\overset{\displaystyle \|}{RNH_2}} + OH° \qquad (4\text{-}33)$$

or it can exchange an oxygen for a hydrogen atom (equation 4-34):

$$RH + HO_2° \rightarrow RO° + H_2O \qquad (4\text{-}34)$$

These two latter types of reactions lead to different molecules which may greatly affect the chemical or biological system under study.

INTERACTION OF RADICALS WITH A SINGLE SOLUTE

When $OH°$ and $H°$ radicals are formed in irradiated aqueous solutions, a certain proportion of these will recombine to form H_2O or will combine in the molecular reactions to form H_2 and H_2O_2. The others will diffuse until they react with another molecule. The number and relative reactivity of the solute molecules compared with the $OH°$ and $H°$ radicals will determine the relative probability of reaction with the radicals; that is, a free radical will most likely react with the first available reactive form. If the concentration of solute molecules is very low, the radicals will most often react with each other, so that the "yield" (extent of reaction of solute molecules) will be

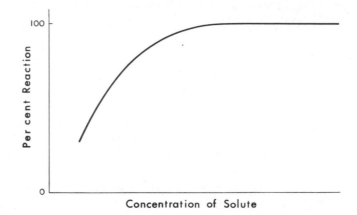

Fig. 4.5. Effect of concentration of solute in a simple aqueous system on the extent of the solute reaction following exposure to a specific dose of radiation.

low (Figure 4.5). As the concentration of solute molecules increases, the probability of an $OH°$ or $H°$ radical reacting with the solute molecule increases. Therefore, the "yield" increases. Once a certain solute concentration is exceeded, however, the yield stays constant. Beyond this value, every available radical reacts with the solute. None are lost by recombination.

Before the solute can react with the radicals, it must be present in a concentration at least as great as that of the radicals in the columns surrounding the path of the ionizing particle (see Figure 4.2). In order to compete effectively for the radicals, the solute concentration should be many times greater than the radical concentration. Electrons with an energy of 60 kv produce in water a radical concentration along the track of the electron of about 10^{-6} M. A maximum yield of a solute reaction is reached at a concentration of about 10^{-4} M. Thus, if there are 100 times as many solute molecules present along the path of the ionizing particle as there are radicals formed, almost all available radicals will react with solute molecules rather than with each other.

Where alpha particles are used to irradiate water, the $OH°$ radicals exist in the core of the track at about one molar concentration, while the $H°$ radicals in the surrounding cylinder are at about 10^{-2} M. Therefore, several times molar concentration of a solute would be needed for effective competition with the radicals. Some effect can be noted from solute reactions with radicals formed along the lower LET delta (secondary) rays from the alpha particles, but the interaction between solute molecules and free radicals is very much less for alpha particles than for gamma radiation at the same solute concentration.

INTERACTION OF FREE RADICALS WITH SEVERAL SOLUTES

When more than one type of solute molecule is present in a solution, the free radicals will react most readily with those that are largest in size, those that are most numerous, and those that are most reactive. The relative importance of these factors depends on the specific solutes involved. An example (from Lea, 1955) will clarify this.

In a solution of a dinucleotide (alloxazin-adenine) and a protein (specific protein component of a D-amino acid oxidase), the probability of a radical reacting on collision with the dinucleotide is about 2.5 times as great as the probability of its reacting on collision with the protein; that is, given an equal number of collisions, more radical reactions will occur with the dinucleotide than with the protein. The dinucleotide is more reactive. However, the protein is considerably larger than the dinucleotide. Therefore, if the two are present in equal concentration, the chance of a radical colliding with a protein molecule is much greater than the chance of its colliding with a dinucleotide molecule. This more than makes up for the differing reactivities. When a solution containing equal molar concentrations of the two types of molecules is irradiated, almost all of the radicals react with the protein and almost none with the dinucleotide. The protein "protects" the dinucleotide from radiation. If the concentration of dinucleotide is increased 10 times or more, without changing that of the protein, some free radicals will react with the dinucleotide molecules instead of the protein.

DIRECT VS. INDIRECT EFFECT IN AQUEOUS SOLUTIONS

The indirect effect of radiation on the solute in an aqueous system involves changes in the solute molecules secondary to the formation of free radicals from the water molecules. Beyond a certain minimum concentration, all of the available free radicals react with solute molecules. One of the characteristics of the indirect effect of radiation is a *linear* relationship between dose and effect, which is *independent* of the concentration (above the critical value). This is indicated in Figure 4.6.

Note that as the radiation dose is doubled the number of ion pairs formed, the number of free radicals formed, and the number of solute molecules affected are all doubled. As the concentration of the solute molecules is increased, a smaller *proportion* of them will be changed, but the same total number will be affected. Therefore, in dilute solutions this may represent a large percentage change, while in concentrated solutions it may represent a small percentage change.

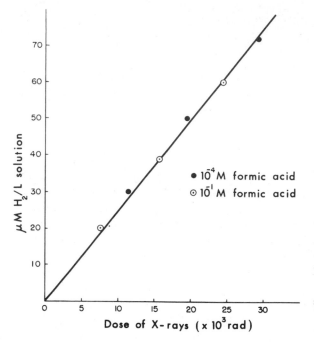

Fig. 4.6. Indirect effect of radiation on formic acid solution, illustrating the independence of absolute yield on concentration of solute. (From Fricke *et al.* [1938], courtesy American Institute of Physics.)

Normally, in dilute solutions of small molecules, the radiation dose that will cause a considerable proportion of the solute to react with free radicals will only suffice to ionize *directly* a negligible proportion of the solute molecules. Thus, the direct action of the radiation on the solute molecules is small. However, the fraction of the total reactions which are related to the direct effect can be increased in several ways. If material is irradiated dry, the water molecules have been removed so that there will be only direct interactions with the molecules of the material. If a solution is frozen, the mobility of the radicals which are produced in the water molecules is decreased. This will decrease the possibility of indirect action and result in a greater proportion of the interactions being of the direct type. Also, chemical protectors can be introduced into the system which will compete successfully for the OH° and H° radicals formed. This will reduce the indirect effect on the molecule which is being studied.

The dose required to produce a chemical change in a given proportion of the molecules of a substance, by direct action, is inversely proportional to the molecular weight of the substance, assuming that the ionic yield is constant. (The larger molecules are more likely to be in the path of the radiation.) If the ionic yield is approximately 1, it requires 10^6 R (1 million roentgens) to change half of the molecules of molecular weight 10^6 by direct interaction. The direct effect is not very important in considerations of simple chemical systems, but is of importance in macromolecular and biological systems because of the presence of many large molecules.

REACTIONS IN AQUEOUS INORGANIC SOLUTIONS

In simple inorganic solutions, the primary reactions are reduction of cations and anions by $H°$ radicals and oxidation of cations and anions by $OH°$ and $HO_2°$ radicals and by H_2O_2.

The general reactions for cations are

$$M^{+N} + H° \rightarrow M^{+(N-1)} + H^+ \tag{4-35}$$

$$M^{+N} + OH° \rightarrow M^{+(N+1)} + OH^- \tag{4-36}$$

The general reactions for anions are

$$A^{-N} + OH° \rightarrow A^{-(N-1)} + OH^- \tag{4-37}$$

$$A^{-N} + H° \rightarrow A^{-(N+1)} + H^+ \tag{4-38}$$

In the presence of O_2, the $H°$ radicals will be converted to $HO_2°$ radicals and will oxidize rather than reduce materials. The net effect in a system containing oxygen, then, is oxidation of the molecules. The extent of the reaction depends on the concentration of the solute, the amount of oxygen in the solution, and the presence of additional materials in solution. Since most inorganic molecules are small, the indirect effect predominates.

REACTIONS IN AQUEOUS ORGANIC SOLUTIONS

When an aqueous organic solution is irradiated, the usual "indirect" reaction on the organic molecule is the removal of either a H atom or an entire radical group (such as the —CH_3 "methyl" group) from the molecule. The general reaction for removal of a hydrogen atom is

$$RH + OH° \rightarrow R° + H_2O \tag{4-39}$$

Since most organic molecules have several hydrogen atoms, a number of different organic radicals ($R°$) can be formed. Although breaks occur most often between carbon and hydrogen atoms resulting in the loss of the hydrogen from the molecule, carbon-carbon (C-C), carbon-oxygen (C-O), carbon-nitrogen (C-N), or carbon-sulfur (C-S) cleavages are not uncommon. The significant feature is not the removal of the hydrogen atom or the group, or rupture of the molecule, but rather the behavior of the remaining radical. A chemically different molecule will result unless the organic radical which has been formed by loss of a hydrogen atom captures another hydrogen atom and incorporates it at exactly the same position.

Other reactions which will produce a different molecular species are typified by the process of condensation, where two radicals combine,

$$\boxed{R° + S° \rightarrow RS} \tag{4-40}$$

by the process of rearrangement, where two radicals combine and then split into two different forms,

$$\boxed{R° + S° \rightarrow RS \rightarrow T + U} \tag{4-41}$$

and by the process of combination of a radical with molecular oxygen.

$$\boxed{R° + O_2 \rightarrow RO_2°} \tag{4-42}$$

The $RO_2°$ radical produced (equation 4-42) may pick up a hydrogen atom.

$$\boxed{RO_2° + R'H \rightarrow RO_2H + R'°} \tag{4-43}$$

The resulting molecule, RO_2H, may be stable, but is a *different* molecular species than the original RH and may exert a different effect chemically and biologically. Also, another radical ($R'°$) has been formed.

Several representative reactions of organic materials will be cited to clarify the general reactions.

1. Most saturated hydrocarbons probably undergo a hydrogen extraction, and are converted to alcohols in a two-step process.

$$\boxed{CH_4 + OH° \rightarrow CH_3° + H_2O} \tag{4-44}$$

or

$$CH_4 + H° \rightarrow CH_3° + H_2 \qquad (4\text{-}45)$$

$$CH_3° + OH° \rightarrow CH_3OH \qquad (4\text{-}46)$$

2. Acetic acid most frequently loses a hydrogen atom attached to the methyl group, and two of the resulting radicals condense to form succinic acid.

$$CH_3COOH + H° \rightarrow CH_2COOH° + H_2 \qquad (4\text{-}47)$$

or

$$CH_3COOH + OH° \rightarrow CH_2COOH° + H_2O \qquad (4\text{-}48)$$

$$2\ CH_2COOH° \rightarrow HOOCCH_2—CH_2COOH \qquad (4\text{-}49)$$

In the presence of oxygen, a rearrangement occurs with the radical form to give glycolic acid and glyoxylic acid.

$$CH_2COOH° + O_2 \rightarrow O_2CH_2COOH° \qquad (4\text{-}50)$$

$$2\ O_2CH_2COOH° \rightarrow 2\ OCH_2COOH° + O_2 \qquad (4\text{-}51)$$

$$2\ OCH_2COOH° \rightarrow CH_2OHCOOH + CHOCOOH \qquad (4\text{-}52)$$

3. Irradiation of aqueous methanol or ethanol will lead to the production of aldehydes and glycols by similar steps.

$$CH_3OH \rightarrow CH_2O + H_2 \qquad (4\text{-}53)$$

$$2\ CH_3OH \rightarrow CH_2OHCH_2OH + H_2 \qquad (4\text{-}54)$$

A few generalizations can be made with respect to radiation effects on organic molecules.

1. Aromatic structures (contain an unsaturated ring) are more resistant than are aliphatic structures (straight chains of carbon atoms). The unsaturated bonds of the ring permit resonance, that is, a distribution of energy over the entire ring. With less energy concentrated at one spot, there is less probability of removing either a hydrogen atom or other groups.
2. A ring structure stabilizes a side chain. A chain (aliphatic structure) which has a ring attached is less radiosensitive than a straight aliphatic structure. For example, the formation of crosslinks in a 12 carbon chain molecule is decreased by attaching an aromatic ring to one of the carbons; the protective effect is greatest when the ring is on one of the center carbons. This suggests a migration of the energy to the ring where it can be dissipated. The migration distance is less than 12 carbon atoms.
3. Energy which is absorbed in any portion of a complex molecule can react at the site where it is absorbed; it can shift within the molecule and react elsewhere; or it can be thermalized (given off as heat) and thus produce no chemical change in the molecule.
4. The transfer of energy from radiation is more efficient when it occurs within molecules than when it takes place between molecules. This generalization applies to energy transferred by ionization and excitation.

MACROMOLECULAR REACTIONS

As a result of the chemical changes which have been described, very large molecules which are common in biological systems may undergo a variety of structural changes which lead to altered function.

"Degradation," or breaking into smaller units, has been shown to occur when these large molecules are irradiated. Moreover, in molecules which contain a series of identical or similar repeating units, the breaks usually occur in the same bond. This results in a number of smaller molecules, of similar chemical composition to the original molecule, but of potentially different functional ability. Since the chance is slight that radiation would always hit the same bond, these findings have suggested that energy which is absorbed any place in the molecule can be transmitted down the molecular chain to the weakest bond. For example, the haemocyanins (giant protein molecules in the blood of certain arthropods and molluscs) can be split by irradiation into identical subunits. These units are held together by hydrogen bonds in the intact molecule.

"Crosslinking" is another common structural change. A long molecule which is somewhat flexible in structure can undergo intramolecular crosslinking (becoming attached to itself) when a chemically active locus is produced on it and when this spot can come in contact with another reactive area. Likewise, intermolecular crosslinking (attachment between different molecules) can occur, as indicated in Figure 4.7. If the crosslinking is extensive, not only are the molecules incapable of normal function but they may no

Intramolecular   Intermolecular

Fig. 4.7. Crosslinking of macromolecules caused by radiation.

longer be soluble in the system. The resulting "gel" represents an entirely different physical-chemical state than did the original solution. One effect of radiation on the nucleic acid, deoxyribonucleic acid (DNA), is an example of crosslinking.

Many macromolecules are normally held in a rigid configuration by intramolecular crosslinking bonds; that is, specific chemical groups are linked together, frequently by hydrogen atoms, to form a three-dimensional structure. The hydrogen bonds are among the weakest in the molecule and, thus, are the first to be broken by radiation. Such structural changes can lead to severe alterations in the biochemical properties of the molecule.

RADIATION EFFECTS ON MOLECULES OF IMPORTANCE IN BIOLOGICAL SYSTEMS

This section will consider briefly the major radiation-induced changes in the molecules which are of particular importance to the structure and function of a living cell. The emphasis will be on chemical and structural changes in these molecules and not on metabolic processes in which they are involved. It is not intended to provide a comprehensive coverage of all of the varied, and often controversial, experiments involving radiation effects on biologically important macromolecules. This section will indicate some of the types of effects that have been consistently observed and the changes that have been proposed to explain them.

Most of the macromolecules of biochemical importance have a three-dimensional complex structure. Because of this complexity, it is often difficult to recognize or measure chemical changes in these molecules directly. Many of the macromolecules, however, have biological activities or physical-chemical properties which can be quantitated, and which can thus serve as a measure of the structural integrity of the molecule. For example, a decrease in the solubility of a particular macromolecule may suggest an increase in crosslinking; a decrease in the ability of an enzyme system to catalyze a reaction may suggest a chemical change in the reactive group of the enzyme. Moreover, it is often possible to measure functional changes in molecules when they are not a part of a living system. It is not always correct to extrapolate these findings to a living situation where the molecule may be in a different chemical form and may be surrounded by other molecules with differing radiosensitivities and protective capacities.

PROTEINS

Proteins are intimately involved with almost all cellular functions. Among the proteins are many of the structural elements of the cell, enzymes which catalyze most of the essential chemical reactions, many of the hormones which regulate metabolic processes, and the antibodies which are produced to counteract harmful agents. "Simple" proteins are chains of amino acids; "conjugated" proteins contain an organic chemical moiety in addition to the amino acids. For example, nucleoproteins contain nucleic acids plus amino acid chains, and glycoproteins contain carbohydrates in addition to the amino acids.

The general formula for amino acids can be written as

$$\begin{array}{c} R \\ | \\ NH_2CH-COOH \end{array}$$

Amino acids are characterized by an amino (-NH$_2$) group, a carboxyl (-COOH) group, and a "side chain" designated as R, which is characteristic of the specific amino acid. When a protein is formed, the amino groups react with neighboring carboxyl groups to form peptide bonds. Thus, the primary structure of a protein, the peptide chain, is formed as indicated in Figure 4.8.

Secondary and tertiary structures are formed in most proteins by a coiling of the peptide chains and bending or twisting of the coiled chains to form a complex, somewhat rigid, three-dimensional configuration. Disulfide

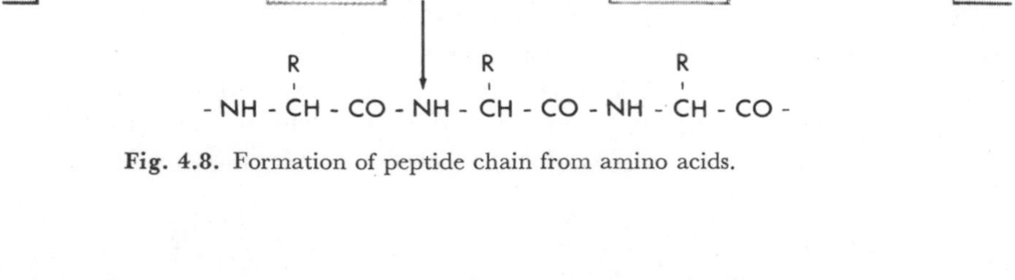

Fig. 4.8. Formation of peptide chain from amino acids.

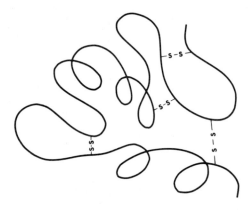

Fig. 4.9. Schematic structure of a protein molecule showing disulfide bonds which maintain the secondary and tertiary structure.

bonds, hydrogen bonds, and nonbonded interactions between side chain groups of the amino acids are important in maintaining the secondary and tertiary structure, as illustrated in Figure 4.9.

Radiation effect on proteins. The amino group is the most radiosensitive portion of an amino acid. However, in the formation of a protein, this group is linked to a carboxyl group and is, therefore, not easily removed from the molecule. Similarly, the carboxyl group is no longer available for a reaction. Thus, the "side chains" (R) are the more radiosensitive portions of a protein molecule. The specific changes that occur in the side chains depend on their chemical composition. A hydrogen atom may be removed or a break may occur between almost any of the atoms in the chain.

Table 4.2 contains some G values for amino acid residues of irradiated bovine serum albumin.[6] The relatively small range in G values indicates that

Table 4.2. G Values for Change in Some of the Side Chains of the Amino Acid Residues (R) of Bovine Serum Albumin Which Was Irradiated with 2 Mev Electrons

Amino Acids	G Value[a]
Cystine	12.0
Tyrosine	7.7
Histidine	7.7
Arginine	5.5
Amino groups of lysine	4.9

$$^{a}G = \frac{\text{residues changed per 100 ev per protein molecule} \times 100}{\text{per cent content of residue in protein}}$$

[6]Bacq and Alexander (1961) contains a more detailed account of the experiments by P. Alexander and L. D. G. Hamilton on irradiation of bovine serum albumin.

energy absorbed in the protein is not funneled into a few sensitive residues, but is distributed over all of the amino acids.

Loss of function of a protein by irradiation is not usually due to breaking peptide bonds or otherwise disrupting the primary skeletal structure of the peptide chain. It may result from a change in a critical side chain or from a break in the hydrogen or disulfide bonds which maintain the secondary and tertiary structures. Such a break can lead to a partial unfolding of the tightly-coiled peptide chains which, in turn, can result in a disorganization of the internal structure, a distortion of necessary spatial relationships of side chain groups, or an exposure of amino acid groups resulting in a change in chemical activity.

The hydrogen bonds of the secondary and tertiary structure are weak bonds. A number of them will be temporarily broken in the vicinity of an ionization because the sudden introduction of a charge disrupts electrical dipoles. One primary ionization, then, can alter the structure of the molecule and lead to extensive change in the overall chemical reactivity.

The behavior of irradiated bovine serum albumin is typical of the physical and chemical changes following irradiation of the globular proteins. The sedimentation constant of the protein increases following radiation without a corresponding alteration in molecular weight. This suggests a change in the shape of the protein. Furthermore, there is also an increase in the chemical reactivity, which may be interpreted as an opening or unfolding of the molecule. Each molecule of bovine serum albumin contains 17 disulfide bonds from the amino acid cystine. These are normally unreactive, mainly because they are located internally — that is, buried in the molecule. When the protein is irradiated, each primary ionization in a molecule makes four of these bonds available for reaction. The energy which is involved could, potentially, alter three amino acid residues in the molecule. However, the similarity of G values, as indicated in Table 4.2, makes it unlikely that cystine groups would be selectively affected by the radiation. On the average, only one of five molecules of albumin which have received a primary ionization will undergo a reaction at a disulfide bond. Much of the energy will be expended on changes in other amino acid components. Accordingly, it seems most probable that the radiation has affected the three-dimensional configuration of the molecule, causing it to unfold and make the disulfide bonds available for reactions.

These changes in bovine serum albumin have been attributed largely to direct action of the radiation on the protein molecules. Similar changes can be produced by indirect transfer of energy from free radicals formed in water. The indirect action is apparently much less efficient than the direct action. Most proteins can be inactivated by between 50 and 200 ev deposited directly in the molecule. In contrast, somewhere between 10 and 200 OH° radicals must be produced from adjacent water molecules for a typical protein molecule to be inactivated by indirect action. This represents considerably more total energy than required for the direct action. Actually, this is a probability effect. Only one of the OH° radicals is needed to initiate the change leading to inactivation; the others are wasted by interaction at unimportant sites.

Experiments of other types have suggested that the indirect effect may be very important under certain circumstances. A decrease in solubility, perhaps from aggregation of protein molecules, is produced by indirect action.[7] This may be caused by an unfolding of the molecule which will facilitate aggregation processes and thus lead to an increase in molecular weight. As the dose of radiation increases, the amount of aggregation increases as does the size of the aggregates. With a sufficiently large dose, precipitation occurs.

Radiation has reportedly produced other changes in proteins, such as an increase or a decrease in viscosity, and changes in refractive index, optical rotation, and electrical conductivity. The data are contradictory or ambiguous. Future work will undoubtedly clarify their existence and significance.

Radiation effect on enzymes. The effects on enzymes, proteins which catalyze specific chemical reactions, can best be studied by measurements of enzyme activity. All enzymes can be inactivated when irradiated in solution, although the radiation doses necessary to inactivate different enzymes vary greatly. Some typical G values for inactivation of enzymes in dilute aqueous solution are given in Table 4.3. If the substrate of a specific enzyme is present in the solution, the inactivation is partially prevented. Since substrate is present in cells, it is probable that the G values for enzymes irradiated *in vivo* are lower than those presented in Table 4.3.

Table 4.3.[a] RADIATION INACTIVATION OF ENZYMES IN DILUTE AQUEOUS SOLUTION

Enzyme	G Value[b]
Carboxypeptidase	0.55
D-amino acid oxidase	0.31
Ribonuclease	0.09
Alcohol dehydrogenase	0.06
Catalase	0.009

[a]Modified from Bacq and Alexander (1961).
[b]Number of molecules inactivated per 100 ev of energy deposited in the solution.

At one time it was thought that enzymes containing sulfydryl groups (-SH) were much more sensitive than nonsulfydryl-containing enzymes.

[7]This is considered to be due to the *indirect* action of radiation, since it follows the dilution law (see Figure 4.6 and text discussion).

Since sulfur-containing enzymes are involved in most essential cell processes, several authors proposed that the effects observed in biological systems after irradiation were due to effects on these enzymes. This theory has not been substantiated. Most evidence indicates that sulfydryl-containing enzymes are probably no more radiosensitive than other enzymes.[8]

In vivo sensitivity of proteins. The proteins of living systems have shown a variety of responses to moderate doses of radiation. There has been a suggestion of a slight to moderate lowering of catalase[9] activity in various tissues, but the magnitude of this effect is not sufficient for it to be a major factor in lethality. The activity of another enzyme, ATPase,[10] has been shown to increase slightly in some tissues following irradiation. It appears, however, that this may not be due to any effect of energy absorbed in or near the molecule itself, but rather to changes in tissues at some distance from those under study. Increased excretion of taurine, a product of protein degradation, follows radiation exposure. How much, if any, of this is due to direct protein breakdown rather than generalized cell breakdown is unknown. It does not appear that alteration of protein structure or function is a major primary factor in the radiation effects on living systems.

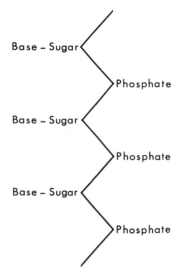

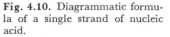

NUCLEIC ACIDS

Nucleic acids are macromolecules which are considered to be the carriers of genetic information. They are composed of nitrogenous bases called *purines* and *pyrimidines*. These bases are linked to pentoses, or 5 carbon sugars, making nucleosides. The nucleosides are linked by phosphate groups as indicated in Figure 4.10 to make the DNA chain. The basic repeating unit of this chain is the nucleotide which is composed of a nucleoside and a phosphate group.

Fig. 4.10. Diagrammatic formula of a single strand of nucleic acid.

[8]Some recent work (Augenstein, L. G., and Grist, L. L., in IAEA, 1962) has reopened this question. These authors feel that subionization, excitation energies of 4.5 to 5 ev, possibly correlated with disruption of S-S bonds, may account for 4% or more of the direct inactivation produced by ionizing radiation.

[9]Catalase is an enzyme that degrades hydrogen peroxide. It is present in varying amounts in most biological systems.

[10]ATPase is an enzyme that hydrolyzes adenosine triphosphate (ATP), the molecule which provides the energy for most biological systems.

There are two general types of nucleic acids. The first, ribonucleic acid (RNA), has D-ribose as its sugar constituent; adenine, guanine, cytosine, and uracil as bases; and is found primarily in the cytoplasm of cells. The second, deoxyribonucleic acid (DNA), has 2-deoxy-D-ribose as its sugar; adenine, guanine, cytosine, and thymine as bases; and is a major constituent of chromosomes. According to the Watson-Crick model, DNA exists as a double-stranded helix; that is, two strands, each of which have the base-sugar-phosphate configuration indicated in Figure 4.10 are linked by hydrogen bonds between the bases, and the entire structure is coiled into a helical configuration. Proteins are associated with the DNA molecules in chromosomes, forming nucleoproteins. The nature of the nucleic acid-protein association is not clear.[11]

Radiation effects on nucleic acids. A number of different kinds of damage which radiation may produce in DNA are listed in Table 4.4. All of these types of damage have been studied, and all have been shown to occur after irradiation.

Table 4.4.[a] Types of Damage to DNA Molecules by Radiation

---

1. Change of a base (for example, deamination)
2. Loss of a base
3. Hydrogen bond breakage between chains
4. Single-strand fracture
5. Double-strand fracture (both chains simultaneously)
6. Crosslinking within the helix
7. Crosslinking to other DNA molecules
8. Crosslinking to protein

---

[a]Modified from Stacey, K. A. "DNA and The Effects of Radiation" from Ebert and Howard (1963).

The change or loss of a purine or pyrimidine base can be supposed to occur by the mechanisms which were described for similar changes in smaller molecules. About 75% of the OH° radicals produced in a DNA solution probably react in this way. Pyrimidine bases are more radiosensitive than purine bases, and thymine appears to be the most sensitive of the pyrimidines. Since the order and composition of the bases on a DNA molecule are thought to determine the genetic code which is carried by the molecule, an alteration in the base sequence can potentially give rise to a change in genetic composi-

---

[11]It has been suggested that protein molecules may form a "backbone" on which DNA components are hooked in precise order. Many of these protein-DNA complexes combine to form the "strands" of a chromosome.

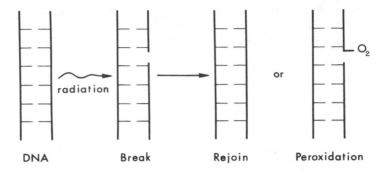

DNA   Break   Rejoin   Peroxidation

Fig. 4.11. Single strand break in DNA following irradiation can lead to rejoining or, in the presence of O_2, to a peroxidized end which cannot rejoin.

tion, or mutation. This, in turn, may produce gross biochemical or physical alterations in an entire cell or organism.

Rupture of the hydrogen bonds which link base pairs in DNA has been considered by some investigators to be an important effect of irradiation. Values of G between 1.09 and 38 have been estimated for this effect. Other authors believe that there is no evidence for the effective rupture of hydrogen bonds by irradiation. They postulate that any of these bonds that are broken will tend to reform immediately. Irreversible damage, according to these authors, will only occur when a very large number of hydrogen bonds are all broken at the same time.

Single-strand fracture or cleavage between a sugar and a phosphate group is often difficult to detect following irradiation, since the rigid structural configuration may limit movement and thus provide no opportunity for the broken portions of the strand to separate. Rejoining would usually occur between the broken portions. However, in the presence of oxygen, if at least one end of a broken strand is in a reactive state, this end may become peroxidized and thus be unable to rejoin (Figure 4.11). Recent studies with the transforming principle[12] have strongly suggested the presence of single-strand breaks in DNA.

Double-chain breaks in irradiated DNA molecules have been demonstrated by several techniques. A main chain double break with separation of the sections will only occur if there is a break in each of the two strands less than about 5 nucleotide units apart. This can occur either when two single breaks come into juxtaposition[13] or when a single particle with high LET produces a break in both strands. The latter requires about 600 ev. Similar-

[12]This is a single-strand portion of DNA which will join another portion of DNA and incorporate its genetic constitution into the transformed portion.
[13]Calculations have indicated that approximately one of these double breaks will occur for every 70 random single breaks.

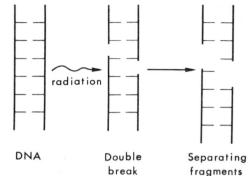

Fig. 4.12. Double chain break in DNA. For separation of fragments to occur, breaks must be less than 5 nucleotide units apart.

DNA   Double   Separating
             break   fragments

ly, when a cluster of ionizations is formed by a radiation with low average LET, a double break may result (see Figure 4.12).

The formation of crosslinks either within the helix of a DNA molecule, between two DNA molecules, or between a DNA molecule and a protein can probably be attributed to the formation of reactive sites at the break. Two such reactive sites will join if contact is made between them. Figure 4.13 illustrates the formation of a crosslink between two DNA molecules.

Crosslinking within the helix of a DNA molecule may occur between the bases. Ultraviolet radiation has been shown to produce thymine dimerization in which two thymine bases join together to form a dimer linkage that is more stable than a hydrogen bond. Recent evidence has suggested that similar dimers are formed by ionizing radiation as well.

The crosslinking process is thought to be primarily a direct effect of the radiation, while double-chain breaks are largely indirect. As the DNA concentration in irradiated solutions is progressively decreased starting from a concentrated solution, there is, first, an increase in the amount of crosslinking. This occurs because the DNA is swollen with water at the intermediate concentrations. Consequently, the molecular movement increases and the active

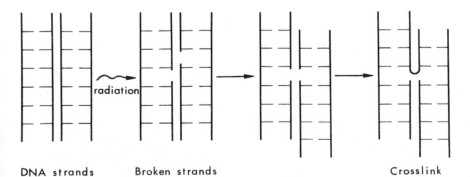

DNA strands   Broken strands   Crosslink

Fig. 4.13. Crosslinking between two irradiated molecules. For diagrammatic purposes, the relative position of the two strands is shifted in the third illustration.

ends are more likely to come together. With an excess of water in dilute solutions of DNA, however, the indirect effect predominates and double-chain breaks are produced.

DNA complexed with proteins is considerably less sensitive to base destruction or chain breakage by radiation than is noncomplexed DNA. Part of this "protection" by the protein results from competition for the radicals, but a large fraction is also due to the structure of the DNA-protein complex. Perhaps, as has been suggested by some authors, the protein may be wrapped around the nucleic acid and thus have a "shielding effect" on the DNA. The temperature, water content, and other conditions of specific experiments have a considerable influence on the degree of protection afforded the DNA by the protein.

Considerably less information is available regarding radiation effects on RNA. It is likely that the changes produced are similar to those in DNA.

In vivo sensitivity of nucleic acids. There is ample evidence that DNA in a cell is damaged by irradiation. Details of mechanisms of the various effects are not always clear. Radiation action on normal processes of synthesis and degradation of DNA is also not fully understood, and, thus, such effects may be confused with direct effects on the molecules.

Bases are undoubtedly altered or damaged in DNA irradiated *in vivo*, as indicated by subsequent genetic changes, and crosslinking has been indicated in a number of experiments. A variety of other changes have been reported which are similar to those observed following *in vitro* irradiation of DNA.

LIPIDS

Lipids include the fats, waxes, and other substances which are generally soluble in organic solvents such as alcohol, acetone, ether, or benzene. "Simple lipids" constitute one of the largest groups of the lipid class. These are esters of fatty acids and an alcohol. The general structure of a simple lipid is illustrated in Figure 4.14.

$$CH_2OOCR$$
$$|$$
$$CHOOCR'$$
$$|$$
$$CH_2OOCR''$$

Fig. 4.14. Chemical structure of a simple lipid which is the ester of glycerol and three unspecified fatty acid chains (R, R', and R").

Saturated fatty acids are, in general, straight chains with the formula $C_nH_{2n+1}COOH$. Unsaturated fatty acids are also known which have a double bond between one or more pairs of carbon atoms. Other lipids include waxes in which glycerol is replaced by a long-chain alcohol; phospholipids which contain nitrogen and phosphoruss; phingolipids and cerebrosides, most of which contain sphingosine; and lipoproteins which are cholesterol esters and phospholipids conjugated with proteins. Lipids are important in cells as nutrients and as structural elements.

Radiation effects on lipids. The most important reactions of irradiated lipids are those of the fatty acid components, especially the unsaturated fatty acids with two or more double bonds. The chemical configuration of one of the more sensitive of these, linoleic acid, is indicated in the top portion of Figure 4.15. When a solution of linoleic acid is irradiated, the point of attack of free radicals is the carbon between the double bonds. The hydrogen is removed and a resonating structure is formed, as shown in Figure 4.15. In the presence of oxygen, peroxyl radicals RO_2° (see equation 4-10) are readily produced and organic peroxides result.

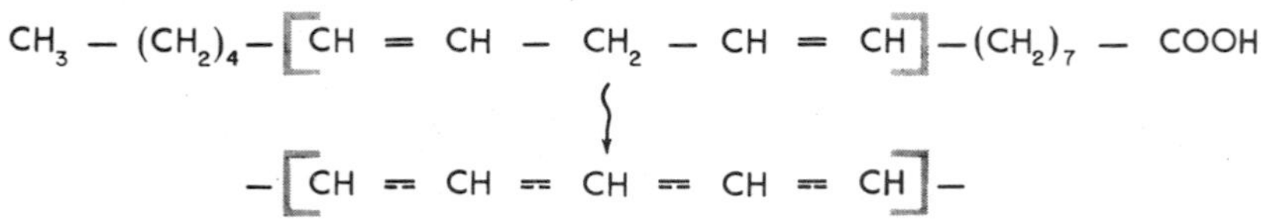

$$CH_3 - (CH_2)_4 - \left[CH = CH - CH_2 - CH = CH \right] - (CH_2)_7 - COOH$$

$$- \left[CH = CH = CH = CH = CH \right] -$$

Fig. 4.15. Resonant structure produced in linoleic acid by irradiation.

Free radical chain reactions often occur during the formation of organic peroxides. These lead to a very great increase in the number of molecules affected. A radical is formed which reacts with oxygen to form the peroxyl radical. The peroxyl radical removes a hydrogen atom from another fatty acid, leaving a radical which can react with oxygen, and so on. These reaction steps are

$$R^\circ + O_2 \rightarrow ROO^\circ \tag{4-55}$$

$$ROO^\circ + R'H \rightarrow ROOH + R'^\circ \tag{4-56}$$

$$R'^\circ + O_2 \rightarrow R'OO^\circ \tag{4-57}$$

It has been estimated that each of the free radical chain reactions may have a range of 0.0001 to 0.001 cm and may produce organic peroxides in between 20 and 1000 molecules.

In vivo irradiation of lipids. It appears that lipids are converted to organic peroxides when irradiated *in vivo*, although isolation of organic peroxides from biological systems has been difficult. Modern techniques, such as electron spin resonance, will permit the identification and quantification of organic peroxide radicals from irradiated tissue.

CARBOHYDRATES

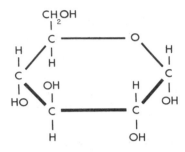

Fig. 4.16. Structure of D-glucose.

Carbohydrates are compounds composed of carbon, hydrogen, and oxygen. Certain members of the class also contain nitrogen or sulfur. Some of the simplest carbohydrates are the 5 and 6 carbon sugars called *monosaccharides*. Some of these are constituents of nucleic acids, and others serve as respiratory substrates in the cell. Carbohydrates generally exist, in solution, in the form of a ring which is in equilibrium with an open chain form. The ring form of glucose, a common simple sugar, is shown in Figure 4.16. Two or more monosaccharides may be joined to form disaccharides (2), trisaccharides (3), or polysaccharides (many monosaccharide units) by condensation. Very large branching molecules may be formed in this way, as is shown in Figure 4.17. Some polysaccharides, such as starches and glycogen, are carbohydrate reserves. Others, such as cellulose, are structural units in plants or animals.

Radiation effects on carbohydrates. Reports of radiation studies on polysaccharides are few in number and far from conclusive. Studies on starch and dry cellulose indicated that the primary direct action of radiation on polysaccharides is probably degradation or chain breaking.

A decrease in viscosity occurs in x-irradiated synovial fluid. This is attributed to depolymerization of the polysaccharide portion of the hyaluronic acid-protein complex. It has been assumed that this is an indirect action of the radiation because of the low concentration and high hydration of the hyaluronic acid polymer. No increase in viscosity has been measured at either high- or low-radiation doses, which suggests that no polymerization occurs. Following small doses of radiation, the endothelial layers of blood vessels become more permeable. This is suggestive of a change in hyaluronic acid.

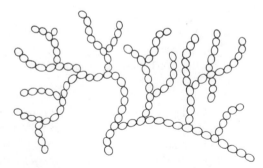

Fig. 4.17. Schematic representation of glycogen molecule.

Irradiation of sugars in dilute solutions, in the absence of oxygen, results in polymer production. Some authors have indicated that this may possibly be attributable to crosslinking. Extremely high doses will also produce a variety of degradation products from sugars. Such an effect is probably of minor importance in *in vivo* systems because of the high doses needed.

GENERAL REFERENCES

Alexander, P., and Charlesby, A. Energy Transfer in Macromolecules Exposed to Ionizing Radiations. *Nature*, **173:** 578–579 (1954).

Bacq, Z. M., and Alexander, P. *Fundamentals of Radiobiology*, Chapters 5, 6, 7, 8, 13, 2d ed. Pergamon Press, New York (1961).

Ebert, M., and Howard, A. (eds.). *Radiation Effects in Physics, Chemistry and Biology*. North-Holland Publishing Co., Amsterdam (1963).

Errera, M., and Forssberg, A. *Mechanisms in Radiobiology*, Vol. I, Chapters 2 & 3. Academic Press, Inc., New York (1960).

Feinstein, R. N. (ed.). Implications of Organic Peroxides in Radiobiology. *Radiation Res.*, Suppl. 3 (1963).

International Atomic Energy Agency. *Biological Effects of Ionizing Radiation at the Molecular Level*. IAEA, Vienna (1962).

Lea, D. E. *Actions of Radiations on Living Cells*, 2d ed. Cambridge University Press, Cambridge (1955).

White, A., Handler, P., and Smith, E. L. *Principles of Biochemistry*, 3d ed. McGraw-Hill Book Co., Inc., New York (1964).

ADDITIONAL REFERENCES CITED

Blois, M. S., Jr., Brown, H. W., Lemmon, R. M., Lindblom, R. O., and Weissbluth, M. (eds.). *Free Radicals in Biological Systems*. Academic Press, Inc., New York (1961).

Brinkman, R., Lamberts, H. B., and Quideveld, J. Contributions to the Study of Immediate and Early X-ray Reactions with Regard to Chemoprotection. II. Irradiation and Chemoprotection of Fresh Synovia as a Model of Mucopolysaccharide Depolymerization. *Intern. J. Radiation Biol.*, **3:** 279–283 (1961).

Collinson, E., Dainton, F. S., and McNaughton, G. S. The Effect of Acrylamide on the X- and γ-Ray Yields of Hydrogen Peroxide from De-aerated Water. *Trans. Faraday Soc.*, **53:** 357–362 (1957).

Fricke, H., Hart, E. J., and Smith, H. P. Chemical Reactions of Organic Compounds with X-ray Activated Water. *J. Chem. Phys.*, **6:** 229–240 (1938).

5 EFFECTS OF RADIATION ON THE CELL

Ionizing radiation acts on a biological system by altering the molecules of which it is composed. The initial chemical change is rarely detected directly and may be repaired almost immediately. Some of the molecular changes may, however, occur within important structures, and a small initial change may eventually result in alterations which are readily recognizable. The transition between a chemical change in a system and the biological manifestation of this change is complicated and often obscure. This chapter will describe the effects of radiation on the major structures and functions of a "generalized" cell with reference, when possible, to the underlying molecular alterations. For the purposes of orientation, the discussion will first review current concepts of the structure and function of the cell, with emphasis on those characteristics of the cell which are related to radiation damage.

THE CELL

The cell theory, first proposed in 1838 by Schleiden and Schwann, states that all living systems are composed of either small units called cells, or of cell products. Schleiden and Schwann did not recognize the complexity of the organization, structure, and function of the cells, however, and these are still areas of intensive investigation with the most advanced methods of instrumentation. The electron microscope, for example, provides, through its high degree of resolution, a powerful tool for cytologic study. It is revealing structural interrelationships that were previously unknown when the light microscope provided the only source of magnification.

The cell is the smallest independent unit of life. (Viruses are thought to be living organisms which do not possess a cellular structure, but they depend upon living cells for their continued existence.) To maintain life and to retain a certain level of organization, a cell must use energy and maintain itself in a condition called the *steady state*.[1] This requires a continuous supply of energy and an organization which is capable of absorbing and channeling the energy into the sites where it is needed. In addition, most cells must be able to reproduce themselves and to perform work.

Absorption and utilization of energy by a cell is a complex chain of events. Plant systems use and store the energy of light, beginning with photosynthesis, to build carbohydrates, proteins, and fats. Animals cannot use light energy directly, but must obtain their energy-supplying food and substances by eating plants or plant products (or other animals which, in turn, utilized plants in some form). The cells of both plants and animals utilize the energy of these substances by breaking them down to simpler substances in a series of enzyme-catalyzed oxidations which are, in turn, coupled with a process of phosphorylation. In this process, inorganic phosphate is chemically added to adenosine diphosphate (ADP) to form "energy-rich" adenosine triphosphate (ATP). When needed, the reverse reaction $(ATP \rightarrow ADP + P)$ serves as a convenient source of energy, which is made available by transphosphorylation and used for carrying out the work of the cell.

Most cellular activities appear to occur in or on specific structures or portions of the cell. Some cellular constituents have been well characterized chemically and structurally; some are still being investigated with the modern techniques of electron microscopy and cytochemistry. The major structures of a "generalized" cell are indicated in Figure 5.1.

Every cell is bounded by a cell membrane which is usually referred to as the *plasma membrane* in animal cells. This limiting structure is primarily a lipoprotein complex in which phospholipids are sandwiched between two layers of protein molecules, forming a strong, flexible, semipermeable structure. Some materials pass through the membrane by simple diffusion; some materials are blocked; others move through against a concentration gradient — from lower to higher concentration — by means of an energy-requiring mechanism called *active transport*. The cell membrane is of vital importance in cell function.

A continuous intracellular membrane network forms an interconnecting series of channels inside most cells. This complex is called the *endoplasmic reticulum*. Between the membranes of the endoplasmic reticulum, in the

---

[1]The "steady state" as used here is *not* the same as a chemical equilibrium. An equilibrium condition means that a system has achieved the lowest possible free energy and the highest possible disorganization and does not require the addition of energy to maintain this state. The structural and functional organization of a cell requires that it contain more than the minimum amount of free energy.

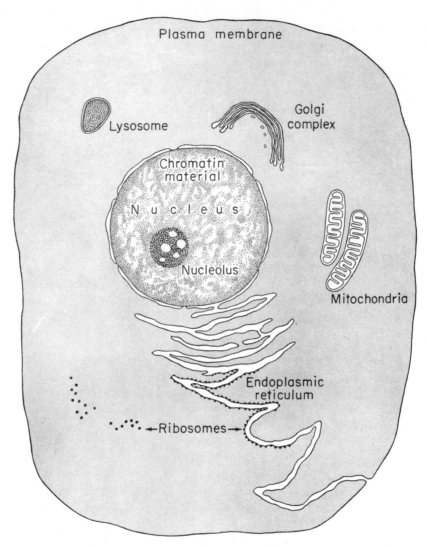

Fig. 5.1. The major structures and features of a generalized cell which will be considered in this chapter and which are probably most related to the radiation effects on the cell.

cytoplasm of the cell, are located inclusions such as ribosomes, mitochondria, lysosomes, Golgi complex, vacuoles, and secretory granules. The number, size, and shape of these inclusions vary according to the type and activity of the individual cell.

Along the cytoplasmic side of some portions of the endoplasmic reticulum one can observe small granules called *ribosomes*. They may also be seen in the cytoplasm of the cell unassociated with the endoplasmic reticulum. Ribosomes are believed to be the site of protein synthesis within the cell.

Mitochondria are also the site of important enzyme functions. Associated with their complex membrane structure are the respiratory enzymes which couple the oxidation and phosphorylation reactions resulting in ATP formation. The oxidative enzymes appear to be an integral part of the insoluble lipoprotein of the inner, folded mitochondrial membrane. The electron transfer between components of the oxidation system is thought to occur by a direct flow of electrons along the structure of the membrane. How the energy is transferred from the oxidation system to phosphorylate ADP to ATP is not well understood.

Lysosomes contain a number of the digestive or degradative enzymes of the cell. Lysosomes are postulated to break when a cell ages or degenerates, thus releasing the enzymes to digest or autolyze the cell.

The largest cellular inclusion, the nucleus, is ordinarily bounded by a double membrane, the outer layer of which appears to be continuous with the membrane of the endoplasmic reticulum. Within the nucleus are one or more nucleoli and the chromosomes of the cell. The nucleoli are visible only during interphase and prophase, that is, between and at the beginning of cell divisions.

Present evidence indicates that chromosomes are multiple-stranded structures. In the interphase period they are in a thin, coiled, filamentous form. At mitotic prophase, they become shortened and thickened and can be seen to be longitudinally divided into two equivalent portions called *chromatids*, each of which, presumably, is composed of many strands. The two sister chromatids formed from a single chromosome are connected at the centromere, a constriction in the chromosome. During metaphase, the chromosomes are oriented on the equatorial plane of the cell, and, at anaphase, the sister chromatids are pulled to opposite ends or poles of the cell. During telophase, the cell separates into its daughter cells and the chromosomes return to an interphase configuration.

Interphase is not a "resting" phase, as had been believed on the basis of morphological examination. Functional studies have indicated that this period can be further divided into three subperiods: a postmitotic nonsynthetic period (G_1), a synthetic period (S) during which duplication of DNA occurs, and a premitotic nonsynthetic period (G_2).

Chromosomes consist primarily of deoxyribonucleic acid (DNA) associated structurally with proteins (see Chapter 4). DNA is conceived to consist of a linear sequence of only four different nucleotides. The particular order in which these four nucleotides are strung together provides a huge amount of "information" — the information which is passed on to succeeding generations of cells when the DNA replicates itself in cell division and the information that is transcribed into messenger ribonucleic acid (RNA) copies. The messenger RNA copies move from the nucleus to the surface of the ribosomes where they provide the templates or "blueprints" for the translation of the information into polypeptide chains from which proteins are made.

Most somatic cells of an organism contain a number of chromosomes which is characteristic of the species. This number of chromosomes is maintained from division to division by the longitudinal separation of each chromosome into two chromatids prior to division. The reproductive cells of an organism undergo a meiotic (reduction) division. As a result, each cell contains only half of the somatic number of chromosomes. At fertilization, the parent cells fuse to form a cell which will, by repeated divisions and differentiation, form a new organism.

The many additional structures within the cell such as the Golgi apparatus, secretory granules, and secretory vacuoles do not, at present, appear to be affected to any major degree by ionizing radiation.

RADIATION EFFECTS ON CELL STRUCTURE AND FUNCTION

As was indicated in Chapter 4, all molecules can be altered by a sufficiently large amount of ionizing radiation. It is not surprising that massive doses of radiation can break and destroy the structural components of a cell and produce cell death. Exposure to smaller amounts of radiation can also result in cell death, but by less obvious mechanisms. For example, it will be pointed out in the following sections that radiation results in such widely varied effects as increased permeability of membranes, gross structural chromosome changes, and subtle chemical change in the structure of DNA molecules.

The effects of radiation on major cell functions and related cell structures and the gross effects on chromosomes and on mitosis will be discussed in the following sections. Genetic changes — changes in chromosomes or DNA which are not physically visible but which lead to mutations — will be discussed in Chapter 6.

RADIATION EFFECTS ON MEMBRANES

An intact membrane structure is essential for cell integrity. Large doses of radiation (3000 to 5000 rads) produce dramatic changes in the membrane systems of the cell. Most often, the plasma membranes are ruptured and the membranous structures of the endoplasmic reticulum are dilated. The nuclear membranes seem to be somewhat more resistant to physical altera-

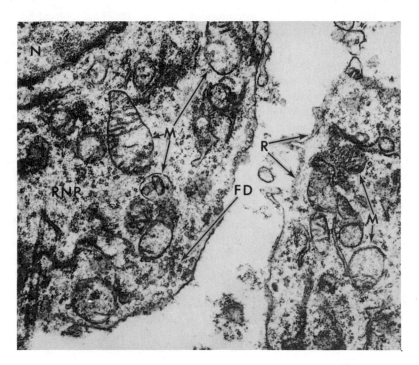

Fig. 5.2. Electron micrograph of portions of two cells of a G tumor fixed 24 hours after irradiation with 3000 rads. The limiting plasma membrane in one cell is ruptured (R). The mitochondria (M) in both cells are disorganized; some are swollen, some have a few cristae, others are empty. The ribosome particles (RNP) have coalesced. Very few intact vesicles of the endoplasmic reticulum are present. A portion of the nucleus (N) is seen in one cell, the limiting membrane of which is apparently intact. A number of disintegrated cytoplasmic components are lying within the empty space. These might be from neighboring cells which were injured. Note the fluid droplets (FD) lined up on the plasma membrane: (Courtesy A. Goldfeder [1963] and Laval Medical.)

tion than are the membranes associated with cytoplasmic structures. Mitochondria often appear swollen and the internal membranes (cristae) are disorganized, as illustrated in Figure 5.2. It has been suggested by Goldfeder (1963) that some of these changes may be due to altered permeability of the mitochondrial membranes. The intracellular fluid droplets (FD) in Figure 5.2 are also interpreted by Goldfeder as due to a change in permeability of the plasma membrane, permitting the extracellular fluid to enter the intracellular space. Additional evidence for altered permeability was described by the same author. She measured the relative concentration of an enzyme, catalase, in the mitochondrial and cell sap (nonparticulate portion of the cytoplasm) fractions of certain tumor cells. Following a large dose of radiation, an increased proportion of the enzyme was in the cell sap. She suggests that a change in the permeability or fragility of the mitochondrial membranous system permits a "leakage" or shift of the enzyme within the intracellular space.

Evidence from these and other experiments has been used to support the *Enzyme-Release Hypothesis* as proposed by Bacq and Alexander (1961). These authors have stressed the significance of intracellular barriers in maintaining the integrity and viability of the cell. They cite particularly the presence of proteolytic enzymes and nucleic acid attacking enzymes in the lysosomes of the cell and the fact that these enzymes are normally released when a cell dies to act on and digest cellular constituents. They suggest that radiation may act by breaking down the lysosomes[2] or other membranous barriers, thus allowing an interaction of enzymes and substrates that are ordinarily kept separate in an intact, viable cell.

In addition to the gross rupture of membranes, more subtle permeability changes occur which may presumably be due to alterations in the protein-lipid structure of the membrane. Among these are changes in transfer rates of single ions such as Na^+, K^+, and Ca^{++}, which are important to cellular function.

No experimental evidence, however, conclusively demonstrates that either permeability changes or the resulting "enzyme release" are more than partially responsible for cellular changes which result from moderate doses of radiation.

RADIATION EFFECTS ON ENERGY METABOLISM

Phosphorylation or ATP production is reduced in certain cells following moderate doses of radiation. This decrease appears to result from an uncoupling of the phosphorylation mechanisms from those of oxidation, since oxidation proceeds without phosphorylation following radiation exposure. Some defect occurs in either the electron transport system of the mitochondria or, more likely, in associated mechanisms of ATP formation. This effect is not universal, however. It has been observed in spleen mitochondrial fractions when they are irradiated *in vivo*, yet spleen mitochondria do not show this change when they are irradiated *in vitro*. Nor are such changes seen following irradiation of mitochondrial preparations from a radioresistant tissue such as liver.

Uncoupling phosphorylation from oxidation has been considered by some investigators to be a primary effect of radiation which can result in the death of the cell. Others suggest that it is associated with cell death and is secondary to other cellular changes.

---

[2]These lysosomes have been termed *suicide bags* by a number of authors because of their ability to destroy or "lyse" the cell. It must be noted that doses in excess of 1000 rads are required before a leakage of enzymes occurs in lysosomes irradiated *in vitro*.

RADIATION EFFECTS ON ENZYMES

The activities of many enzymes have been studied following irradiation of *in vitro* and *in vivo* systems. In general, high doses of radiation are required to inactivate enzymes *in vitro* (see Chapter 4). In most living systems in which alterations in enzymatic activity have been demonstrated, the radiation doses required are considerably higher than those which produce mutations or chromosome damage or which prevent cell growth or cell division.

Furthermore, in some experiments, measurements have been made hours or even days after the radiation exposure. At these times, cells are dying or have already died and been eliminated from the experimental system. It is often possible to explain changes observed in enzyme systems in terms of the altered nature of the cell population — a preponderance of radioresistant survivors or a change in the viability of remaining cells. Thus, alteration in the quantity or activity of existing enzymes does not appear to be an important initial effect of radiation, although it may be a secondary indication of other types of radiation damage.

RADIATION EFFECTS ON SYNTHETIC PROCESSES

A decrease in the rate of synthesis of DNA following radiation has been reported in experiments which measure the rate of cellular incorporation of radioisotope-labeled precursors of DNA. This effect may be partly a result of radiation decreasing the concentration or activity levels of the enzymes which regulate the synthesis of DNA, although, as indicated above, enzymatic activity does not appear to be a particularly radiosensitive process. Moreover, since it is well known that radiation exposure delays mitotic activity, decreased DNA synthesis may result from the delay rather than be a cause of it. In addition, the decrease in DNA incorporation may reflect a changing population of cells.

The synthesis of RNA may be delayed or depressed by irradiation. Figure 5.3 illustrates such an effect. The incorporation of carbon-14-labeled uracil (RNA precursor) was measured in irradiated lettuce seeds which had not yet germinated. It appeared that RNA synthesis was delayed by radiation and that, in addition, the total synthesis was depressed by the higher dose. Protein synthesis (measured by carbon-14-labeled leucine) was also delayed and, with the larger dose, was depressed in these seeds. It has been suggested that such effects may be due, indirectly, to some alteration in DNA molecules which makes them unable to function as a template for RNA production.

Similarly, x-irradiation of the bacterium *E. coli* can inhibit the formation of the enzyme beta-galactosidase under certain conditions. This enzyme will

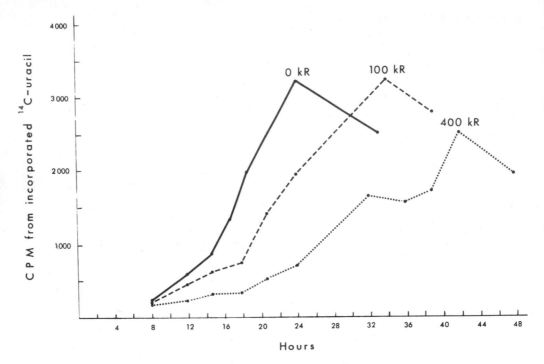

Fig. 5.3. Delay and depression in RNA synthesis in ungerminated lettuce seeds exposed to gamma radiation from cobalt-60, as measured by incorporation of carbon-14 labeled uracil. Abscissa is time after planting. (Courtesy B.P. Stone.)

be synthesized by the normal cell when it is suitably stimulated. If radiation is given before the stimulus is applied, formation of the enzyme is suppressed. This effect has been interpreted as caused by radiation damage to the DNA, so that it cannot serve as a template for the synthesis of a specific messenger RNA. Without that RNA, the beta-galactosidase cannot be synthesized. If the radiation is given after the stimulus has had a chance to function, sufficient messenger RNA will have been formed, and the enzyme synthesis will proceed despite the radiation.

This effect of radiation in rendering DNA molecules incapable of serving as a template for the production of normal messenger RNA has been suggested by Kaplan (1963) as a primary cause of death following moderate amounts of radiation. The end result is that some enzymes and other proteins cannot be formed, and the cell can no longer maintain its normal capacity for sustained cell division. It undergoes a "reproductive death."

Recently, an interesting technique has been used to show that very small alterations in the DNA molecules can influence the effects of radiation. Synthetic purines and pyrimidines can be prepared which differ from natural purines and pyrimidines only in a substitution of one chemical group (see

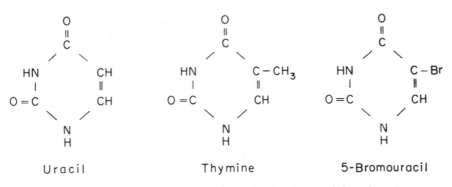

Fig. 5.4. Structural comparison of uracil, thymine, and brominated analogue of uracil. 5-bromouracil may be incorporated into DNA in place of thymine.

Figure 5.4). These synthetic analogues are incorporated into DNA molecules. For example, a uracil molecule which contains a substituted bromine atom at the carbon-5 position will resemble thymine sufficiently to be incorporated into DNA in the place of thymine. Thus, DNA which is produced in the presence of 5-bromouracil will contain this analogue in place of thymine in certain positions of its structure. Such cells are much more radiosensitive than those with nonsubstituted DNA, especially when loss of reproductive integrity is studied and when death appears to result from eventual inability to divide. Such experimental results suggest that DNA may be a very important site of radiation damage.[3]

RADIATION EFFECTS ON CHROMOSOMES

For many years it has been recognized that structural aberrations can be induced in chromosomes by radiation. These changes have been described and classified in great detail. More recently the aberrations have been studied from a biochemical point of view in attempts to determine the mechanisms involved. The more common aberrations are described in the following sections.

Structural aberrations. Structural changes in chromosomes can be produced by irradiation of cells at any stage of their mitotic cycle. They can most easily be seen microscopically when the cells are in the metaphase or anaphase stages of division; that is, when the chromosomes are short, com-

---

[3]If changes in membrane permeability, for example, were responsible for loss of reproductive integrity and death, the substitution of a DNA-incorporated analogue would not be expected to alter the radiosensitivity of the cell unless the altered DNA was serving as a template for the production of more sensitive cytoplasmic constituents.

pact, and easily visible as discrete structures. For the purposes of this discussion, the aberrations may be viewed as the result of breaks in chromosomes
or chromatids although it has been suggested that actual "breakage" may
not be the primary process.

Aberrations have been classified according to the portions of the chromosome which appear to have been involved in the initial alteration. Thus, a
"chromosome-type aberration" involves both chromatids of a chromosome at
identical loci and acts as if it is the result of a single "break" in the chromosome
before it has replicated into chromatids.
Under this situation, both chromatids
of a chromosome are affected identical-
ly. A "chromatid-type aberration"
appears to be the result of a change in
an individual chromatid after replication has taken place. These are produced by irradiation of chromosomes
which are either visibly separated into
chromatids or act as if they were
separated. "Subchromatid aberrations"
involve changes in only part of individual chromatids.

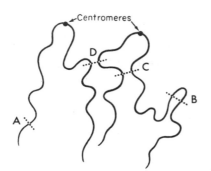

Fig. 5.5. Lesions leading to chromosome-type aberrations. (A) *Single break*
in one arm of a chromosome. (B) *Intra-
arm intrachange* of ends of two breaks
which occur in the same arm of a
chromosome. (C) *Interarm intrachange*
between broken ends from each arm
of a chromosome. (D) *Interchange*
between broken ends from two different
chromosomes.

Chromosome-type   aberrations.
Figure 5.5 illustrates the types of lesions
which may lead to chromosome-type
aberrations. The chromosomes are
drawn as thin, filamentous, twisted
structures as they probably exist in
interphase when these aberrations are produced. Figure 5.6 diagrams the
"breaks" as they would appear in interphase, if one can imagine the chromosomes to be contracted and straightened. The figure also illustrates, diagrammatically, the appearance of these aberrations at metaphase and anaphase.
Figure 5.7 shows some of these changes as they appear in photomicrographs
of *Tradescantia* microspores.

If a single break is produced in a chromosome, as in Figure 5.5 at (A),
the ends may rejoin in the process called *restitution*. In such cases, no lesion
will be visible in the chromosome, and, most likely, complete recovery will be
achieved. It has been determined that a very high percentage of these single
breaks do restitute. Estimates are as high as 99% restitution.

If the break does not heal, the end portion will remain as a chromosome
fragment. During cell division, this fragment will separate into two chromatid fragments which may be visible at metaphase. Since they do not possess
a centromere, these pieces will not move to the poles of the cell at anaphase,
but will remain at the equatorial plane, as indicated in Figure 5.6. They

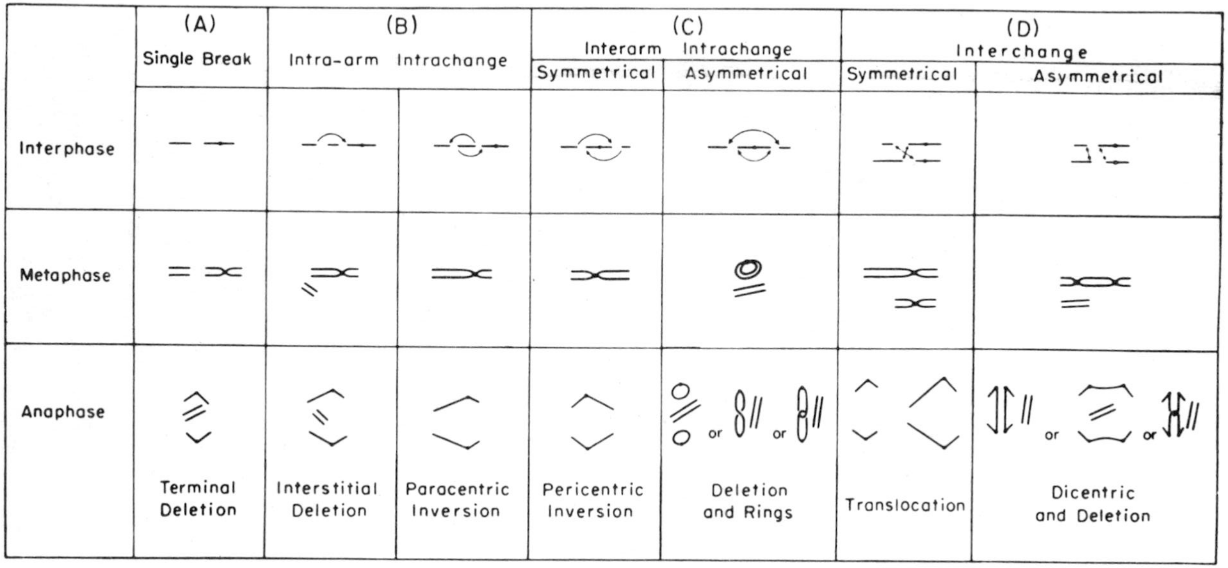

Fig. 5.6. Some chromosome-type aberrations. The lesions which occur in interphase are diagrammed (see also Figure 5.5) and some possible types of interactions between the broken ends are indicated. The diagrams for interphase are unrealistic, since the chromosomes are not in a rod-shaped configuration at that stage. The metaphase and anaphase diagrams show the result of the interactions and, diagrammatically, the appearance of the aberrations at these stages. Note that lesions of types (B), (C), and (D) all involve two breaks.

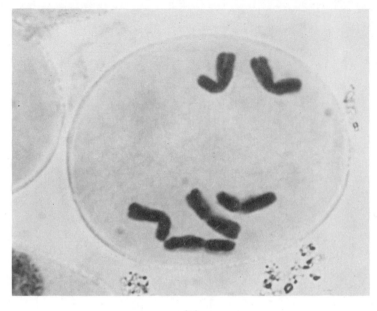

(A)

Fig. 5.7. Chromosomes of *Tradescantia* microspores in metaphase illustrating irradiation-induced chromosome-type aberrations. (A) Normal. (B) Dicentrics (1) from asymmetrical interchanges. (C) Deletions (2)

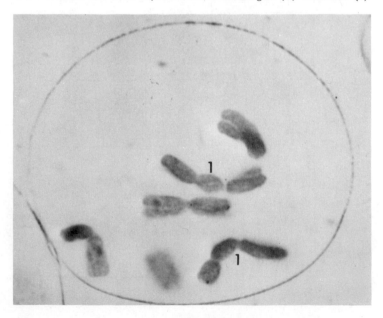

(B)

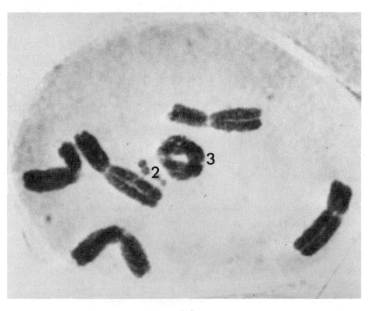

(C)

and ring (3) from asymmetrical interarm intrachange. (D) Translocation (4) from symmetrical interchange. (Courtesy H. Luippold.)

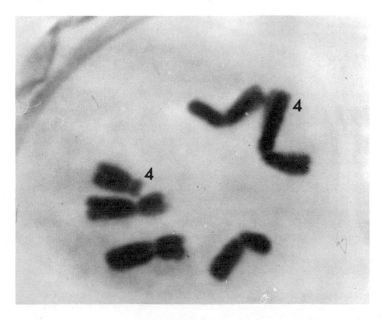

(D)

may both be incorporated into one of the daughter cells or may be lost. This type of chromosome aberration is called a *terminal deletion*.

If two breaks occur in the same arm of a chromosome, as in Figure 5-5 at (B), proper healing or restitution may occur. However, if the breaks are close together, an end from one break may join an end from the other break to form an intra-arm intrachange. As illustrated in Figure 5.6, the portion between the two breaks may remain free, as a fragment, and the terminal broken portion may join the rest of the chromosome. This results in an interstitial deletion or removal of an internal part of a chromosome. If the deletion is very small, it is usually called a *dot deletion*.[4]

All four of the broken ends of the chromosome may rejoin, however, but with the central portion inverted, as illustrated. This is referred to as a *paracentric inversion*. Such a change will usually not be visible microscopically[5] but may influence the function of the chromosome, as a result of a change in the order of the bases of the DNA.

When two breaks occur in the same chromosome, but in separate arms (C), lesions may be produced which are referred to as *interarm intrachanges*. Some of the more common of these are indicated in Figure 5.6. If the two fragments rejoin to the center portion, but at opposite ends, a "symmetrical intrachange" occurs which results in a pericentric inversion. This may be detectable if the two fragments are very unequal in length or it may not be visible. If union occurs between the two broken ends of the centromere-containing portion (asymmetrical intrachange), "ring" formation may result (Figure 5.7C). When the ring chromatids separate at anaphase, several different ring structures may result as illustrated, depending on the relative intertwining of the chromatid rings.[6]

Interchanges are the result of union of broken ends from two different chromosomes (D). If the two portions containing centromeres unite (an asymmetrical interchange), a dicentric is formed which contains two centromeres within a single chromosome structure (see Figure 5.7B). At anaphase, there may be a dicentric chromatid at each pole, each chromatid may stretch between the two poles, forming a "bridge," or the dicentrics may be hooked together. The acentric fragments (no centromere) may go to one daughter cell, be split between them, or be lost.

[4]Many of the dot deletion fragments are about 1μ or less in length. Since this is approximately the estimated size of the minor spiral or "relic coils" of chromosomes, it has been suggested that these dot deletions represent the deletion of 1 coil.

[5]Specific bands can be identified in certain chromosomes, such as those in the salivary glands of *Drosophila*. Inversions can be recognized in such situations. In addition, inversions can be identified by the presence of certain loops at meiosis, when chromosomes are paired.

[6]This can be demonstrated by the reader as follows: Join the ends of a strip of paper to form a single ring (ring chromosome). Cut the paper in half, lengthwise (chromosome divided into 2 chromatids). Two rings will result. Now, join the ends of another strip of paper, but first twist the paper once, so that a twisted loop is formed. When this is cut, a single large ring will appear. Next, join the ends of another strip of paper, but first twist it twice. This time, two interlocking rings result.

In a symmetrical interchange, each acentric fragment joins a centromere-containing portion to form a translocation (Figure 5.7D). If the fragments are of similar lengths, this exchange may be difficult to see; but if the fragments are of different lengths, the aberration may be obvious.

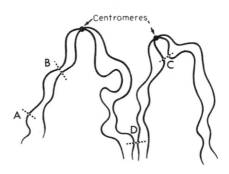

Chromatid-type aberrations. A second major group of aberrations are those which appear to arise from "breaks" in chromatids. Figure 5.8 illustrates some of the lesions which apparently result in the most common chromatid aberrations. Figure 5.9 diagrams these aberrations as they might appear in prophase, if the chromosomes were contracted and straightened, and

Fig. 5.8. Lesions leading to chromatid-type aberrations. (A) *Single break* in one chromatid. (B) *Sister union* of breaks in two sister chromatids which occur a the same location. (C) *Interarm intrachange* of ends of two breaks which occur in the same chromatid arm. (D) *Interchange* between chromatid breaks in two different chromosomes.

as they actually appear at metaphase and anaphase. Figure 5.10 shows some of the chromatid lesions in irradiated *Tradescantia* microspores.

If a single chromatid is broken (A), the free end may rejoin (restitution) or may remain as a single fragment or terminal deletion. If both chromatids

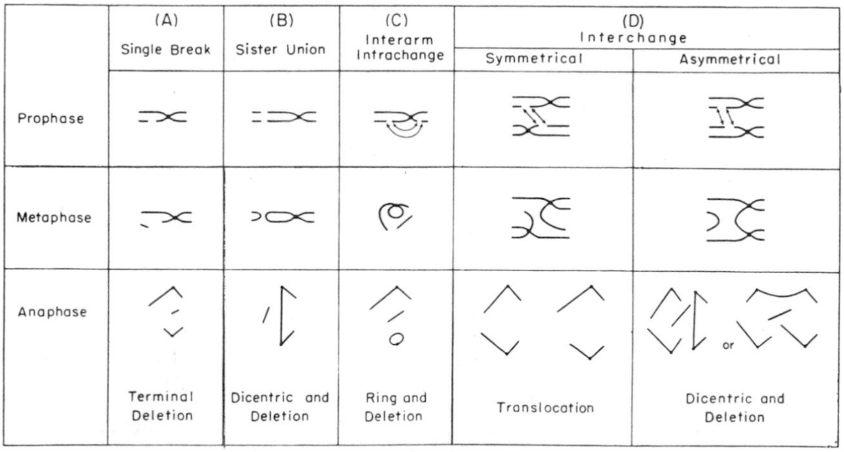

	(A) Single Break	(B) Sister Union	(C) Interarm Intrachange	(D) Interchange	
				Symmetrical	Asymmetrical
Prophase	⟩⟩⟨	⟩⟩⟨	⟩⟨	⟩⟨	⟩⟨
Metaphase	⟩⟨	⟩⟨	⟨⊙⟩	⟩⟨	⟩⟨
Anaphase	⋀ ⋁	⟋ ⟩	⋀ ○	⋀ ⋁	⋀ ⋁ or ⋀ ⋁
	Terminal Deletion	Dicentric and Deletion	Ring and Deletion	Translocation	Dicentric and Deletion

Fig. 5.9. Some chromatid-type aberrations. These lesions are shown diagrammatically as they may be produced by radiation delivered in late interphase (G_2) or early prophase. Also indicated are some possible interactions of the broken ends and the result of these interactions as seen in metaphase and anaphase. Lesions of types (B), (C), and (D) all involve two breaks.

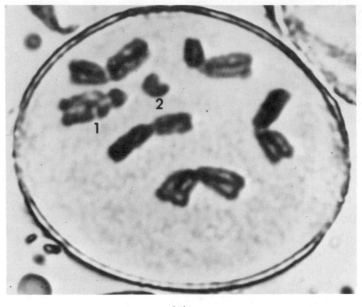

(A)

Fig. 5.10. Chromosomes of *Tradescantia* microspores illustrating irradiation-induced chromatid-type aberrations: (A) Dicentric (1) and deletion (2) from sister-union, (B) Dicentric (3) and deletion (4) from

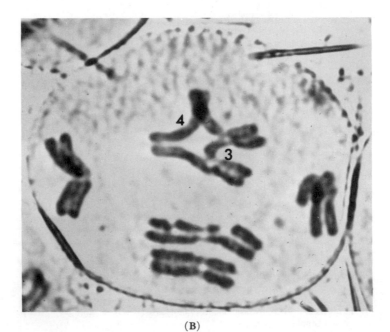

(B)

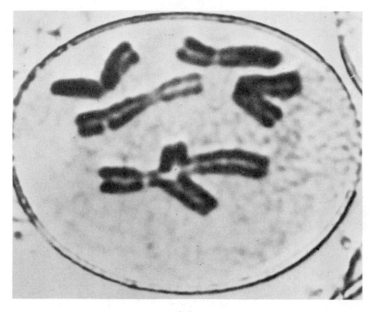

(C)

asymmetrical interchange, (C) Translocation from symmetrical interchange, (D) Dicentric and deletions resulting in anaphase bridge.
(Courtesy H. Luippold.)

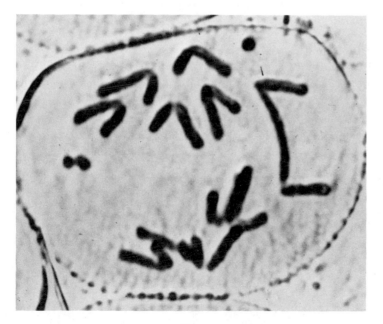

(D)

are broken at the same locus (B), the two proximal (toward the centromere) ends may join forming a sister union (see Figure 5.10A). This dicentric will usually form a bridge at anaphase. There may be one or two fragments remaining, depending on whether or not the two distal (away from centromere) portions of the chromatid join. Figure 5.9 illustrates the formation of one fragment.

A chromatid break in each arm of a chromosome (C) can lead to an interarm intrachange. As a result, at anaphase, there is one normal chromatid, one ring chromatid, and one deletion, as illustrated.

If breaks in the chromatids of different chromosomes are involved (D), interchanges occur, as illustrated in Figure 5.9. If the proximal end of one broken chromosome joins the distal end of the other, and vice versa, the interchange is termed *symmetrical*. The resulting translocation may be seen at metaphase before the chromatids separate (Figure 5.10C) but may not be visible at anaphase, depending on the relative lengths of the fragments.

If the two proximal broken ends join in an asymmetrical interchange, a dicentric and deletion will be formed (Figure 5.10B). At anaphase there will be two normal chromatids, one dicentric, and one or two deletions.

Many other types of chromosome and chromatid aberrations are also seen. It is not necessary in any of the multiple-break aberrations that all of the broken ends join other ends. In the formation of a chromosome ring, for example, two sets of fragments may be present. Those are the two terminal ends which have remained separated rather than joining to form one fragment. In addition, more than two breaks may be produced in chromosomes or chromatids, and various combinations of unions may occur.

Subchromatid-type aberrations. A third class of visible chromosome changes is that of subchromatid aberrations. These are seen less frequently than the chromosome and chromatid effects and involve portions of chromatids. They are formed when the cells are irradiated in late prophase or early metaphase and frequently take the form of partial bridges, where only part of the chromatid is stretched between the poles at anaphase.

Gaps. Under certain conditions, when irradiated chromosomes are examined, there are "gaps" in the chromosomes. A fragment appears to be lined up directly with the remainder of the chromosome, but there is no obvious connection between the pieces. Originally, such "gaps" were classified as breaks, but it now appears that they are not true discontinuities, since they do not lead to acentric fragments at anaphase. Various authors have suggested that they represent partial breaks, discontinuities in staining, a localized despiralization of the chromatids, or a loss or depolymerization of DNA. They are apparently reversible, since they appear in relatively high numbers shortly after exposure, but are infrequent at later time periods (see Figure 5.11).

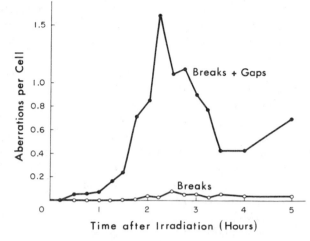

Fig. 5.11. Frequency of "real" chromatid breaks and "conventional" chromatid breaks (that is, "real" breaks + gaps) scored at metaphase after about 50 R. This distinction has been proposed by S. H. Revell. (From Revell [1959], courtesy Royal Society London.)

Chromosome stickiness. When cells are irradiated just as they enter division, there is apparently some change in the surface properties of the chromosomes which cause them to adhere to each other when they happen to touch. This "stickiness" has been attributed to a partial dissociation of the nucleoproteins and an alteration in their pattern of organization. There is some evidence to suggest that such changes may be of the subchromatid aberration type and involve exchanges between these subunits.

When the chromosomes stick to each other in this way, they appear clumped at metaphase and are often unable to separate completely at anaphase. This may lead to the formation of an anaphase bridge.

Chromosome "stickiness" appears to be reversible. Cells which have been irradiated in interphase or early prophase and which have been delayed in entering mitosis do not show this type of effect. This is presumably because there has been time for repair to occur during the delay.

RELATION BETWEEN ABERRATION STRUCTURE AND THE MITOTIC OR MEIOTIC CYCLES

Chromosomes are sensitive to aberration induction at all stages of the mitotic or meiotic cycle. However, certain types of aberrations are produced by irradiation of certain stages, and the relative efficiency of production differs at different stages.

Irradiation of cells in the G_1 (presynthetic) portion of interphase results in the appearance of chromosome-type aberrations at the next metaphase. During the synthetic period (S), either chromosome or chromatid types are produced. Irradiation in G_2 (postsynthetic) leads to chromatid aberrations. Thus, it seems that once the DNA has duplicated, the chromosomes act as paired chromatids, even though a physical separation of chromatids[7] is not apparent. Irradiation of mitotic prophase cells results in subchromatid aberrations such as stickiness. In meiosis, subchromatid aberrations may be produced by irradiation as late as the first metaphase division. Aberrations are produced in cells which are in metaphase or beyond, and, indeed, such cells are extremely sensitive to such effects. However, the changes cannot be scored until the *next* division of the cell.

The cells which are least susceptible to structural alterations are those in the presynthetic G_1 period. The G_2 cells are more sensitive and S cells vary in sensitivity depending on the extent of the DNA duplication that has taken place in the cell. Cells in division are most susceptible. Furthermore, cells in meiotic division (especially in diplotene and metaphase I stages) are more sensitive than those in mitotic division and have been estimated to be 10 times as sensitive as resting cells.

Cells in metaphase or anaphase are about three times as sensitive as those in prophase. It has been suggested that restitution of breaks is less likely when the chromosomes are in movement and that gaps produced in metaphase do not have a chance to heal before the stress of anaphase separation is imposed on the chromosomes. Accordingly, breaks may result from these gaps.

RELATION BETWEEN ABERRATIONS AND "HITS"

Certain of the aberrations which have been described are apparently produced by the passage of a single particle or photon, or are the result of a

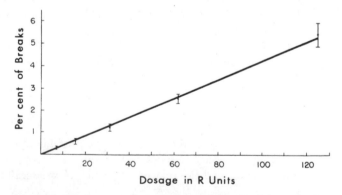

Fig. 5.12. Relation between x-ray dosage and the production of single chromosomal breaks in grasshopper neuroblasts. (Courtesy J. G. Carlson [1949] and National Academy of Sciences.)

[7]It must be remembered that an increase in nuclear histone also occurs during the S phase. This increase in DNA-associated protein might be of great importance in the production of aberrations.

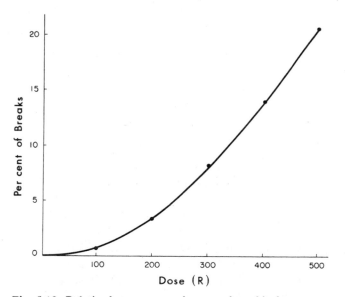

Fig. 5.13. Relation between x-ray dosage and two-hit chromosome breaks in *Tradescantia* microspores. (Modified from Sax [1949], courtesy University of Texas Press.)

single break. They are usually called *one-hit aberrations*. Simple deletions, certain of the dot deletions, and sister unions[8] are examples of these. Other aberrations such as rings or dicentrics require an exchange of parts between chromosomes or chromosome arms (or chromatids) and presumably require two breaks. These are referred to as *two-hit aberrations*.

When the total number of breaks is low, the incidence of single breaks has been found to increase linearly with the dose of radiation (see Figure 5.12). The rate at which radiation is delivered does not appreciably influence the number of breaks formed. Moreover, no minimum dose or threshold exists for this effect.

In contrast, the frequency of two-hit aberrations from x- or gamma radiation increases more rapidly than a single power of the radiation dose[9] (see Figure 5.13), and the number of two-hit aberrations is dependent on the dose rate. If the duration of exposure is varied, and the total dose of radiation is kept constant, more two-hit aberrations are scored following a short-time,

---

[8]If the two breaks in spiraled interphase chromosomes are sufficiently close together, they can presumably be produced by the passage of a single particle. They are thus classed as one-hit effects.

[9]Actually, the incidence of two-hit aberrations is usually found to increase as the dose to the 1.5 power, 1.7 power, or 1.8 power. The discrepancy between this and a simple dose squared relationship has been attributed to the presence of a mixture of one- and two-hit events in the low dose range and to an effect from a limited number of sites for breakage, which is important at high doses. (See Evans in Wolff, 1963.)

high-intensity exposure than following a long, low-intensity exposure. This suggests that the two breaks must be produced within a certain period of time and that recovery can occur.

There appear to be different kinds of breaks, some of which will heal within a few minutes and some of which stay open several hours. Certain authors, such as Read (1959), have suggested that the rapidly repaired breaks involve ionic bonds formed by divalent ions of calcium and magnesium. Repair of these breaks is not dependent on oxidative metabolism. The breaks which repair more slowly, however, seem to be dependent on the presence of both oxidative metabolism and protein synthesis.

Densely ionizing radiations, such as alpha particles, produce the two-hit aberrations in numbers proportional to the dose of radiation. This suggests that an individual alpha particle forms multiple breaks which are sufficiently closely spaced to permit interaction of the broken ends and, thus, the formation of a two-hit aberration. In contrast, the sparsely ionizing radiations, such as x-rays or gamma radiation, produce a somewhat randomly distributed pattern of single breaks. Two of these single breaks must occur close together spatially, within a certain interval of time, for a two-hit aberration to result.

SIGNIFICANCE OF CHROMOSOME ABERRATIONS

As has been indicated earlier, chromosomes are the primary information system of a cell. The sequence of bases on the DNA molecules appears to constitute the code for this information system which directs the formation of all other cellular constituents.

If a portion of a chromosome is broken, inverted, or exchanged, the cell may quite probably survive for a considerable period of time, since it still has a full quantity of DNA. If, however, the linear base sequence has been altered at a critical point, certain information will not be available to the cell. For example, it might be unable to form certain structural elements or enzymes when they are needed. As a result, the function of the cell might eventually be impaired. If division occurs in a cell which contains fragments or exchanges, some of the genetic information will not be transmitted to the daughter cells. This may be immediately lethal or may lead to disfunction or death at a later period. Many of the more extensive aberrations are lethal to the cell as soon as it attempts to enter division.

It is apparent that the chromosome effects which have been described are potentially lethal to a cell, either immediately after exposure to radiation or at some later period.[10] Less obvious alterations in chromosomes can pro-

[10]It is well known that many cells survive without certain of their chromosomal components. However, it must be remembered that such cells are missing some of their genetic information and, thus, are presumably less "fit" for survival than are cells with a full genetic complement.

duce alterations in cell function or synthetic activity as was discussed earlier in this chapter or may lead to inheritable alterations as will be described in Chapter 6.

RADIATION EFFECTS ON CELL DIVISION

Radiation influences the process of cell division. Some of the radiation effects on mitosis have been clearly demonstrated by Carlson[11] with the use of hanging drop preparations of grasshopper neuroblasts. This technique permits visualization of individual cells throughout their mitotic cycle.

Examination of irradiated neuroblasts has revealed the presence in these cells of a critical period between late and very late prophase. At 38°C, this period occurs approximately 5 minutes before the nuclear membrane disappears, when the chromosomes are almost at the end of prophase contraction, the nucleoli are disappearing, and the cytoplasmic viscosity is falling rapidly. If a cell is exposed to a small dose of radiation after this period, mitosis proceeds. If radiation is given before this time, mitosis stops or the cell reverts to an earlier prophase or interphase configuration. The specific time of irradiation prior to the critical phase is important. For example, if exposures as low as 8–16 R are given during middle or late prophase, mitosis will stop for a short time. If low-dose irradiation is done during interphase or early prophase, the mitotic process progresses more slowly as the cell nears the critical stage, but cell division does not stop.

As the radiation exposure is increased, the delay is longer. With 250 R or more to prophase cells, delay is followed by a reversion or simulated reversion to a stage in which the chromatin resembles that present in interphase nuclei. Exposure of cells beyond the critical phase to 250 R or more may cause a very slight delay in the remaining mitotic stages. The lag is due, at least in part, to a delay in the breakdown of the nuclear membrane and to chromosome "stickiness" which prolongs the anaphase period.

Comprehensive studies[12] of eggs of *Arbacia* and other marine invertebrates have indicated a delay in the first cleavage when irradiation is given to the unfertilized egg, the sperm, or the newly fertilized egg. Maximum delay is measured when radiation is delivered during the period between the fusion of the pronuclei, that is, egg and sperm nuclei, and early prophase. Treatment during very late prophase, metaphase, anaphase, or telophase does not retard division.

Recent studies[13] on tissue culture preparations have indicated that the mitotic process in mammalian cells can be modified by radiation given dur-

[11]Carlson in Hollaender (1954).

[12]Work of Henshaw and co-workers and Yamashita *et al.* quoted by Carlson in Hollaender (1954).

[13]The following examples of mitotic effects on mammalian cells are reported by Lesher (1963).

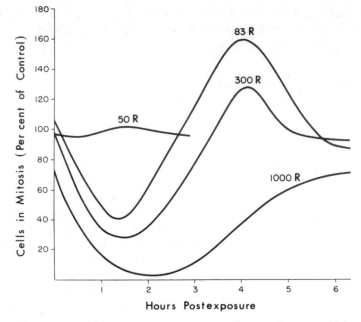

Fig. 5.14. Mitotic effects of gamma rays from radium on chick fibroblasts *in vitro*. (Modified from Canti & Spear [1929], courtesy Royal Society London.)

ing any of the stages of the mitotic cycle, including interphase. The specific response which is measured depends on the type of cell, the normal mitotic rate, the dose of radiation, the dose rate, and the exposure conditions. The results of many experiments may be summarized as follows.

Most cells which are already in division at the time of irradiation complete the division; those that are about to enter division may be stopped in the G_2 stage. Synthesis of DNA may be partially or completely inhibited in cells in S phase. Cells in G_1 may be held back from entering the synthetic phase.

Relatively low doses of irradiation block the cells in G_2. For example, exposures as low as 50 R to human kidney cells in tissue culture result in mitotic delay. Another type of cultured cell, the mouse L-cell, was delayed approximately 12 hours in G_2 by 200 R and about 19 hours by 5000 R.

As has been indicated, radiation will delay cells which are irradiated in G_2 and will cause prophase cells to delay or to revert back to the G_2 configuration. If the less sensitive G_1 and S phases have not been delayed, cells pile up in G_2 as a result of the normal progression of the G_1 and S cells into G_2, the holding of G_2 cells in that stage, and the reversion of prophase cells back into G_2. When the inhibited cells recover, more cells enter mitosis than would be seen in a nonirradiated sample. This "increase" in mitotic cells has given rise to the notion that irradiation can enhance the mitotic rate. Radiation has not truly increased mitosis,[14] but, rather, it has altered the timing of some of the cells and has thus resulted in an increased number of mitotic figures at a specific time interval after exposure. Figure 5.14 illustrates such an effect following irradiation of chick fibroblasts. As indicated, the 1000 R exposure has apparently inhibited cells in all stages of the cell cycle. There is no pile up of mitotic cells at any time after this high dose of radiation.

PRIMARY SITE OF
RADIATION DAMAGE

A variety of morphological and functional changes have been described which occur in irradiated cells. The bulk of both direct and inferential evidence suggests that cell nuclei are a major site of the radiation damage leading to cell death. For example, Patt[15] has calculated that a million rads are required to inactivate certain enzyme systems in cells. Doses of a thousand roentgens or more are usually required to damage membranes. In contrast, chromosomal aberrations and mutations are produced by low radiation doses. Since only a few hundred roentgens are needed to produce a high degree of lethality in most mammalian cells in tissue culture, it seems most logical that the nuclear changes produced by the low doses are responsible for the cell death.

It is pertinent to mention several additional experiments which relate to this problem. Very precisely focused microbeams[16] of alpha particles, protons, x-rays, or other radiations have been used to irradiate specific areas of individual cells. In most of these experiments much larger doses are required to produce cell death or mitotic delay when the cytoplasm is irradiated than when the cell nucleus is irradiated.

Another series of studies by Ord and Danielli (1956) has involved transplantation and irradiation of parts of amoeba. An irradiated nucleus combined with nonirradiated cytoplasm is more than twice as sensitive as a nonirradiated nucleus combined with irradiated cytoplasm, when the criterion of sensitivity is death of the resulting cell. When death does occur, however, it takes place much more rapidly in the latter combination. It is likely that different mechanisms of death may be involved in the two situations.

Sparrow has found marked correlations in plants between cell sensitivity and many nuclear parameters such as chromosome number, nuclear volume,

---

[14]However, the suggestion has been entertained that duodenal crypt cells which are irradiated daily will compensate for radiation damage by a reduction in the duration of the G stage. This results in a decrease in the duration of the total mitotic cycle.

[15]Cited by Gray (1954).

[16]For example, see Zirkle (1957).

interphase chromosome volume, and DNA content (see Chapter 8). These factors support the case for the nucleus as a major site of radiation damage.

Alterations in membrane permeability and integrity cannot be disregarded. Cytoplasmic inclusions, such as mitochondria, are important, especially in relation to the potential of a damaged cell for repair. Changes in cytoplasmic structures may also contribute to radiation-induced lethality, especially in cells which are not in active division. Moreover, precise relationships between cytoplasmic factors and nuclear factors are not clear.

It seems most probable that cell death following irradiation cannot be attributed to any one mechanism. More likely, any one of several changes can be lethal to the cell, depending on the type of cell studied, the activity of the cell at the time of exposure, and the exposure conditions.

GENERAL REFERENCES

Bacq, Z. M., and Alexander, P. *Fundamentals of Radiobiology*, Chapters 9 & 10, 2d ed. Pergamon Press, New York (1961).

Errera, M., and Forssberg, A. *Mechanisms in Radiobiology*, Vol. I, Chapters 4, 5, 6. Academic Press, Inc., New York (1960).

Evans, H. J. Chromosome Aberrations Induced by Ionizing Radiations. *Intern. Rev. Cytol.*, **13**: 221–321 (1962).

Hollaender, A. *Radiation Biology*, Vol. I (2), Chapters 9, 10, 11. McGraw-Hill Book Co., Inc., New York (1954).

Kaplan, H. S. Biochemical Basis of Reproductive Death in Irradiated Cells. *Am. J. Roentgenol., Radium Therapy Nucl. Med.*, **90**: 907–916 (1963).

Lea, D. E. *Actions of Radiations on Living Cells*, 2d ed. Cambridge University Press, Cambridge (1955).

Sax, K. The Effect of Ionizing Radiation on Chromosomes. *Quart. Rev. Biol.*, **32**: 15–26 (1957).

Wolff, S., and Luippold, M. E. Metabolism and Chromosome-Break Rejoining. *Science*, **122**: 231–232 (1955).

ADDITIONAL REFERENCES CITED

Canti, R. G., and Spear, F. G. The Effect of Gamma Irradiation on Cell Division in Tissue Culture *In Vitro*, Part II. *Proc. Roy. Soc. (London) Ser. B*, **105**: 93–98 (1929).

Carlson, J. G. An Analysis of X-ray Induced Single Breaks in Neuroblast Chromosomes of the Grasshopper (*Chortophaga Viridifasciata*). *Proc. Natl. Acad. Sci. U. S.*, **27**: 42–47 (1941).

Goldfeder, A. Radiosensitivity at the Subcellular Level. *Laval Med.*, **34:** 12–43 (1963).

Gray, L. H. Some Characteristics of Biological Damage Induced by Ionizing Radiations. *Radiation Res.*, **1:** 189–213 (1954)

Lesher, S. Radiosensitivity of Rapidly Dividing Cells. *Laval Med.*, **34:** 53–56 (1963).

Ord, M. M., and Danielli, J. F. The Site of Damage in Amoebae Exposed to X-Rays. *Quart. J. Microscop. Sci.*, **97:** 29–37 (1956).

Read, J. *Radiation Biology of* Vicia faba *in Relation to the General Problem.* Charles C Thomas, Springfield, Illinois (1959).

Revell, S. H. The Accurate Estimation of Chromatid Breakage, and Its Relevance to a New Interpretation of Chromatid Aberrations Induced by Ionizing Radiations. *Proc. Roy. Soc. (London) Ser. B*, **150:** 563–589 (1959).

Sax, K. An Analysis of X-Ray Induced Chromosomal Aberrations in Tradescantia. *Genetics*, **25:** 41–68 (1940).

Wolff, S. (ed.). *Radiation-Induced Chromosome Aberrations.* Columbia University Press, New York (1963).

Zirkle, R. E. Partial-Cell Irradiation. *Advan. Biol. Med. Phys.*, **5:** 103–146 (1957).

RADIATION
GENETICS

The science of genetics is concerned with the study of heredity. The specific linear base sequences of the nucleic acids in a cell determine the activities of the cell and the characteristics of the individual. These base sequences are carried in the chromosomes and are transmitted to the next generation when the cell divides.

A change in any specific linear sequence changes the information which is passed on. This is a mutation. Such a broad definition of mutations includes gross structural changes in chromosomes, such as the chromosome and chromatid aberrations discussed in Chapter 5. In this chapter, the term mutation will be limited to a more restricted meaning, namely, the alterations that are not visible as aberrations of the chromosomes, but are recognized by a change in phenotype or survival of the progeny.

The postmeiotic sex cells of each organism contain a specific number of chromosomes, designated the haploid number (n). In man, $n = 23$. Somatic cells of higher organisms are usually diploid $(2n)$. If the diploid cells of an individual contain identical genes at a given locus of a given chromosome pair, the individual is homozygous at that locus; if the genes on the two chromosomes differ at the locus, the individual is heterozygous. Dominant mutations are expressed in the immediate descendants of the mutated cell; that is, they are revealed in the first generation offspring of an individual in whose germ cells the mutation has occurred (genetic mutation), or in the daughter cells of dividing somatic cells in which a mutation has occurred (somatic mutation). Recessive mutations are only expressed if both chromosomes of the cell contain the mutation, or if there is no dominant gene to mask the recessive characteristic.

GENETIC MUTATIONS

Although the biological importance of genes and gene mutations has been recognized for many years, the study of mutations was limited for a long time by the low natural frequency of their occurrence and by the difficulty of increasing their occurrence by artificial means. Prior to 1927, there were indications that radiation might produce gene mutations; however, the experimental results were not clear. In that year, Dr. H. J. Muller, then at the University of Texas, reported the occurrence of "true gene mutations" in a high proportion of sperm from irradiated fruit flies (*Drosophila melanogaster*).

In a paper entitled "Artificial Transmutation of the Gene," Muller (1927) described the production of several hundred mutant flies by treatment of sperm with relatively heavy doses of x-rays. He found that his heaviest x-ray treatment produced detectable mutations in a particular chromosome (the X chromosome) in about one-seventh of the offspring that hatched from eggs fertilized by the treated sperm. At the time of his report, many of these mutants had been followed through four or more generations, most were stable in their inheritance and "behaved in the manner typical of the (Mendelian) chromosomal mutant genes found in organisms generally."

Recessive lethals, semilethals (mutants having a viability between about 0.1 and 0.5 of some standard viability), and visible mutations were recorded. In addition, Muller was able, for the first time, to obtain evidence for the occurrence of dominant lethal mutations. Since the zygotes receiving dominant lethals never develop to maturity, such lethals cannot be detected individually. Their number in Muller's experiments was so great that they could be detected indirectly by egg counts and by effects on the sex ratio of the offspring.

While Muller was engaged in his earliest successful experiments with *Drosophila*, quite independent studies of radiation-induced mutations in plants were being done by Stadler in Missouri and by Goodspeed and Olson in California. The latter authors irradiated mature plants (containing numerous flower buds) of a species of *Nicotiana*, permitted self-pollination of irradiated buds, and observed "variant" plants in the next generation. In one group of plants studied, 136 out of 168 were some type of variant. For example, plants were dwarf, low, or tall; leaves were large, small, smooth, wavy, corrugated, or different in color; flowers were small, long, fluted, notched, light pink or reddish purple. Many plants exhibited some reduction in fertility. Second generation seedlings were also variants.

Stadler used x-irradiation to produce mutations in maize and barley. He found that the percentage of mosaics[1] in x-rayed corn was more than 20 times as great as in the untreated series. He interpreted these results as due to chromosome aberrations rather than to gene mutations.[2] Stadler also irradiated germinating barley seeds and observed mutants among the progeny in the next generation. Most aberrant seedlings showed chlorophyll defects: "white" (colorless from emergence), "virescent" (colorless from emergence, but later developing a pale green color), or "yellowing" (green at emergence, but becoming yellow at about one week). Older seedlings showed a variety of variegation patterns such as "banding" (transverse white bands) or striping (two distinct patterns), as well as morphologic abnormalities such as "tapering."

Following these pioneering studies of Muller, Stadler, and others, ionizing radiation was used to induce lethal and phenotypically visible point mutations in a wide range of types of living organisms. The relatively simple microorganisms have been of great interest, since most are haploid. Therefore, a "recessive" mutation is not hidden[3] by an intact site on a duplicate chromosome.

Many different types of mutations have been studied in irradiated bacteria. A typical experiment is as follows, using E. coli. Wild-type E. coli will grow in a medium composed of only salts, water, and a carbon source. Strains of this bacteria are known which require vitamins or amino acids, in addition. When the requiring strains are plated on the minimal medium (without the additional growth requirement), they will not grow. Any colonies which do grow are generally considered to be the result of a mutation to the original type. Radiation of the bacteria increases the number of mutants. Such a selective system is extremely valuable for studying mutation rates, since only the mutated forms are observed. Therefore, large numbers of bacteria can be irradiated without the requirement that a colony from each irradiated cell must be examined in order to detect mutated forms. In many other experiments, very large numbers of individuals have been studied in similar ways for a wide variety of phenotypic mutants.

Viruses are another useful group of microorganisms for radiation studies. For example, there are strains of tobacco mosaic virus which normally grow

[1]Nutrient material in the corn seed (endosperm) has a distinctive visible color or consistency which is genetically controlled. If a plant with a recessive endosperm character is pollinated by the corresponding dominant, heterozygous seeds are produced. Occasionally, mosaic individuals result in which a portion of the endosperm shows the recessive character. These can be attributed to chromosome deficiencies at the site of the dominant gene or to mutations of the dominant gene, permitting the recessive character to be revealed.

[2]A majority of the mosaics for two specific chromosomes (I and V) involved closely linked characters. The linked genes were lost together, indicating to Stadler that at least these mosaics were due to the loss of a portion of the chromosome.

[3]Some bacteria such as E. coli contain multiple identical chromosomes. A recessive mutation on one of these requires several generations for expression.

on a variety of strains of tobacco plants. If the viruses which are introduced into the host plants have previously been irradiated, some of them may react differently on some of the plants. Careful purification of the viruses contained in the lesions may permit isolation of an altered form of the virus which can be studied further.

The mutations which radiation produces in microorganisms, in the mold *Neurospora*, and, most recently, in the small flowering plant *Arabidopsis* have been of particular interest to biologists for the information they provide on gene-enzyme relationships. They are also, of course, of interest to those biologists who are primarily concerned with the biological effects of radiation. For example, *Arabidopsis* is a convenient organism for studies of the types of mutations which are produced by radiation in higher plants. It has a short life cycle and can be easily grown under aseptically controlled conditions. Over 190 mutant forms of this tiny plant have already been identified.

Insects such as *Drosophila*, silk worms, and the wasp *Habrobracon* have been used to study radiation mutagenesis. Recently, large-scale studies with mice have been started. As examples of the results of the many experiments in this field, some of the *Drosophila* and mouse results follow.

MUTATION STUDIES IN DROSOPHILA

TECHNIQUES

Muller and other workers have measured recessive lethal mutations in most of their studies of radiation effects in *Drosophila*. These mutations cause the affected organism to die before the completion of its development. Recessive lethals on the X chromosome are especially useful for study since they are carried without adverse effect in the heterozygous state in females (mutation is on only one of the two X chromosomes); in males (only one X chromosome), their presence always results in death. Furthermore, one can be sure that the parental males used to start an experiment do not contain such a lethal.

A carefully constructed experimental technique is of prime importance in these studies. For his early work, Muller used the ClB method. This makes use of *Drosophila* which contain three specific characteristics on the X chromosome. A large inversion is incorporated into the X chromosome which suppresses recombination between it and the X chromosome of standard gene arrangement. This crossover suppressor (C) maintains the integrity of the X chromosomes. The same X chromosome also contains a recessive lethal mutation (1), insuring that all males which inherit the chromosome will die, and a dominant visible marker, Bar (B), which reduces the width of the eyes of all flies in which it is found. The B gene permits identification of females containing the ClB chromosome.

The use of the ClB method for detecting radiation-induced mutations is as follows (Figure 6.1). A female containing one ClB chromosome[4] is mated to an irradiated wild-type male. The irradiated X chromosome of the male will be inherited by the females in the first generation; all of these females must carry any mutation induced in the X chromosome of the irradiated male. Some of these F_1 females will be wild-type, some will have Bar eyes (indicating a ClB chromosome). When the ClB females are mated to their brothers, the offspring in the F_2 generation will contain wild-type females, Bar females, and wild-type males. There will be no Bar males (ClB/Y) since the ClB chromosome is lethal in males. If a recessive lethal mutation is pro-

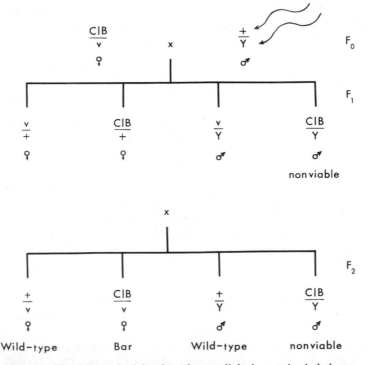

Fig. 6.1. The ClB method for detecting sex-linked recessive-lethal mutations in *Drosophila*. If the irradiated X chromosome of the F_0 male contains a lethal mutation, the wild-type males in F_2 are nonviable (see text).

[4]Often the other X chromosome of these females contains a recessive visible mutation, such as vermillion (v) eye.

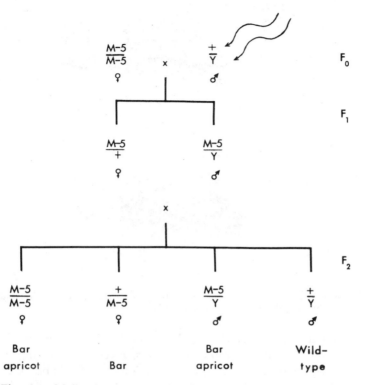

Fig. 6.2. Muller-5 method for studying sex-linked recessive-lethal mutations in *Drosophila*. The X chromosome (M-5) contains a dominant Bar marker, a recessive apricot marker, and a crossover suppressor. Bar females and wild-type males in the F_2 generation contain the same irradiated X chromosome.

duced by the irradiation, the expected "wild-type" males will also be non-viable, and there will be no males at all in F_2.

Experimental details must be such that, in any one vial, all the F_2 male flies come from a single sperm of an irradiated male in the F_0 generation. Then, the *occurrence* of a recessive-lethal mutation induced by the radiation can be detected by the presence of a vial of F_2 flies which contain no viable males. The *number* of recessive lethal mutations can be estimated by the number of containers of F_2 flies which contain no viable males.

A second technique for detecting sex-linked, recessive lethal mutations is the Muller-5 (M-5) method (Figure 6.2). This is similar in principle to the ClB method except that instead of the lethal mutation, a recessive visible marker (eye color mutation apricot) is located on the X chromosome. The dominant Bar eye marker and a crossover suppressor are included, as in the ClB method. Females which are homozygous for this M-5 chromosome are mated to irradiated wild-type males. The F_1 females contain one M-5

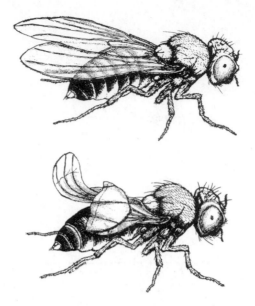

Fig. 6.3. An example of a mutation produced by radiation in *Drosophila.* Top: Normal fly. Bottom: Curly wing mutant fly.

chromosome and one irradiated X chromosome; F_1 males contain a single M-5 chromosome. An F_1 brother-sister cross gives the following in the F_2 generation: M-5 females (Bar, apricot), M-5 males (Bar, apricot), Bar females containing an irradiated X chromosome, wild-type males containing the same irradiated X chromosome. Absence of the latter group indicates a recessive lethal mutation induced by the radiation. This mutation is not lost from the population and can be further characterized with the use of the Bar females which contain the same mutation.

Similar methods are available for detecting sex-linked visible mutations, recessive lethal mutations on other chromosomes, and recessive visible mutations. An example of a typical visible mutation is shown in Figure 6.3. An additional method of study is the specific-locus technique. Irradiated males are mated with females which are homozygous for a number of recessive visible mutations. If there is no mutation, the offspring appear normal since they are heterozygous for the recessive genes. (All have the normal genes from the male to mask the recessive genes from the female.) If irradiation has produced a mutation at one of the gene loci in the male, the offspring will show a mutant phenotype. This may be the phenotype of the female, or it may be another different phenotype, depending on the mutation which was produced.

EXPERIMENTAL RESULTS

Between 1928 and 1950, many studies utilized the ClB and Muller-5 techniques to detect recessive lethal mutations from irradiated *Drosophila*

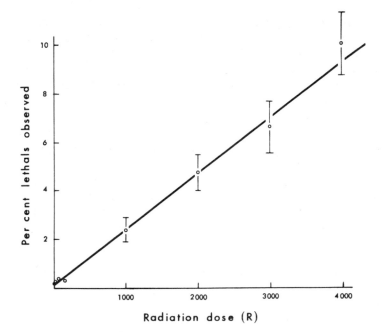

Fig. 6.4. Relationship between mutation rate and radiation dose to *Drosophila* spermatozoa. (Data from Spencer and Stern [1948].)

spermatozoa. A number of points emerged from these experiments. It appeared that radiation-induced mutations add to those which arise spontaneously; that is, radiation does not produce a multiplication or magnification of the spontaneous frequency. Rather, the total mutation frequency is made up of two independent components: (1) The spontaneous mutations which would have arisen without radiation, and (2) the radiation-induced mutations. Most of the experiments also indicated that the mutation frequency is substantially linear with increasing dose. For example, Figure 6.4 indicates the data from the comprehensive studies of Spencer and Stern (1948). Those workers used the Muller-5 method to detect recessive lethal mutations from radiation exposures extending from 25 R to 4000 R. Even over this very wide dose range, the relationship appears approximately linear.[5] There is no

---

[5]As has been pointed out by Muller (1959), it may be assumed that this linearity applies primarily to point mutations. He indicates that most recessive lethals are associated with no visible structural change below 3000 R. However, at higher exposures (for example, 4000 R) many structural changes are visible; these probably contribute to the number of recessive lethals. Since structural changes induced in spermatozoa (of the multibreak types) have a frequency which is approximately proportional to the 1.5 power (see Chapter 5), one might expect that the number of recessive lethals from higher doses should be greater than predicted from a linear extrapolation from low-dose data in which most mutations result from point defects. As indicated in Figure 6.4, such an increase was not found by Spencer and Stern. Muller suggests that the expected increase at high doses may be balanced and masked by a decrease in frequency caused by the elimination of offspring via dominant-lethal chromosome aberrations.

indication of a threshold for the induction of mutations or of a reduction in the mutagenic effectiveness of the radiation at very low doses.

The results of other experiments suggested that the dose rate did not influence mutation rate and that fractionation of the total dose over several days did not change the mutation rate from that resulting from a single acute exposure. All of these observations seemed to point to the conclusion that the production of a point mutation in *Drosophila* spermatozoa resulted from a single interaction of radiation with genetic material, and that such a lesion was "fixed" immediately and could not be altered or repaired by subsequent treatment.

More recently, *Drosophila* geneticists have expanded their interest to study the mutation frequencies which result from irradiation of germ cells at different stages of development and under varying conditions. Muller (1959) has summarized (Table 6.1) the relative number of recessive lethal mutations

Table 6.1. RELATIVE FREQUENCIES OF RECESSIVE LETHAL MUTATIONS IN *Drosophila* PRODUCED BY A GIVEN DOSE OF RADIATION[a]

Stage	Relative Number Mutations Produced
Male	
Spermatogonia and early spermatocytes	1
Meiotic division stages	8
Spermatids	12
Spermatozoa (more than 1 day pre-ejaculation)	3
Spermatozoa (within 1 day of ejaculation)	$4-4\frac{1}{2}$
Spermatozoa (within inseminated females)	5–6
Female	
Oogonia and early oocytes	1
Late oocytes (last 3–4 days)	2–3
Postfertilization	
Immediate (within 10 to 25 minutes after egg laying)	< 1
Cleavage	≤ 12

[a]After Muller (1959).

which are produced in different germ cell stages. It is apparent that, in the male, the spermatogonia are the least sensitive germ cells, and the spermatids are the most sensitive. The fertilized egg is sensitive during cleavage. It has also been determined that more mutations are produced in spermatogonia when the dose rate is high than when the radiation is given at a low dose rate.

Although the mutation rate can vary considerably, depending on the experimental conditions, most estimates for *Drosophila* fall between 1.5×10^{-8} and 8×10^{-8}/locus/R. Another way of measuring mutation production is by the "doubling dose" or the dose of radiation which induces the same number of mutations as arise spontaneously in one generation. Muller has considered this approach and suggests a value of about 350 R for either recessive-lethal or specific-locus point mutations induced in oogonia, and a value of 75 R for such mutations induced in spermatozoa.

In most *Drosophila* experiments, it has been apparent that both induced and spontaneous mutations are overwhelmingly detrimental in character. However, in certain situations where homogeneous material has been irradiated, the heterozygosity which resulted from the radiation has apparently resulted in a higher likelihood of survival. It is doubtful that such results can be called "beneficial" in respect to human populations.

MUTATION STUDIES IN MAMMALS

The experiments with *Drosophila* spermatozoa prior to 1950 established some "principles" of radiation genetics. It was not really known, however, whether these would also apply under other experimental conditions, in other germ cell stages, and to other species. Accordingly, a comprehensive radiation genetics study with mice was started at Oak Ridge National Laboratory under the direction of Dr. William L. Russell. This "megamouse" project has shown that there are very significant differences in the genetic responses of *Drosophila* and mice to irradiation.

Of particular interest in this mouse work has been the comparison of sensitivities of the different stages of spermatogenic cells. The sensitivity of spermatogonia is especially important in mammals, since mammalian spermatogenesis is a continuous process; that is, the stem cells of the spermatogenic series, the spermatogonia, divide at regular intervals. Some of the daughter cells divide and differentiate to form mature sperm; other daughter cells remain as spermatogonia to initiate the next cycle. Thus, cells which are beyond the spermatogonial stage at the time of irradiation may carry a mutation, but, at most, only a limited number or cluster of mutated sperm will result from each. In contrast, mutated spermatogonia may be stem cells for many sperm over the entire reproductive life of the individual. The first sperm produced after irradiation will have been irradiated as postspermatogonial cells. All subsequent gametes will result from cells irradiated as spermatogonia.

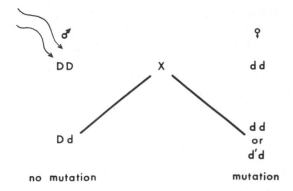

Fig. 6.5. Specific-locus method for determining genetic mutations. The irradiated male is homozygous for a dominant character (DD); the female is homozygous for a recessive character (dd). Nonmutated offspring will have the dominant phenotype (D); offspring with a mutation at that locus will have a recessive phenotype (d or d'). (Modified from Russell [1963a], courtesy American Philosophical Society.)

TECHNIQUES

A specific-locus method has been used by Russell to study mutation rates. This method may be illustrated as follows (Figure 6.5). An animal homozygous at a particular locus for the normal gene, for example, dark coat color (DD), is irradiated and mated to one that is homozygous at the same locus for a visible recessive mutation, in this case, dilute coat color (dd). If no mutation occurs at this locus, the offspring (Dd) will have dark coats since the gene for dark coat (D) is dominant to that of dilute color (d). If, however, the specific gene (D) from the germ cell of the irradiated mouse mutates to the gene for dilute color (d) or a similar gene (d') at the same locus, then the offspring will be dilute colored.

Actually, Dr. Russell uses mice which contain seven such marker genes for phenotypically visible features such as coat color, spotting and mottling of coat, and ear and tail defects. Mutations in any one of the seven loci can be scored in a single experiment. The mutations can be recognized at a glance and are determined in first generation offspring. Without such a specific-locus method, the detection of all visible mutations at unmarked loci necessitates a thorough examination of all animals (frequently including dissection). Sex-linked lethals can be recognized only by large-scale tests of the progeny.

The mutations used by Russell range from those with slight visible effects to those which are lethal in the homozygous state. In addition, two of the marker genes which give rise to short ear (se) and dilution of coat color (d) are linked; that is, they are presumed to be closely located so that a mutation in both genes probably represents a deficiency, involving the region of both genes, rather than separate point mutations.

EXPERIMENTAL RESULTS

Using the specific-locus method, Dr. Russell has examined the rate of mutation production in mice with a variety of dose rates, fractionation patterns, and cell stages. Perhaps the most surprising finding is the dependence of mutation rate on the dose rate of the radiation.

It was pointed out in an earlier section of this chapter that the mutation rate induced in *Drosophila* spermatozoa is independent of dose rate. Similarly, when mouse spermatozoa are irradiated, the mutation rate does not depend on dose rate. However, when spermatogonia are irradiated at a high exposure rate (90 R/min), the number of mutations produced by a particular dose is about three times as high as when the same total dose is delivered at a low exposure rate (0.8 R/min). Reducing the exposure rate further, to 0.009 R/min or 0.001 R/min, does not further lower the number of mutations. Figure 6.6 illustrates this dose-rate effect. It will be noted that the number of mutations decreases after 1000 R (90 R/min). It has been suggested that this is not really a decrease in the rate of mutation production. Perhaps

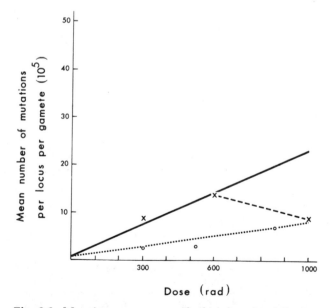

Fig. 6.6. Mutation rates at specific loci in mice following irradiation of spermatogonia. $\times$ = 90 R/min. (Solid line shows extrapolated rate, dotted line observed rate.) O = 0.009 R/min. (Modified from Russell [1963b], courtesy Pergamon Press.)

those cells in which a mutation has occurred are also those which are most easily killed by the 1000 R exposure.

When a total exposure of 1000 R is given in two fractions of 600 R and 400 R, separated by more than 15 weeks, the mutation frequency is not significantly different from the line fitted to the 300 R and 600 R points (Figure 6.6). Since this 15 week period is sufficient to allow restoration of normal spermatogenesis, the experiment supports the view that cell selection is at least partially responsible for the lower mutation rate at 1000 R than at 600 R. Other fractionation experiments[6] have also supported this view.

One surprising result has been obtained with fractionation. If the total exposure of 1000 R is administered in two fractions, 24 hours apart, the mutation rate per roentgen is very high. This suggests that mutation production may depend on some other cellular factors (possibly stage of cell division), which are themselves influenced by radiation; that is, radiation can alter the distribution of spermatogonial cells in various stages of mitosis. If a particular stage is especially sensitive to mutagenesis, a second radiation exposure given at a time when many cells are in this stage may produce mutations in an abnormally high proportion of cells. This notion has yet to be substantiated.

An even greater dose-rate effect has been found when female mice are irradiated. Following 400 R given at a high rate (90 R/min), the mutation rate in oocytes is very high (19.26 mutations/locus/gamete $\times$ 10^5); at 0.8 R/min, or at 0.009 R/min the same exposure results in 4.8 or 0.7 mutations/locus/gamete $\times$ 10^5, respectively. At this lowest dose rate, the mutation frequency has not been demonstrated to be different from that of non-irradiated females.

Russell has summarized this data on dose-rate dependence as indicated in Table 6.2. He has proposed a model which will account for the observed dose-rate effect[7] and will also allow the mutation frequency at low dose rates to be linearly related to dose. According to this model, there is a dose-rate dependent repair process[8] which is inactivated or saturated by a certain quantity of radiation. A particular radiation exposure delivered at a low dose rate interferes less with the repair mechanism than the same exposure at a high dose rate. Therefore, at a low rate there is greater repair of the initial injury and fewer observable mutations than at a high rate. Some of the initial injury in spermatogonia is not repaired even if the dose rate is very low. As indicated previously, lowering the exposure rate below 0.8 R/min does

[6]See Russell, W. L. (1963b).

[7]Although one might expect that the dose-rate effect indicates multiple-hit phenomena in mutation production, Russell gives many reasons for believing that this is not the case. For example, dose-response curves do not fit the typical (concave upward) curve for 2-hit aberrations; in spermatozoa, where 2-break aberrations occur with the greatest frequency, there is no dose-rate effect; in spermatogonia and oocytes, which have few translocations, there is a marked dose-rate effect; fractionation of exposure (90 R/min) does not decrease the mutations compared with a single exposure.

[8]There is strong evidence in *Paramecium* and other organisms that premutational damage induced by ionizing radiation is repaired under certain conditions.

Table 6.2. Specific-Locus Mutation Frequencies at Various Dose Rates[a]

Dose Rate (R/min)	Mutation Frequency	
	Spermatogonia	Oocytes
90	high	very high
9	intermediate	—
0.8	low	intermediate
0.009	low	very low
0.001	low	—

[a]From Russell, W. B. (1963a).

not decrease the mutation rate. This suggests that there is no threshold situation for mutation production. In oocytes, however, the mutation production from radiation delivered at a low dose rate is not distinguishable from the control rate. This suggests that a large proportion of the injury is repaired at the low dose rates. Spermatozoa do not show a dose-rate effect. The repair mechanism may be different or nonexistent in these cells. It is not unlikely that spermatozoa with a low metabolism and hypoxic state may be inefficient at repair.

SPECTRUM OF MUTATION RATES

In Russell's data, there is wide variation in the mutation rates at each of the seven specific loci. For mutations induced in spermatogonia, for example, there can be a thirtyfold difference in mutation rates induced at different loci. There is less variation in mutation rates among the loci when post-spermatogonial stages are irradiated. Oocytes from irradiated females have still another spectrum of mutation rates for the seven loci.

The occurrence of deficiencies (as detected by the two linked loci) also depends on the cell type irradiated. They are rarely found in sperm which were irradiated as spermatogonia, common in sperm irradiated in post-spermatogonial stages, and of intermediate frequency in irradiated oocytes.

MUTATIONS IN MAN

It is obvious from the preceding sections that it is not possible to generalize and assign a specific mutation rate to a specific dose of radiation. Gene loci differ markedly in their mutability. Mitotic stage, cell type, sex, species, and dose rate all influence the rate of mutation production. It is, therefore, difficult to extrapolate the existing animal data to man in order to predict the mutation rates that can be expected from various types of radiation exposure. Direct evidence of radiation-induced mutation in man is also lacking. The largest group of humans which are available for study are the descendants of those exposed to radiation at Hiroshima and Nagasaki. While there has been no detectable effect on the frequency of prenatal or neonatal deaths or on the frequency of malformations, this does not mean that there have been no hereditary effects produced by the irradiation.[9] The number of exposed parents was small and the dosages were so low that it would have been surprising if an increase in mutation had been detected as yet. There has not been sufficient time for the several generations needed to reveal recessive damage.

It is logical to expect that radiation exposure will increase the mutation rate in humans. Based largely on the mouse experiments, it has been estimated that the doubling dose (that which will double the spontaneous mutation rate) for man probably lies between 10 and 100 rads. For acute irradiation, the probable value is between 15 and 30 rads; for chronic irradiation, it is probably around 100 rads. These figures may be changed as new information becomes available.

SOMATIC MUTATIONS

Mutations which are produced in cells which are not germ cells are usually termed *somatic mutations*. These may affect the individual in which the mutation has occurred, but the mutant genes are not passed on to offspring in sexual reproduction.

Somatic mutations occur spontaneously; radiation increases the incidence. An interesting demonstration of somatic mutations comes from radiation studies on flowering plants. As a simple illustration, assume that a flower bud contains cells which are heterozygous for a flower color gene; that is, they have one dominant gene for red color and one recessive gene for white color. The flower will ordinarily be red. If, in a certain cell, the dominant gene is mutated,[10] either to the gene for white color or to another form over which white is dominant, the portion of the flower which forms from that cell will be white. The recessive gene for white color is "unmasked" in the original cell where the mutation occurred and its daughter cells pro-

---

[9] One survey of these individuals did detect a significant shift in the sex ratio of immediate progeny.

[10] Similarly, the dominant gene might be lost if a chromosome break results in a deletion of that portion of the chromosome. Such a change might not be distinguishable from the point mutation described above.

Fig. 6.7. "Red" dahlia bloom in which approximately half of the flower petals are white. The plant was exposed to gamma irradiation administered at a rate of 118 R per day in the Brookhaven gamma field. (Courtesy Brookhaven National Laboratory.)

duced by subsequent cell division. The white area may be a few cells, a large streak, or even a major portion of the flower. This depends on the stage of flower bud when the mutation occurred and on the portions of the flower arising from the affected cell. Figure 6.7 illustrates the color change in a dahlia bloom from a plant that had been exposed to chronic irradiation. Approximately half of the individual flowers (petals) are white[11] instead of red.

Somatic mutations occur in animals, also, although their detection is sometimes more difficult than with the color change in plants. As early as

[11]Such a color change in other plants (for example, carnation) may not be the result of somatic mutations, but simply of damage to the outer layer of a chimera, thus allowing the underlying cell layer to express its phenotypic potential.

1929, Patterson reported somatic mutations in *Drosophila* which had been x-irradiated as eggs or larvae. In one series of experiments, he used flies which were heterozygous for certain of the sex-linked genes for eye color. For example, in a cross between a normal red-eyed female and a white-eyed mutant male, irradiation of the eggs or larvae during different stages of development resulted in a certain proportion of the flies showing white patches on their eyes. Treatment of eggs during the first few hours of their development resulted in large white areas in the eyes of adults. If radiation was given during the mid-larval period, the white area consisted of a medium-sized spot; if x-ray was given during the late larval stages, the patch was small. If pupal stages were irradiated, no changes in eye color were produced. From these results, it was concluded that a mutation of the dominant (red) gene was induced in a single cell of the eye rudiment, permitting the expression of the recessive white gene. If the mutation occurred early, the patch would be large because the mutated cell had many cell descendants. At later stages, there would be fewer cell descendants and, therefore, the patch would be smaller. More recently, the suggestion has been made that somatic mutations may contribute to some of the long-term effects of irradiation in mammals, such as the formation of some tumors. This subject is discussed in Chapter 12.

One further aspect of the production of mutations by radiation should be mentioned. Inasmuch as radiation from both terrestrial and cosmic sources has impinged on living organisms throughout their history, it is likely that a considerable number of mutations has resulted. It has been pointed out that mutations are generally considered to be detrimental. At some stages, however, such genetic changes may have provided a positive contribution to the evolutionary process.

GENERAL REFERENCES

Errera, M., and Forssberg, A. *Mechanisms in Radiobiology*, Vol. I, Chapter 6. Academic Press, Inc., New York (1960).

Hollaender, A. *Radiation Biology*, Vol. I (2), Chapter 12. McGraw-Hill Book Co., Inc., New York (1954).

Muller, H. J. Advances in Radiation Mutagenesis Through Studies on *Drosophila*. *Progr. Nucl. Energy, Ser. VI*, **2**: 146–160 (1959).

Russell, W. L. Evidence from Mice Concerning the Nature of the Mutation Process, pp. 257–264 in Genetics Today. *Proc. Intern. Congr. Genet., 11th, The Hague, 1963.* Pergamon Press, New York (1964).

Russell, W. L. The Nature of the Dose-Rate Effect of Radiation on Mutation in Mice, in Mechanisms of the Dose Rate Effect of Radiation at the Genetic and Cellular Levels. *Suppl. Japan J. Genet.*, **40**: 128–140 (1965).

Russell, W. L. Studies in Mammalian Radiation Genetics. *Nucleonics*, **23:** 53–56 (1965).

Srb, A. M., Owen, R. D., and Edgar, R. S. *General Genetics*, 2d ed. W. H. Freeman, San Francisco (1965).

ADDITIONAL REFERENCES CITED

Goodspeed, T. H., and Olson, A. R. The Production of Variation in Nicotiana Species by X-ray Treatment of Sex Cells. *Proc. Natl. Acad. Sci. U. S.*, **14:** 66–69 (1928).

Muller, H. J. Artificial Transmutation of the Gene. *Science*, **66:** 84-87 (1927).

Patterson, J. T. X-rays and Somatic Mutations. *J. Heredity*, **20:** 261–267 (1929).

Russell, W. L. Genetic Hazards of Radiation. *Proc. Am. Phil. Soc.*, **107:** 11–17 (1963a).

Russell, W. L. The Effect of Radiation Dose Rate and Fractionation on Mutation in Mice, pp. 205–217 in Sobels, F. H. (ed.) *Repair from Genetic Radiation Damage*. Pergamon Press, New York (1963b).

Spencer, W. P., and Stern, C. Experiments to Test the Validity of the Linear R-Dose Mutation Frequency Relation in *Drosophila* at Low Dosage. *Genetics*, **33:** 43–74 (1948).

Stadler, L. J. Mutations in Barley Induced by X-rays and Radium. *Science*, **68:** 186–187 (1928a).

Stadler, L. J. Genetic Effects of X-rays in Maize. *Proc. Natl. Acad. Sci. U. S.*, **14:** 69–75 (1928b).

7 RADIATION EFFECTS ON MICROORGANISMS AND INDEPENDENT CELL SYSTEMS

The changes which occur in an irradiated cell are partly the response of the cell itself to the radiation and partly the response of the cell to changes in surrounding cells. The latter influence can be recognized when one considers an organism composed of many cells arranged in a very complex structural organization and in which the behavior of the entire organism is regulated by interactions between the cells and tissues (see Chapter 9). In contrast to the higher organisms, there are some biological systems in which the cellular components are relatively independent of each other. This is true in a suspension of microorganisms such as bacteria or yeast or, to a somewhat lesser degree, in a culture of mammalian cells. By studying these systems, one can examine the radiation effects on individual cells or on populations of cells.

TARGET THEORY

Early in the history of radiation biology it was noted that there was a direct relationship between the number of microorganisms which were killed by a radiation exposure and the radiation dose they had received. In order to put this relationship into mathematical terms, the target theory was formulated. Although the theory was proposed by Crowther, it was developed and expanded by D. E. Lea[1] and is usually associated with him.

[1]See Lea (1955) for details of the theory as it has been expanded and modified.

The target theory is strictly a model which is considered to be applicable when the biological effect meets certain criteria in its relation to dose. The theory states that the production of ionization in or very near some particular molecule or structure (target) is responsible for the measured effect. The production of an effective event in the target is often called a *hit*. The target may be a whole cell, part of a cell, or a critical molecule. Generally, the system studied is a cell population in which the measured effect may be cell death or inability to grow or divide. In the simplest form of the target theory, one hit is sufficient to produce the measured effect in the associated organism.

As a first approximation, the simple target theory assumes that each event occurs at random in an irradiated system. There is, therefore, a statistical chance that any particular target (and its associated organism) will be hit. With very small radiation doses, the number of targets hit and affected will be directly proportional to the amount of radiation; that is, if the dose is doubled, twice as many targets will be hit. Therefore, twice as many organisms will be affected. With increasing doses, some of the events will occur within targets that have already been hit. These hits will be wasted, so the number of effective hits will decrease with increasing exposures (Figure 7.1).

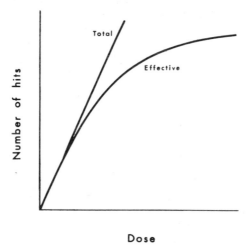

Fig. 7.1. Relationship between dose and number of hits according to the simple target theory. The number of "effective" hits deviates from the total number as the number of duplications increases. Note that the initial portion of the effective hit curve is approximately linear.

The relationship demonstrated in Figure 7.1 is usually expressed in reverse when the target theory is applied to a cell population in which the effect measured is lethality. The fraction of the population which survives is plotted against the radiation dose (see Figure 7.2A). The numbers of organisms killed by successive increments of dose are not equal, but each dose increment kills the same proportion of the number of organisms that have survived until then. Thus, the number of viable organisms decreases in a geometric progression. One may say, therefore, that the survival curve is

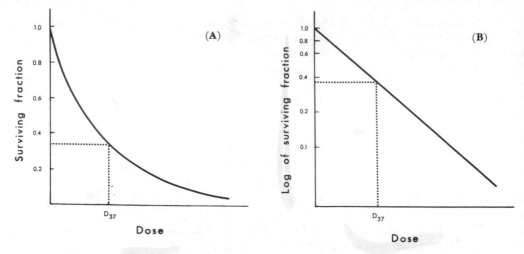

Fig. 7.2. Relationship between dose and surviving fraction of organisms on linear scale (A) and logarithmic scale (B).

exponential. If the surviving fraction is plotted on a log scale (Figure 7.2B), one obtains a straight line.

The numerical relationship of the simple target theory can be expressed as follows: If $N_0 =$ number of organisms initially present and $N =$ number of organisms surviving after a dose D, then if an increment of dose (dD) is added, N will be decreased by dN in an amount which is proportional to the number present, N.

$$-\frac{dN}{dD} = kN \tag{7-1}$$

or

$$\frac{dN}{N} = -kdD \tag{7-2}$$

This integrates to

$$\ln\left(\frac{N}{N_0}\right) = -kD \tag{7-3}$$

or

$$\frac{N}{N_0} = e^{-kD} \tag{7-4}$$

If the fraction of surviving cells is S, then[2]

$$S = e^{-kD}$$ (7-5)

This can be written in the logarithmic form as

$$\ln S = -kD$$

Notice that the slope of the survival curve (Figure 7.2B) is $-k$, the proportionality constant.

If it is assumed that the distribution of hits follows the Poisson distribution,[3] then it can be shown that the proportionality constant k is equal to $1/D_0$, where D_0 is the dose that gives an average of one hit per target. When dose D has been given such that $D/D_0 = 1$, then since $e^{-1} = 0.37$, D_0 can be called the 37%, or the e^{-1}, dose,

$$S = e^{-D/D_0} = e^{-1} = 0.37$$

Therefore, when there has been an average of one hit per target (that is, the number of hits equals the number of targets), 37% of the original number of organisms will still survive.

In practice, this relationship is used in reverse; that is, if a survival curve is plotted, the dose required to reduce the number of individuals in the population to 37% can be determined directly from the graph. This is the D_0, or D_{37} dose. It is often used to compare the radiation sensitivity of different systems which behave kinetically according to the simple target theory. Some authors prefer to use $1/D_0 = k$, the sensitivity constant. Note that the dimensions of D_0 are in units of dose, and those of k are in reciprocal units of dose (since kD or D/D_0 must be dimensionless). Note also that D_0 varies in an inverse way with radiation sensitivity, whereas k is always directionally the same.

Many assumptions in the simple target theory are oversimplifications. For example, it assumes that the degree of effect is not influenced by the dose rate, and that the experimental conditions during irradiation or immediately after will be unimportant. In many experimental situations, these assumptions are not fully justified. The model has, however, been extremely valuable as a means of describing the kinetics of the response of certain organisms to radiation, and with modifications has given some idea of the mechanisms of radiation action in a wide variety of populations composed of single cells.

[2]Notice that S is used here as a fraction. In some books S is used for the number of cells surviving a dose and S_0 for the original number.
[3]The Poisson distribution describes all random processes whose probability of occurrence is small and constant. It is discussed in many statistics books, or see Evans (1955).

TARGET SIZE

One interesting application of the simple target theory has been its use for estimating target size, based on the D_0 or k value. The calculation is made as follows: One roentgen is approximately equal to 2×10^{12} ionizations per gram of tissue (or target). Therefore, if D_0 is the dose that will result in 1 hit per target, then $D_0 \times 2 \times 10^{12}$ is the number of *ionizations* per gram which will result in an average of one hit per target. This is the number of targets per gram. The number of targets per gram is also the reciprocal of the target volume in grams. Therefore,

$$D_0 \times 2 \times 10^{12} = \frac{1}{\text{target volume}} \qquad (7\text{-}6)$$

If one assumes that the target is spherical, the target radius or diameter can be calculated from the volume. Table 7.1 shows the results of such a calculation, using radiation data on the inactivation of the S13 bacteriophage by gamma, x-, and alpha radiation. The target diameters in the first row (I) are calculated from equation 7-6, using the D_0 values for the various radiations. The simple target theory, on which this calculation is based, assumes that ionization is distributed randomly. This is not strictly true. If a correction is applied for the distribution of the ionization in clusters (II) and for the presence of delta rays (III), it can be seen that the values calculated for the three

Table 7.1. ESTIMATION OF DIAMETER OF S13 BACTERIOPHAGE FROM RADIATION DATA[a]

	Radiation		
	γ	x (1.5 A)	α(4 Mev)
Inactivation dose (R $\times 10^{-5}$)	5.8	9.9	35
Calculated target diameter (mμ)			
I	11.2	10.0	5.9
II	16.3	16.2	24.4
III	15.5	15.9	16.3

[a]Data from Lea (1955). Method I calculates from simple relationship in equation 7-6; method II takes into account the distribution of ionization in clusters; method III takes delta rays into account. The diameter of the unhydrated virus has been measured by other techniques and found to be about 16 mμ.

Table 7.2. Nonapplicability of the Simple Spherical Target Theory to the *Vaccinia* Virus (mean unhydrated virus diameter = 200 mμ)[a]

	Radiation		
	γ	x (1.5 A)	α(5 Mev)
Inactivation dose (R $\times$ 10^{-6})	0.080	0.104	0.211
Calculated target diameter (mμ)	31	41	70

[a]From Lea (1955).

radiations are quite similar. Moreover, the values are all close to 16 mμ. This is known from other techniques to be about the diameter of the bacteriophage. It appears that the entire phage is the radiation target and that a single event (hit) within the phage will result in its inactivation.

When the simple target theory is applied to the calculation of target size of large viruses, it is apparent that the target volume does not coincide with the volume of the virus. For example, in Table 7.2 are shown the inactivation doses and calculated target diameters for the *Vaccinia* virus, which is known from other techniques to have a mean unhydrated virus diameter of 200 mμ. Correction factors have been applied to these calculations for clustering and delta tracks. It is apparent that the target is only a small portion of the total virus, amounting to less than 1% of the total volume.

Survival data on larger cells such as bacteria or yeast give target volumes which are also much smaller than the total cell size. In addition, the survival curves obtained from radiation experiments on these larger organisms are often not of the simple exponential shape. In order to explain differences in the shapes of the survival curves, modifications and extensions of the target theory have been suggested. Two theories which are most often considered are generally known as the *multitarget* and *multihit* theories.

MULTITARGET THEORY

According to the multitarget theory, certain organisms contain more than one "target." In order to inactivate the entire unit, each of the targets must receive a hit. This can be expressed mathematically.

The probability of a target not being hit is equivalent to the percentage of surviving organisms in equation 7-5.

$$S = e^{-kD}$$ (7-7)

The probability of a target being hit is one (total probability of being hit or not being hit) minus the probability of not being hit. This is then

$$(1 - e^{-kD})$$

The probability of all the targets (n in number) being hit is the probability of any one being hit raised to the n power, or

$$(1 - e^{-kD})^n$$

if the probability for each is equal. The probability of the unit surviving is, then, the probability that not all of the targets will be hit, or

$$S = 1 - (1 - e^{-kD})^n \tag{7-8}$$

For large values of D, equation 7-8 is approximately[4]

$$S = ne^{-kD} \tag{7-9}$$

This can be written in the logarithmic form instead of as an exponential equation.

$$\ln S = \ln n - kD \tag{7-10}$$

When such a relationship is plotted on semilog paper (Figure 7.3), the result is a straight line at doses where the simplification from equation 7-8 to 7-9 can be made. The slope of the straight portion is $-k$, the sensitivity constant. The intercept of the straight portion extrapolated to the vertical axis is n which, according to equation 7-10, represents the number of targets in this model. Often n is referred to as the *extrapolation number*.

MULTIHIT THEORY

The multihit theory has been invoked to interpret certain types of data. This theory postulates that some systems contain a single target which must be hit m times in order to inactivate the system. Those units will survive which receive $m - 1$ or fewer hits.

[4]In the expansion of $(1 - e^{-kD})^n$, all terms beyond the first two (1, and $-ne^{-kD}$) involve powers of e^{-kD}. If D is large, these additional terms become negligible relative to the first two.

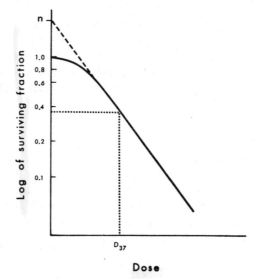

Fig. 7.3. Logarithmic plot of relationship expressed by multitarget theory.

It can be shown by application of the Poisson distribution that the probability of receiving

$$0 \text{ hits is } e^{-kD}$$
$$1 \text{ hit is } kDe^{-kD}$$
$$2 \text{ hits is } \frac{(kD)^2}{2}e^{-kD}$$

$$\qquad \qquad \cdot \qquad \qquad \cdot$$
$$\qquad \qquad \cdot \qquad \qquad \cdot$$
$$\qquad \qquad \cdot \qquad \qquad \cdot$$

$$m - 1 \text{ hits is } \frac{(kD)^{m-1}}{(m-1)!}e^{-kD}$$

The surviving units are all those receiving less than m hits.

$$S = e^{-kD} \left(1 + \frac{kD}{1!} + \frac{(kD)^2}{2!} + \cdots + \frac{(kD)^{m-1}}{(m-1)!} \right) \qquad (7\text{-}11)$$

or

$$S = e^{-kD} \sum_{i=0}^{m-1} \frac{(kD)^i}{i!} \qquad (7\text{-}12)$$

This is not an exponential relationship although it may markedly resemble the curve in Figure 7.3; that is, when plotted on semilog paper, there is a reasonably linear portion of the curve with slope of $-k$ which extrapolates to some value (m) on the vertical axis. This value of m may be interpreted as the number of hits required for inactivation of a target. The value obtained for k may be used as a measure of relative radiosensitivity.

It would be of theoretical interest to know whether a given set of experimental data fit the multitarget or multihit theory. However, such a distinction is seldom possible.

MULTITARGET MULTIHIT THEORY

A combination theory has been suggested, called the *multitarget multihit theory*. According to this model, each irradiated unit contains n targets, each of which must be inactivated by m events. The general formula for surviving fraction, according to this theory, is

$$S = 1 - \left(1 - e^{-kD} \sum_{i=0}^{m-1} \frac{(kD)^i}{i!}\right)^n \qquad (7\text{-}13)$$

While this theory may make some sense intuitively, it is far too complex for effective use in interpreting data.

THE BIOLOGICAL TARGET

The identity of the "target" is not always obvious. When small organisms are studied, and the target size is calculated to be about the same size as the organism, an event any place within the organism will produce the measured effect. The whole organism can be viewed as the target. The actual inactivation process, however, may involve either the direct interaction of the radiation with a critical molecule or an indirect effect from changes in a nonessential molecule (such as water) which is close enough to a critical molecule to transfer the energy.

When a target is small relative to the total volume of a cell, it is possible to conceive of several possible target sites — chromosomes, mitochondria, membranes, etc. There is considerable evidence (see Chapter 5) that the nucleus and, more particularly, the chromosomes may be the primary site of radiation damage in a cell. It would not be surprising, then, if the chromosomes constituted the major portion of the target. A series of experiments by Munson and Bridges (1966) have demonstrated that inactivation of *E. coli* B/r grown in minimal medium is due to the inactivation of discrete units (targets?) within the organism, and that these units segregate from one another at cell division. The authors present evidence that this target is the bacterial "chromatid." They also point out, however, that under certain experimental conditions there may be some mechanism for the repair of lethal chromosomal damage. Under such conditions, it is quite conceivable that nonchromosomal damage may contribute significantly to the lethal effect. There may also be additional targets.

In larger, more complex cells, it is more difficult to associate the observed effect with a particular structure. One can speculate whether there is a discrete target (or even a target process such as DNA synthesis), or one can use the kinetics predicted by the target theory to quantitate and describe the effectiveness of radiations without attempting to correlate effect with a particular target.

INTERPRETATION OF EXPERIMENTAL RESULTS IN TERMS OF THE TARGET THEORIES

Certain biological systems appear to respond to radiation according to kinetics predicted by the simple target theory; that is, the survival data can be fitted to a curve with an exponential shape and there is no appreciable dose-rate effect. One may consider these systems as if they contained some target region or sensitive site in which one ionization results directly or indirectly in an inactivation. The rate of inactivation of the sensitive element may be represented by k, the negative of the slope of the curve when it is plotted on semilog paper.

Other systems yield survival curves of the types predicted by the more complex target theories. Survival data are approximately exponential beyond a threshold or shoulder value. When such data are plotted on semilog paper, the straight portion of the curve can be extrapolated to an intercept on the vertical axis which is some value greater than one. Curves of this type can be characterized by the two parameters, n and k. The slope of the straight portion of the curve is k and the intercept of the straight portion extrapolated to the vertical axis is n.

As with the interpretation of the simple target theory, k may represent the rate of inactivation of some sensitive element. It is significant that the k values obtained for experiments on a given organism are not always constant but appear to depend on the experimental conditions. Under carefully controlled conditions in which each parameter is individually varied, the change in k values which result may give some insight into the mechanisms of radiation interaction within the system. For example, Powers (1962) reports that the k values for the survival curves of dried *Bacillus megaterium* spores depend on the gaseous atmosphere in which the spores are irradiated (see Figure 7.4). When irradiation of spores takes place in oxygen, the data

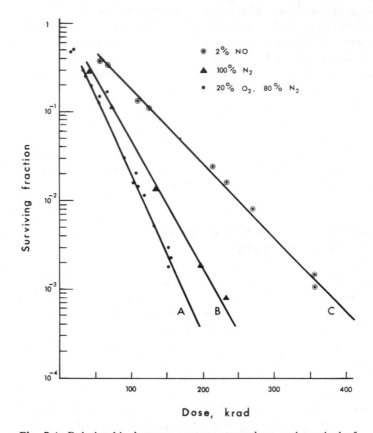

Fig. 7.4. Relationship between gaseous atmosphere and survival of spores of *Bacillus megaterium*. Spores inactivated in situation A are more sensitive than those inactivated in B because of the presence of oxygen in A. Spores in C are most resistant because the presence of nitric oxide has reversed an indirect component of injury, presumably by inactivation of free radicals. (Courtesy Powers [1962] and Taylor & Francis.)

exhibit an exponential relationship ($n = 1.4$, $k = 0.0396$ kR^{-1}); irradiation in nitrogen decreases sensitivity ($n = 0.95$, $k = 0.0308$ kR^{-1}); irradiation in nitric oxide further decreases the radiation sensitivity ($n = 1.26$, $k = 0.0186$ kR^{-1}); but the inactivation is still exponential.

With these values of k it is possible to recognize three ways in which lethality is brought about in these cells by x-rays. In the presence of oxygen, all three are operating. When irradiation is in nitrogen (followed by exposure to oxygen), that kind of damage which is dependent upon oxygen at the time of irradiation is not observed (k is lower). When irradiation is carried out in nitric oxide, another kind of damage is not observed (k is still lower). The cause of this latter damage has been identified as the induction of free radicals that secondarily react with oxygen to cause damage. The damage represented by the lowest k (nitric oxide irradiation) is a third type. Extensive experiments have shown that these modes are independent of each other, showing no interaction, and that each has a unique set of characteristics. The understanding of the mechanisms of radiation damage must be based on understanding at least these three (and perhaps more) components of the overall radiation effect.

The parameter n can be viewed in many ways. It may be considered to represent the number of targets which must be hit for the unit to be inactivated or the number of hits necessary within a single target to produce the measured effect. In certain instances such an explanation may be intuitively satisfying, for example in considering a diploid microorganism with an n number of 2. It is possible to consider that each of two homologous chromosomes must be inactivated by the radiation (two targets). In other cases the n numbers are difficult to interpret in biologically significant terms.

Moreover, changing the conditions of an experiment may cause considerable variation in the n number (with or without a change in k). This is often difficult to relate to a change in the number of some physical targets.

Several authors (see Powers, 1962) have suggested that in some instances the value of n may reflect a variable physiological condition. Suppose, for example, that there is a pool of chemical compounds which can reverse part of the radiation injury after the effect is induced. These compounds might be hydrogen donors. As free radicals are formed, hydrogen atoms are removed from the pool to inactivate the radicals. When the pool is exhausted, the sensitivity of the "target" molecules is revealed and the radiation appears to become more efficient. Inactivation occurs with exponential kinetics relative to dose. In this interpretation, the value of n reflects the pool size. Alterations in culture conditions directly affect the hydrogen ion pool size, and, therefore, result in a shift in the value of n.

Another suggestion which may apply in interpretation of certain experimental data refers to the shoulder on a survival curve. It may imply that individual components of the population are clumped and that every cell within a clump must be inactivated to prevent colony formation by the clump. Then n is the average number of organisms in a group.

Whatever the actual value of n may represent in a particular situation, it seems evident that values of n greater than 1 imply some form of damage accumulation. This may be a number of ionizations required within a system, a number of bonds broken, or an amount of protective substance exhausted. The implication is, therefore, that cells which survive irradiation may have received some degree of sublethal damage which cannot be detected if survival is used as the criterion of effect.

RADIATION EFFECTS ON MICROORGANISMS

Microorganisms of many varieties have proved to be particularly useful for certain types of radiation studies. They can be manipulated more readily than most higher organisms, and are adaptable for experiments which require desiccation, extremes of temperature, or unusual salt concentrations. Moreover, statistically significant numbers of organisms can be handled readily, and each member of the population can be considered as a discrete unit, independent of other members. However, there are certain disadvantages to the use of microorganisms. For example, they are relatively radio-resistant, and the results from most studies cannot be quantitatively extrapolated to higher forms such as man.

Commonly used microorganisms are coliform bacteria, yeast, spores of *Aspergillis*, and viruses such as *Vaccinia* or tobacco mosaic viruses. Generally, lethality has been measured.

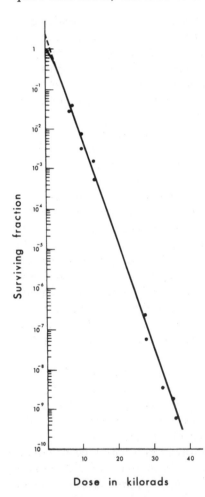

Fig. 7.5. The colony-forming ability of *Serratia marcescens* after exposure to 200 kv x-rays in oxygen. (By Dr. D. L. Dewey, Mount Vernon Hospital, England, courtesy Powers [1962] and Taylor & Francis.)

SURVIVAL STUDIES

Survival in microorganisms is operationally defined as ability of the irradiated organisms to multiply and form visible colonies upon incubation on some suitable growth medium. This only measures ability to reproduce; it does not measure nonreproductive essential metabolic activity. The survival curves for most microorganisms follow the exponential kinetics of either the simple target theory or one of the modifications of the theory (see previous sections). For example, Figure 7.5 illustrates the exponential decrease in colony-forming ability of *Serratia marcescens* over a wide range of radiation doses.

Most microorganisms are relatively radioresistant, especially in the vegetative stages. The D_{37} values are generally in the kiloroentgen range. For example, typical experiments have given D_{37} values in the range of 3 to 8 kR for suspensions of *Escherichia coli*, about 100 kR for *Bacillus mesentericus* spores, and about 400 kR for the tobacco mosaic virus. Specific values and the shapes of the survival curves depend greatly on the experimental conditions.

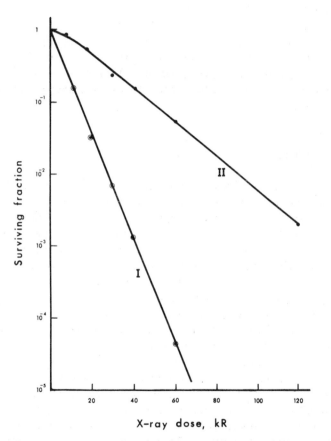

Fig. 7.6. Survival of *E. coli* cultured aerobically in broth and irradiated in O_2-saturated buffer (I) or N_2-saturated buffer (II). (Modified from Hollaender *et al.* [1951], courtesy *Nature*.)

For example, oxygen presence or absence influences the radiosensitivity of microorganisms. This is illustrated in Figure 7.4 and in Figure 7.6 which shows the survival curves for *E. coli* strain B/r cultured aerobically in broth and irradiated in an O_2-saturated buffer (I) or in an N_2-saturated buffer (II). The n and k values calculated by Powers (1962) for the curves in Figure 7.6 are $n = 0.77$ and $k = 1.64 \times 10^{-1}$ kR^{-1} for I, and $n = 1.99$ and $k = 6.22 \times$

10^{-2} kR^{-1} for II. It is of interest to note that in this experiment the oxygen effect,[5] as measured by the ratio of k values (I/II), is 2.64.

Radiosensitivity also depends on the stage of the growth cycle of the microorganisms at the time of irradiation. Under certain growth conditions, *E. coli* B/r cells exposed to 250 kvp x-rays are relatively insensitive during the lag phase (nonproliferative growth). This is followed by a marked increase in sensitivity during the exponential (log) phase of growth. Maximum sensitivity is reached at the end of the exponential phase, and then decreases to the level characteristic of the lag phase.

This apparent variation in sensitivity at different stages may be related to repair ability. Several enzymatic, energy-requiring repair mechanisms have been studied which can only act before the onset of DNA synthesis. Therefore, there is much greater opportunity for repair during the lag phase of nonproliferative growth than during the exponential growth phase (see Haynes, 1964).

METABOLIC STUDIES

A variety of experiments have been designed to examine the effects of radiation on metabolism of microorganisms. In general, the radiation doses required to produce measureable changes in the common catabolic processes are higher than are those necessary to decrease survival significantly.[6]

Many of the examples of altered metabolic function are probably related to changes in cell membranes. For example, yeast can be shown to lose its ability to retain potassium ions following irradiation. This effect has been tentatively attributed to an uncoupling of the oxidative-phosphorylation associated with active transport across the cell membrane. Potassium loss in irradiated yeast will be used to illustrate the parameters associated with the dose-response curves for metabolic changes.

Figure 7.7 shows the effect of radiation on the ability of yeast to retain potassium ions. It can be seen that when oxygen is present during irradiation, the potassium retention decreases exponentially with dose. The magnitude of this effect is approximately the same at 0°C and at 25°C. In a nitrogen environment, considerably higher radiation doses are needed to decrease potassium retention. The curves obtained for inactivation in nitrogen have a broad shoulder (large n), although the slopes of the exponential portions of these curves are similar to the slope obtained for irradiation in oxygen. Moreover, much higher doses are required to produce a comparable effect when irradiation is carried out in nitrogen at 0°C rather than at 25°C.

---

[5]Chapter 11 contains a discussion of the oxygen effect in a variety of organisms. Values of about 2 are commonly obtained for the effect.

[6]Only processes such as DNA synthesis show sensitivities comparable to those seen when cell survival is measured.

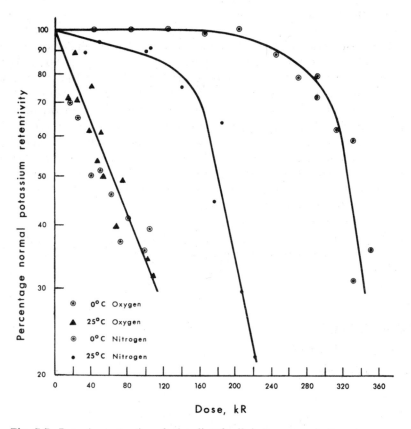

Fig. 7.7. Potassium retention of x-irradiated cells in oxygen and nitrogen at 0°C and at 25°C. (Courtesy Bruce [1958], and Rockefeller University Press.)

The shoulders on the nitrogen curves suggest to A. K. Bruce (see Bruce, 1958) that, among other possibilities, protection may result from the accumulation of reducing compounds which occurs during anaerobic metabolism. The organisms are considered to be relatively insensitive to events producing increased potassium leakage until the reducing compounds are exhausted by reaction with oxidizing products of the irradiation.

It is interesting to note that survival curves for the same strain of yeast, as determined by the same author (Bruce, 1958), follow the kinetics of the simple target theory in both oxygen and nitrogen environments. The D_{37} values (at 25°C) are about 17 kR (oxygen) and about 40 kR (nitrogen). Corresponding doses to reduce potassium retentivity to 37% of control value at 25°C (from Figure 7.7) are about 92 kR (oxygen) and about 195 kR (nitrogen). Thus, based on this comparison, the colony-forming ability of yeast appears to be about five times as sensitive an index of radiation damage as is potassium leakage.

GROWTH STUDIES

An interesting radiation effect on microorganisms has been demonstrated in a series of morphologic studies of bacteria (see Adler *et al.*, 1965).   In some strains of *E. coli*, such as B, failure to produce colonies following irradiation apparently results primarily from damage to division mechanisms.   Growth *per se* is not inhibited.   In these cases, when the bacteria are grown on nutrient media during the postirradiation period, radiation causes the production of giant cells in the form of long, microscopically visible filaments.   The filamentous cells contain nuclei at regular intervals.   They have failed to cleave into individual cells following nuclear replication.   Some authors attribute cell death of the irradiated B strain, at least in part, to an inability to maintain metabolic function in these giant cells.

Other strains of *E. coli*, such as B/r, do not show this filament formation under similar conditions.   The relative radioresistance of this strain is sometimes attributed to slowing of metabolic processes during which repair can occur.

RADIATION EFFECTS
ON MAMMALIAN CELLS
IN TISSUE CULTURE

Recently, a quantitative technique has been developed (Puck and Marcus, 1955) whereby single mammalian cells can be plated in a nutrient mixture and grown into macroscopic, countable colonies.

Survival curves can be determined to measure the reproductive function of mammalian cells in the same way as is done routinely with microorganisms.   This technique also permits microscopic observation of the morphologic alterations which are the direct result of irradiation of the cell and are not the indirect result of damage to surrounding cells or tissues.

Some mammalian cells are more easily cultured than others.   Among the tissues successfully established as culture cell lines are many types of tumor cells, such as the HeLa cells (derived from human cervical carcinoma), cells from a variety of embryonic tissues, and some of the somatic cells of normal humans, hamsters, rabbits, or mice.   Many types of primary cultures have also been studied of cells such as epithelial cells from liver, conjuctiva, and appendix or from fibroblastic cells from skin, spleen, and ovary.

SURVIVAL STUDIES

Figure 7.8 is a typical x-ray survival curve for the reproductive function of mammalian cells (S3 HeLa cells in this experiment) exposed to ionizing

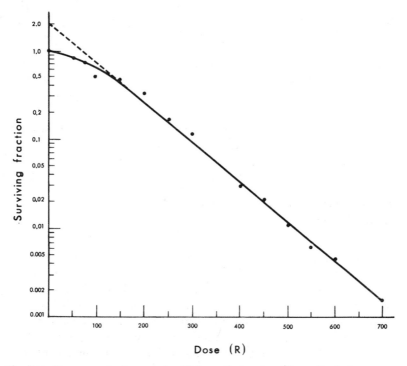

Fig. 7.8. X-ray survival curve for HeLa cells *in vitro*. (From Puck & Marcus [1956], courtesy Rockefeller University Press.)

radiation *in vitro*. The *n* number is 2.0, and the D_{37}[7] is 128 rads. Survival curves for most mammalian cells have a shoulder with an *n* number of around

Table 7.3. VALUES OF *n* AND D_{37} (FROM EXPONENTIAL PORTION OF CURVE) FOR VARIOUS CELL TYPES IN TISSUE CULTURE[a]

Cell strain origin	D_{37} (rads)	*n*
HeLa	128	2.0
Human embryonic lung	221	2.0
Variety of normal human tissues	80	1.3
Hamster	110–155	3.2
Rabbit	125	~1
Normal mouse	115	2
Mouse lymphoma	114	~1

[a]Modified from Puck (1964).

[7]Most estimates of D_{37} for mammalian tissue cultures are computed from the exponential portion of the survival curve. If *n* is greater than 1, the D_{37} is *not* the amount of radiation that will reduce the original population to 37%. It is the amount that will decrease the population to 37% when exponential kinetics are reached. It is equal to $1/k$.

2; the D_{37} doses are generally in the range between 80 and 250 rads. Values for n and for D_{37} for reproductive death of a variety of representative cells are given in Table 7.3. Two aspects of the D_{37} values are particularly significant. First, while there are individual differences in radiation sensitivity among the mammalian cells which have been cultured, these differences are relatively small. They never approach the range that might be expected from the apparent relative sensitivities of the parenchymal cells of various organs (see Chapters 8 and 9). Secondly, the D_{37} values are extremely low compared with values for bacteria, yeast, and protozoa.

MORPHOLOGIC AND METABOLIC EFFECTS IN
TISSUE CULTURE

Puck and Marcus (1956) have described in some detail the radiation effects on cultures of HeLa cells. Many other cells apparently show the same types of changes. With radiation doses below a few hundred rads, there is a decrease in the number of colonies formed. Those colonies that are formed are smaller than in a nonirradiated culture, suggesting a reversible mitotic delay. With increasing doses, up to about 600 rads, there is a decrease in numbers of colonies and in the average number of cells per colony. Many of the irradiated cells which do not form visible colonies may, however, multiply for as many as five or six generations before terminating reproduction. Microcolonies (less than 50 cells) may result. Those cells which lose the ability to divide either immediately or after a limited number of divisions often become enlarged to form giant cells. These *in vitro*-formed giants are similar in appearance to giant cells found in the body after irradiation. They are capable of a high level of metabolic activity as evidenced by rapid utilization of nutrients and ability to synthesize the proteins which bind the cells to glass culture flasks.

Following radiation doses greater than 600 rads, few if any divisions occur, but many giant cells are formed. These may become as large as 0.8 mm in diameter. Fewer giant cells are formed with increasing doses, although between 5 and 20% of the original cells will be well-developed giants after 10,000 rads. The remaining cells disappear. Some giants are still detectable following exposures as high as 10,000 to 20,000 R. Even these are capable of metabolic activity.

The chromosomes of irradiated cells have been studied. Stickiness, breaks, translocations, and bridges have been reported. Genetic changes have also been demonstrated in studies of isolated survivors. Cells have been found which are altered in their nutritional requirements for growth, or show changes in chromosome numbers or in colonial morphology. Genetic damage may also be responsible for the inability of cells to maintain divisions for more than five or six generations.[8]

MITOTIC DELAY IN TISSUE CULTURE

A reversible mitotic delay occurs as a result of low doses of radiation. This is probably partially responsible for the decrease in colony size. Puck (1964) has found in all the cells that he has studied that the delay is greatest when cells are irradiated in the G_2 period (post-DNA synthesis; see Chapter 5). As an indication of the extent of delay, S3 strain HeLa cells which are irradiated while in the middle of the G_2 phase are delayed in reaching mitosis by a time equal to 0.15 hr/rad.

REPAIR OF DAMAGE IN TISSUE CULTURE

Since most survival curves for mammalian cell cultures have n numbers greater than one, it is likely that lethality is the result of an accumulation of damage. This implies that most cells that survive a sublethal dose of radiation have been damaged. The question then arises as to whether this sublethal damage is reparable. Many authors have clearly demonstrated that at least in some situations it is.

Repair can be demonstrated by multiple-dose irradiation as follows. A culture is irradiated. After a suitable time interval, survivors from the first exposure are reirradiated. If they respond to the second irradiation as if they had received no previous irradiation (same n and k), it can be assumed that damage from the initial exposure has been repaired. Various experiments (see, for example, Elkind and Sutton, 1960 or Han *et al.*, 1964) have indicated that repair processes start immediately after irradiation and that in at least some cell lines sublethal damage is substantially, if not completely, repaired in a few hours. Moreover, the systems responsible for the repair are not attenuated by repeated exposures. The ability to display recovery does not require a rapid growth rate since recovery has been demonstrated with two cell lines which have different growth rates. Recovery does not require cell division, since cells are capable of recovery during the division delay which is induced by an exposure. On the contrary, recovery of certain types of damage may be possible only up until some stage of the mitotic cycle, after which the damage is irreparable. Considerably more information must be accumulated before all of the mechanisms of the repair processes are understood.

---

[8] It is difficult to visualize a type of damage which is not manifest for five or six generations and then appears in all cell progeny. This is reminiscent of the phenotypic lag characteristic of mutations in bacteria.

RADIATION EFFECTS ON INDEPENDENT CELL SYSTEMS IN VIVO

Under certain conditions, single-cell techniques can be applied to an analysis of the radiation sensitivity of mammalian cells *in vivo*. Studies with transplantable tumors and with normal bone marrow cells have shown that the radiation sensitivity of these cells *in vivo* (as measured by proliferative ca-

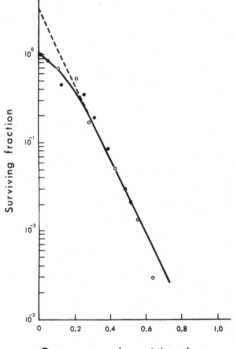

Gamma-ray dose, kilorads

Fig. 7.9. Survival curve for normal mouse bone marrow cells irradiated *in vitro* and injected into recipient mice. Number of colonies (nodules) formed in the spleen is used as a measure of numbers of surviving, proliferating cells. About 10^4 nucleated marrow cells must be injected to produce a colony which probably results from an individual cell. The open and closed circles represent different strains of mice (C57Bl and C3H). The extrapolation number, n, is 2 and D_{37} is 115 ± 8 rads (measured on the exponential portion of curve). (From Till & McCulloch [1961], courtesy Academic Press, Inc.)

pacity) is of about the same magnitude as was found for most mammalian cells *in vitro* (see Figure 7.9). This supports the suggestion derived from *in vitro* work that the radiation sensitivity of a variety of types of mammalian cells is similar and that it is not strongly dependent on the tissue of origin.

RADIATION EFFECTS ON MEDIA

It is possible that decreased survival of cells after irradiation may be related not only to direct changes in the cells but also to indirect effects from the irradiated medium. The latter effect can be studied by irradiating the medium alone before adding the cells. It has been shown that extremely high radiation exposures of the medium can result in decreased survival of microorganisms or tissue culture cells.

At least part of the effect is related to a pH change of the medium. In addition, both hydrogen peroxide and organic peroxides have been detected. Since treatment with catalase will sometimes reverse the inhibition, peroxides apparently contribute to the effect in at least some situations. Certain breakdown products of sugars have also been detected (see Molin and Ehrenberg, 1964). These have not been fully identified, and their significance in cell growth is not clear. The doses required to produce an amount of any of these substances which is sufficient to inhibit reproduction is considerably greater than that required to inhibit cell replication directly.

GENERAL REFERENCES

Elkind, M. M. and Whitmore, G. F. *The Radiobiology of Cultured Mammalian Cells*, Gordon and Breach, New York (1967).

Errera, M., and Forssberg, A. *Mechanisms in Radiobiology*, Vol. I, Chapter 1 (Pt. 2) and Chapter 5. Academic Press, Inc., New York (1960).

Lea, D. E. *Actions of Radiations on Living Cells*, 2d ed. Cambridge University Press, Cambridge (1955).

Powers, E. L. Considerations of Survival Curves and Target Theory. *Phys. Med. Biol.*, **7**: 3–28 (1962).

Setlow, R. B., and Pollard, E. C. *Molecular Biophysics*, Chapter 10. Addison-Wesley Publishing Co., Reading, Massachusetts (1962).

ADDITIONAL REFERENCES CITED

Adler, H. I., and Hardigree, A. A. Postirradiation Growth, Division and Recovery in Bacteria. *Radiation Res.*, **25**: 92–102 (1965).

Bruce, A. K. Response of Potassium Retentivity and Survival of Yeast to Far-Ultraviolet, Near-Ultraviolet and Visible and X-radiation. *J. Gen. Physiol.*, **41**: 693–702 (1958).

Crowther, J. A. Some Considerations Relative to the Action of X-rays on Tissue Cells. *Proc. Roy. Soc. (London) Ser. B*, **96**: 207–211 (1924).

Elkind, M. M., and Sutton, H. Radiation Response of Mammalian Cells Grown in Culture. I. Repair of X-ray Damage in Surviving Chinese Hamster Cells. *Radiation Res.*, **13**: 556–593 (1960).

Evans, R. D. *The Atomic Nucleus*, Chapter 26. McGraw-Hill Book Co., Inc., New York (1955).

Han, A., Miletic, B., Petrovic, D., and Jovic, D. Survival Properties and Repair of Radiation Damage in L-cells After X-irradiation. Intern J. Radiation Biol., **8**: 201–211 (1964).

Haynes, R. H. Role of DNA Repair Mechanisms in Microbial Inactivation and Recovery Phenomena. *Photochem. Photobiol.*, **3**: 429–450 (1964).

Hollaender, A., Stapleton, G. E., and Martin, F. L. X-ray Sensitivity of *E. coli* as Modified by Oxygen Tension. *Nature*, **167**: 103–104 (1951).

Molin, N., and Ehrenberg, L. Anti-bacterial Action of Irradiated Glucose. *Intern. J. Radiation Biol.*, **8**: 223–231 (1964).

Munson, R. J., and Bridges, B. A. Site of Lethal Damage by Ionizing Radiation in *Escherichia coli* B/r Growing Exponentially in Minimal Medium. *Nature*, **210**: 922–925 (1966).

Puck, T. T. Cellular Interpretation of Aspects of the Acute Mammalian Radiation Syndrome, in Harris, R. J. C. (ed.) *Cytogenetics of Cells in Culture*. International Society for Cell Biology. Academic Press, Inc., New York (1964).

Puck, T. T., and Marcus, P. I. A Rapid Method for Viable Cell Titration and Clone Production with *HeLa* Cells in Tissue Culture: The Use of X-irradiated Cells to Supply Conditioning Factors. *Proc. Natl. Acad. Sci., U. S.* **41**: 432–437 (1955).

Puck, T. T., and Marcus, P. I. Action of X-rays on Mammalian Cells. *J. Exptl. Med.*, **103**: 653–666 (1956).

Till, J. E., and McCulloch, E. A. A Direct Measurement of the Radiation Sensitivity of Normal Mouse Bone Marrow Cells. *Radiation Res.*, **14**: 213–222 (1961).

DIFFERENTIAL
CELL SENSITIVITY

8

As early as 1906, the French radiobiologists Bergonié and Tribondeau recognized that different types of mammalian cells differ greatly in their apparent radiosensitivity. On the basis of a series of comprehensive studies of rodent testes, they generalized that cells are radiosensitive if (1) they have a high mitotic rate, (2) they have a long mitotic future (that is, under normal circumstances, they will undergo many cell divisions), and (3) they are of a primitive type. These generalizations have become known as the Law of Bergonié and Tribondeau. Although many exceptions to the "Law" are now known, the criteria still serve as a "rule of thumb" for predicting the relative radiosensitivity of cells.

Comparisons of cell sensitivity can be made in several ways. A higher organism is composed of a variety of tissues which contain cells with varying functional and morphological characteristics. Certain features, such as somatic chromosome number, are common to most of the cells of an organism. Other characteristics, such as mitotic rate or number of mitochondria, can differ widely among the cells of various tissues. The relative radiosensitivities of the cell types within an organism can be studied with reference to the cellular characteristics in which they differ. By this method, one gets an indication of the influence of these factors on radiosensitivity.

Among different species, there can be even greater variation in cellular characteristics. For example, the somatic chromosome number or the chromosome size may be very different. In suitably controlled situations, the influence of certain of these cellular characteristics on radiosensitivity can be determined by interspecies comparisons.

Unicellular organisms are also useful for studies of cell sensitivity. For example, one can study populations of cells of different ploidys or of various mutant types.

Various cellular characteristics which have been shown to correlate with radiosensitivity will be discussed in this chapter. The presentation will be limited to those characteristics which have been most studied, or seem to show the clearest correlations. In addition, a section is included which describes the relative radiosensitivity of various types of mammalian cells.

CRITERIA OF SENSITIVITY

All cells do not respond to radiation in the same way. Some will be unable to divide, but may appear to function normally in other respects; others will not be capable of normal function; still others will die. When comparing the radiosensitivities of different types of cells, it is important to use the same criterion of damage — for example, the dose of radiation which will produce cell death (that is, disintegration and loss of all normal functions) or which will stop division or cause the loss of reproductive integrity. It is most meaningful to compare the dose of radiation which will produce cell death in one type of cell with the dose that will produce the same effect in another cell type, rather than comparing the lethal dose in one cell with the division-inhibiting dose in another.

It is also important to specify the criterion which is used in a particular experiment. If the number of cells of type A decreases following irradiation and the number of cells of type B remains the same, it can be said that cell type A is more radiosensitive than cell type B, *with respect to disappearance*.

The production of chromosome aberrations, disappearance of cells, and mitotic inhibition have often been used to compare the radiosensitivities of mammalian cells *in vivo*. Cell death or disappearance are also very frequently used criteria. There are a number of reasons for this: (1) Many experiments have involved the study of histologic changes in irradiated tissues so that information is available on the morphologic manifestations of cell death or disappearance; (2) cell death can be measured in an individual cell; (3) death is a nonreversible end point; and (4) all cells can be killed. There are disadvantages to the use of this criterion, too. Different cell types "die" or disappear at different rates.

Many of the interspecies comparisons have been done with higher plants. The end points usually selected for comparisons are lethality, severe growth inhibition, or slight growth inhibition of the plants. It is assumed, in these experiments, that the radiosensitivity of the plants is related to the sensitivity of the meristematic cells in the plant shoots or roots. It is, therefore, really the characteristics of these meristematic cells that are compared.

FACTORS INFLUENCING
SENSITIVITY

The cell nucleus is generally considered to be the major site of radiation injury in a wide variety of organisms. In earlier chapters it was shown that DNA may be one of the most likely primary sites for radiation damage in growing cells. It is the present purpose to examine the phenomena at a more macro-level of organization.

Plants, rather than animals, have been used most extensively as test systems for many of the studies on nuclear parameters which affect radiosensitivity. This is partially due to the interests of the individual investigators, and partly to the availability of plants with wide ranges of meristematic nuclear volume, chromosome number, and ploidy. Presumably, the same basic factors control radiosensitivity in both plant and animal cells.

NUCLEAR AND CHROMOSOME VOLUME

The radiosensitivities of many species of plants have been determined by Sparrow and co-workers.[1] They found that the dose necessary to produce a given degree of growth inhibition in a series of species is correlated with the chromosome number and with the average volumes of interphase nuclei of cells in the meristematic regions (regions of active division, such as plant shoots or roots). For example, if one compares a series of plants which have about the same chromosome number, those which have the largest nuclei are most sensitive. In another series of plants, each of which has about the same nuclear volume, those with the fewest chromosomes are most sensitive. Sparrow (1965) has combined these two parameters to obtain a single index of radiosensitivity. Dividing the average nuclear volume of meristematic cells by the number of chromosomes which is characteristic of the species gives a value which has been termed the *Average Interphase Chromosome Volume*. This represents the average volume occupied by each chromosome in the nucleus of a cell in interphase. No allowance is made for nucleoli or for interchromosomal space, but it is anticipated that this omission will not greatly affect the correlation.

When the radiation exposure which produces acute lethality in a group of plants is plotted against the interphase chromosome volumes of these plants (Figure 8.1), a clear correlation is apparent. The species with largest chromosome volumes are the most radiosensitive.

[1]Most of the studies correlating nuclear factors with radiosensitivity have been done by A. H. Sparrow and co-workers at the Brookhaven National Laboratory. (See, for example, Sparrow, 1966.)

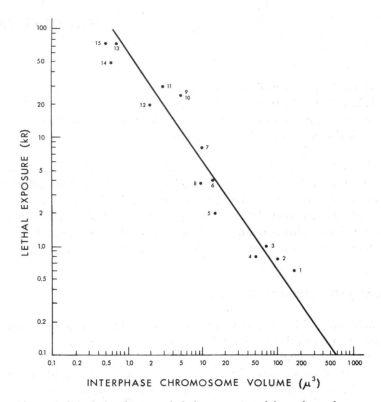

Fig. 8.1. Correlation between lethal exposure and interphase chromosome volumes for 15 herbaceous species, following acute irradiation (plotted with a slope of —1). Each numbered point represents one species which are: 1. *Trillium grandiflorum* 2. *Podophyllum peltatum* 3. *Hyacinthus orientalis* HV. Innocence 4. *Lilium longiflorum* 5. *Chlorophytum elatum* 6. *Zea mays* HV Golden Bantam 7. *Aphanostephus skirrobasis* 8. *Crepis capillaris* 9. *Sedum ternatum* (U643) 10. *Gladiolus* HV. Friendship 11. *Metha spicata* 12. *Sedum oryzifolium* 13. *Sedum tricarpum* 14. *Sedum alfredi* var. *nagasakanium* 15. *Sedum rupifragum*. (Modified from Sparrow (1965), courtesy Williams & Wilkins Co.)

It is possible to calculate the amount of energy which is absorbed per interphase chromosome at a particular exposure for a given plant species. For example (Sparrow, 1966), if one assumes an average value of 32.5 ev/ ion pair and 1.77 ionizations/μ^3 of wet tissue per roentgen, then 1 R = 57.5 ev/μ^3. The interphase chromosome volume of *Trillium grandiflorum* is 145.19 μ^3 and the lethal exposure is 0.6 kR. The energy absorbed per chromosome at this exposure is 145.19 $\times$ 57.5 $\times$ 0.6 = 5009 kev. When the energy absorbed per chromosome at the acute lethal exposure is calculated for the other species of herbaceous plants studied, it is found that all of the values are similar (3711 $\pm$ 434 kev). In fact, the variation in this value between species is very much less than the variation in total exposure for the different species.

A similar correlation between radiosensitivity and interphase chromosome volume has been demonstrated using other end points for the radiation effect. Severe growth inhibition of herbaceous species occurs when about 2330 ($\pm$371) kev is absorbed per interphase chromosome; slight growth inhibition occurs with about 1030 ($\pm$206) kev. Other experiments have been done using chronic irradiation instead of acute, and also with woody species of plants. Similar correlations with the interphase chromosome volumes have been obtained within the groups, although the overall radiosensitivity of woody plants is greater than that of herbaceous plants.

One may conclude from these results that a particular amount of energy must be absorbed within a chromosome in order to produce a given effect. This amount is relatively constant within certain groups of organisms, but is different between these groups. Since the amount of DNA per chromosome also varies widely among species, one may further conclude that the biological damage leading to the given end point is not due to the production of a constant proportion of damaged DNA molecules. Rather the damage is related to the absorption of a relatively constant amount of energy in each chromosome. It can be assumed that the constant amount of energy produces about the same number of lesions of unspecified nature in the genetic structure of an average chromosome of each species. In other words, the degree of difference in radiosensitivity found among species depends to a high degree on the concept of dose, and where and how dose is measured.

In certain groups of plants, chromosome size tends to decrease as chromosome number increases. It is thought that the decrease in radiosensitivity with increasing chromosome number (discussed earlier in this section) is related to the effect of chromosome volume.

PLOIDY

The radiosensitivity of organisms has been demonstrated to relate to ploidy, although the relationship varies greatly from one type of organism to another. In general, haploids (n) are most radiosensitive. In certain systems, such as higher plants and cells in tissue culture, diploids ($2n$) are often shown to be more radiosensitive than polyploids. Sparrow has suggested that the greater resistance of the polyploids in plants may be related to the decrease in average interphase chromosome volume (see previous section) which is often concomitant with increasing ploidy. Other authors have suggested that the resistance of polyploids may be due to the redundancy of genetic information at the duplicate loci of the multiple chromosome sets.

In yeast, the correlation of radiosensitivity with ploidy is the reverse of that usually determined in higher plants. A typical experiment with yeast is

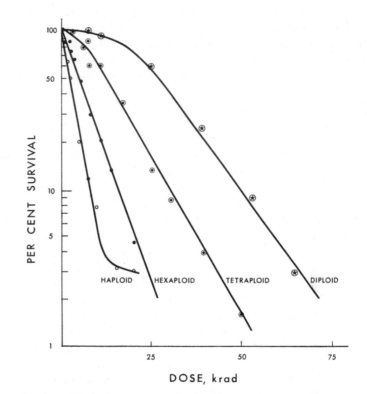

Fig. 8.2. Survival curves of haploid (n), diploid ($2n$), tetraploid ($4n$) and hexaploid ($6n$) yeast (*Saccharomyces cerevisiae*). Criterion for survival was formation of macrocolonies from single cell isolates. 50 kv x-rays were used at room temperature. (Modified from Tobias *et al.* [1958].)

shown in Figure 8.2. With the exception of the high sensitivity of haploid yeast (n), there is an increase in sensitivity with increasing ploidy. This is consistent with the idea that a major part of the induced lethal damage in diploids and cells of higher ploidy (as measured by inability to form colonies) is related to dominant mutational effects. The presence of multiple chromosomes provides more sites for lethal damage. Recessive mutations in polyploids would not be lethal if there are duplicate loci on other chromosomes — that is, genetic redundancy. In haploid cells, there is no genetic redundancy and a recessive mutation is not masked by a nonmutated gene on another chromosome. The decreased survival in haploid cells, then, may result from both recessive- and dominant-type mutations.

It seems likely that there may be additional factors which are not yet recognized which are involved in the relationship between ploidy and radiation sensitivity. The mechanisms which have been proposed do not fully explain all the results which have been observed.

ADDITIONAL NUCLEAR PARAMETERS

Other nuclear factors have been correlated with chromosome changes in irradiated plants and animals. It is likely that these factors can also be correlated with the radiosensitivity of the cell.

For example, cells which have extra nucleoli in their nuclei have been observed to have fewer chromosomal aberrations. Moreover, the nucleolar chromosomes[2] are involved in chromosome interchanges less often than are the other chromosomes. For such an exchange to occur, the portions of the chromosomes must be less than 0.1 μ apart. It may be that the nucleoli, by virtue of their physical size, isolate the nucleolar chromosomes from other chromosomes and decrease the probability of exchanges if breaks occur. The number of chromosomal aberrations appears also to be related to the amount and distribution of heterochromatin. Presumably, there is a greater chance for exchanges to occur at the heterochromatin areas.

The location of the centromeres may also be important in determining the frequency and types of chromosomal aberrations. Assuming random

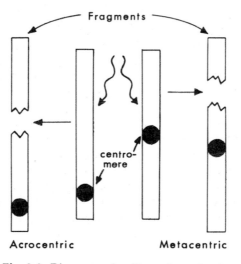

Acrocentric   Metacentric

Fig. 8.3. Diagrammatic illustration showing the influence of centromere position on the amount of chromosomal material lost in an "average" break. Assuming that the "average" break occurs halfway between the centromere and the end of the chromosome, the fragment will be larger if the centromere is located near the end of the chromosome (acrocentric) than if the centromere is in the middle of the chromosome (metacentric).

[2]Nucleolar chromosomes are those chromosomes which have been observed to be physically associated with the nucleoli during the short portion of the mitotic cycle when both are visible.

breakage of chromosomes, the average break will be halfway between the centromere and the end of the chromosome. Therefore, the average deletion will be larger in acrocentric than in metacentric chromosomes (see Figure 8.3).

There is ample evidence that the nuclear factors which have been discussed are important determinants of the radiosensitivity of a cell. They are probably of greatest importance when one is comparing the relative sensitivities of cells of the same tissue in different species (for example, the meristematic cells in different species of plants). When different cells in the same organism are compared, however, there is less variation in interphase chromosome volume, ploidy, etc. Certain other characteristics then become of relatively greater importance.

CYTOPLASMIC FACTORS

Numerous authors have pointed out the significance of certain cytoplasmic characteristics in determining the relative radiosensitivities of mammalian cells. For example, the ratio of nuclear volume to cytoplasmic volume is apparently directly related to radiosensitivity; that is, most cells are radioresistant if they have a large cytoplasmic volume as compared with their nuclear volume. Likewise, the total amount of cytoplasm influences radiosensitivity. Cells with a large volume of cytoplasm are generally more radioresistant than those with little cytoplasm. It has been suggested that the presence of the cytoplasm may help the cell to recover from the initial damage caused by the radiation.

Mammalian cells containing many mitochondria in their cytoplasm are more radioresistant than those with few mitochondria. If there is a causal correlation between these, it may be proposed that, since mitochondria are the chief site of oxidative phosphorylation in the cell, the presence of many such sites may be important in repairing damage caused by radiation.

OTHER CELLULAR FACTORS

Perhaps the feature of mammalian cells which most influences their radiosensitivity is mitotic rate. Since the reports of Bergonié and Tribondeau, many radiobiologists have observed that rapidly dividing cells are killed with relatively low radiation exposures. Cells which are either postmitotic or have a low rate of mitosis are relatively radioresistant, based on the criteria of cell death or disappearance. It is likely that certain types of damage (such as chromosome aberrations) may be produced in cells at the time of irradiation, but that the cell will not die as a result of the damage until it enters division. The apparent resistance of nondividing cells might simply reflect the fact that there has not been an opportunity for the damage to become apparent. There is also evidence that some cellular lesions can be repaired, but that this repair must be accomplished prior to a cellular division. A cell type with a high mitotic rate would have less opportunity for repair prior to mitosis than would a cell type with a low mitotic rate. Damage might also be apparent sooner.

As Bergonié and Tribondeau observed, mammalian cells which are less differentiated (primitive stem cells, for example) are usually more radiosensitive than are highly specialized cells, such as muscle or nerve cells, or mature sperm. Many exceptions, however, have been noted for this last criterion of sensitivity.

At present, none of the theories that have been presented can be used as more than guideposts for predicting probable radiosensitivity. On the basis of these guides and, more especially, on the basis of histologic observations of cellular changes, it is possible to group certain cells into major categories of sensitivity.

CATEGORIES OF
MAMMALIAN CELL SENSITIVITY

On the basis of histologic observations of early cell death, G. Casarett (in Harris, 1963) has divided mammalian parenchymal[3] cells into four general categories of radiosensitivity, as indicated in Table 8.1.

Category I includes the most radiosensitive "vegetative intermitotic" cells which divide regularly, but undergo no differentiation between cell divisions. The basal cells of the epidermis are an example of cells in this group.

Category II is composed of the "differentiating intermitotic" cells which are somewhat less radiosensitive than those in the first category. They divide regularly, but do undergo some maturation or differentiation between divisions. Myelocytes are an example of cells in this category. These are cells of the bone marrow which are precursors to the granulocyte series of white blood cells.

The "reverting postmitotic" cells in category III are relatively radioresistant. Under usual circumstances they do not undergo mitosis, but they are capable of division under the proper stimulus. The parenchymal cells of the liver are in category III. Ordinarily, they do not divide, but removal of a large portion of the liver can provide a stimulus for cell division.

---

[3]Parenchymal cells are the functioning cells of an organ, as opposed to cells of blood vessels and connective tissue.

Table 8.1. Categories of Mammalian Parenchymal Cells in Order of Decreasing Radiosensitivity

---

I. Vegetative Intermitotic Cells.
> divide regularly
> undergo no differentiation between divisions
> most radiosensitive

II. Differentiating Intermitotic Cells.
> divide regularly
> differentiate to some extent between divisions
> relatively radiosensitive

III. Reverting Postmitotic Cells.
> do not divide regularly, but are capable of
> division under proper stimulus
> variably differentiated
> relatively radioresistant

IV. Fixed Postmitotic Cells.
> do not divide
> highly differentiated
> most radioresistant

The "fixed postmitotic" cells in category IV are highly differentiated and have completely lost their ability to divide. These may be long-lived cells such as neurons (nerve cells) which are not replaced after the death of a given cell. Or, they may be short-lived such as the granulocyte series of white blood cells, which are rapidly replaced by cells from category II.

The nonparenchymal cells — the endothelial (lining) cells of small blood vessels and the cells of the connective or supporting tissue of organs — are between categories II and III in radiosensitivity.

SPECIFIC CLASSIFICATIONS OF MAMMALIAN CELL SENSITIVITY

It is of interest to compare specific cells of adult mammals with regard to their radiosensitivities. In most instances, the cells are grouped in the order that one would expect from the generalizations and criteria that have been given in this chapter. There are, however, a number of exceptions. The following classifications are based on the criterion of early cell death, judged by histologic study.

Group 1. Mature lymphocytes (one of the two major classes of circulating white blood cells), erythroblasts (red blood cell precursors), and certain of the spermatogonia (most primitive cell in the spermatogenic series) are undoubtedly the most radiosensitive mammalian cells. Note that the erythroblasts are primitive cells, with a high mitotic rate, a long mitotic future, and a relatively large nuclear/cytoplasmic volume ratio. Accordingly, one would predict that they would be radiosensitive. The small, radiosensitive lymphocytes, however, have a large nuclear/cytoplasmic volume ratio, but are believed to divide only under unusual circumstances, and then perhaps only after enlarging. These cells represent one of the major inconsistencies in the Law of Bergonié and Tribondeau and with the categories of sensitivity presented above. Certain spermatogonia are very radiosensitive when cell death is used as a criterion, as one might expect from their characteristics. In addition, they are extremely sensitive to mitotic inhibition.

Group 2. Granulosa cells (cells surrounding ovum) in developing and mature ovarian follicles are only slightly less sensitive than cells in Group 1, as are myelocytes (in bone marrow), intestinal crypt cells, and germinal cells of the epidermal layer of the skin.

Group 3. Gastric gland cells and endothelial (lining) cells of small blood vessels comprise the next most sensitive group.

Group 4. Cells in this category are only moderately sensitive to radiation. These include osteoblasts (bone forming cells), osteoclasts (bone resorbing cells), chondroblasts (precursors to cartilage cells), granulosa cells of primitive ovarian follicles, spermatocytes, and spermatids. The latter two cell types are in the spermatogenic series.

Group 5. Somewhat more radioresistant are granulocytes (one type of white blood cell), osteocytes (bone cells), sperm, and superficial cells of the gastrointestinal tract.

Group 6. Fairly large doses of radiation are required to produce changes in parenchymal and duct cells of glands, fibroblasts (form intercellular fibrous matrix), endothelial cells of large blood vessels, and erythrocytes (red blood cells).

Group 7. The relatively radioresistant cells in this group include fibrocytes (connective tissue cells), reticular cells (fixed hematopoietic stem cells), chondrocytes (cartilage cells) and phagocytes (scavengers).

Group 8. It is generally considered that the least radiosensitive cells of the adult mammalian body are muscle cells and nerve cells.[4] Note that these are fully differentiated cells which are incapable of division. It must be stressed again that these cells are most radioresistant, based on *morphological* comparison, but that they may rank higher if functional impairment is considered.

[4]Evidence is now accumulating, that certain nerve cells are sensitive, but that it takes a long time for them to show an effect.

GENERAL REFERENCES

Harris, R. J. C. (ed.). *Cellular Basis and Aetiology of Late Somatic Effects of Ionizing Radiation.* Academic Press, Inc., New York (1963).

Sparrow, A. H. Relationship Between Chromosome Volume and Radiation Sensitivity in Plant Cells, pp. 199–222 in *Cellular Radiation Biology.* Williams & Wilkins Co., Baltimore (1965).

Sparrow, A. H. Research Uses of the Gamma Field and Related Radiation Facilities at Brookhaven National Laboratory. *Radiation Botany*, **6**: 377–405 (1966).

Tobias, C. A., Mortimer, R. K., Gunther, R. L., and Welch, G. P. The Action of Penetrating Radiations on Yeast Cells, pp. 420–426 in *Proc. Second U. N. Intern. Conf. Peaceful Uses At. Energy, Geneva*, **22** (1958).

RADIATION EFFECTS ON MAJOR ORGAN SYSTEMS OF MAMMALS

9

It has been observed repeatedly that the various tissues of an animal differ in their response to irradiation. Some show gross histologic damage and are unable to function following low doses of radiation, while some can receive massive doses without apparent injury. Part of this variation in radiosensitivity may be attributed to differences in the sensitivities of the cells which make up the tissues. In addition, apparent tissue sensitivity is influenced by such factors as normal turnover time of the cells, interaction between the various types of cells within a tissue, the ability of the tissue to replace or to repair damaged cells, and the reserve capacity of the tissue.

This chapter is concerned with the histologic and functional changes in the major mammalian organ systems following exposure to radiation. Total body irradiation will be inferred unless other conditions are specified. The direct effects on a tissue are usually similar whether the radiation is given to a portion of the body containing the tissue or to the entire body. Local irradiation experiments will be cited to demonstrate the specific tissue effects of large doses in cases where total-body irradiation of that magnitude would be lethal before the tissue effect could be demonstrated.

The presentation will involve the summarization of many experiments in which slightly different radiation exposures were used. Since details of dosage tend to obscure the overall picture of the radiation effects, doses will be referred to as small (0 to 500 rads), moderate (500 to 1000 rads), large (1000 to 2000 rads), and massive (more than 2000 rads). Within wide limits, the qualitative effects of radiation are not specific for a particular dose. If sufficient radiation is given to produce an effect, a slightly larger dose will probably give qualitatively the same effect. It might be greater, appear sooner, or last longer. Likewise, with a slightly smaller dose, the effect might be less pronounced, appear later, or disappear sooner. Certain specific doses will be cited for illustrative purposes, for clarity, or for emphasis of specific details.

A brief review of the normal histology and function of the tissues is included where such information is considered important to an understanding of the radiosensitivity of the organ.

INTERDEPENDENCE OF
CELLS AND TISSUES

Tissues are composed of parenchymal cells, which carry out the function of the tissue, plus supportive elements (e.g., connective tissue) and the blood vessels which transport metabolites to the parenchymal cells. The apparent radiosensitivities (based on cell disappearance or death) of various categories of parenchymal cells were described in Chapter 8. It was indicated that intermitotic cells (categories I and II) are, in general, radiosensitive and that postmitotic cells (categories III and IV) are usually radioresistant. Most of the vascular and connective tissue cells are intermediate in sensitivity — that is, between parenchymal cells of types II and III. Some generalizations are possible regarding tissue response to radiation, based on the categories of cell sensitivity. There are many exceptions to these generalizations; they are intended only to serve as a guide for the descriptions of specific organ systems which will follow.

If a tissue is composed of radiosensitive parenchymal cells (categories I or II), such as those of the epidermis, mucosa of the gastrointestinal tract, or the blood-forming tissues, then the early radiation effects, particularly after low radiation doses, depend primarily on destruction or mitotic inhibition of parenchymal cells. The disappearance of parenchymal cells may be rapid but the cells are usually replaced within a short time. Accordingly, there may be early damage, but quick recovery.

If the exposure is high enough to produce vascular changes leading to narrowing or occlusion of small blood vessels, there may be a delay in the repair or replacement of parenchymal cells which are dependent on an intact blood supply. In addition, changes initiated in the vascular or connective tissue (vasculo-connective tissue) is often not repaired in a typical fashion, and this damage may actually increase in severity in the months and years following radiation. Therefore, long after irradiation there may be secondary changes in parenchymal cells which are a result of the vascular damage. Such long-term changes will be considered in Chapter 12.

In contrast, in tissues which contain postmitotic parenchymal cells of types III or IV, such as liver, kidney, muscle, or brain, very high doses are required to remove the parenchymal cells, but only moderate doses are needed to damage the vasculature. Thus, the major changes seen in parenchymal cells following moderate doses of irradiation are late effects and are secondary to changes in the vascular tissue. Massive radiation doses are required to produce effects directly in the parenchymal cells of such tissues. It is difficult to assess the indirect contribution of changes in the vasculo-connective tissue following massive radiation exposures.

The radiation response of an organ can also be markedly influenced by the condition of other irradiated or nonirradiated tissues and cells in the body. If certain parenchymal cells are removed but can be replaced by differentiation of more primitive cells or by migration of similar cells from a different tissue, the overall effect may be lessened. For example, bone marrow cells from one part of the body can repopulate irradiated marrow in other parts of the body.

There is some evidence to suggest that radiation injury to some cells and tissues may be partially caused by circulating "toxins." No specific toxins have been demonstrated, but it has been suggested that they may be cell decomposition products. Moreover, a particular function of a cell or tissue may be dependent on specific hormone levels in the blood (for example, adrenal secretions). If radiation produces changes elsewhere in the body which alter this blood hormone level, the function of the dependent cell or tissue might alter. Evidence for such "abscopal" or indirect effects is that certain tissues show changes even when they are shielded from the radiation and other portions of the body are irradiated. Also, certain tissues show a greater effect when the entire body is irradiated than when only the tissue itself is exposed.

GENERALIZED TISSUE CHANGES

Certain characteristic changes occur in many irradiated tissues. Division will be inhibited immediately in cells which normally are mitotically active. Chromosome aberrations may be seen in those cells which are in division, and many cells may show degenerative changes.

Somewhat later, hemorrhage or edema may indicate a change in permeability or a break in the continuity of the vascular system. The population of cells may change as the more radiosensitive degenerate and are removed from the tissue. Radioresistant phagocytic cells may be numerous and active in the removal of cellular debris.

Still later, regeneration begins with the first abortive mitoses leading to degenerative cells. Later, the mitotic activity may be greater than normal, resulting eventually in a more or less complete regeneration of the tissue. Months or years after radiation, there may be secondary changes in parenchymal cells, perhaps associated with vasculo-connective tissue effects.

All of these histologic changes may be accompanied by physiological and functional alterations in the tissue. The specific types of changes, their extent, time of onset, duration, and significance to the organism are dependent on many factors, such as the tissue studied and the dose of radiation used.

EFFECTS ON
SPECIFIC ORGAN SYSTEMS

BLOOD AND HEMATOPOIETIC TISSUE

The circulating blood is composed of a fluid plasma and the formed elements, or blood cells. The mammalian red blood cells, erythrocytes, are nonmotile and contain no nucleus, Golgi apparatus, mitochondria, or centrioles. About 95% of their dry weight is hemoglobin, the oxygen-carrying pigment. Immature erythrocytes, called *reticulocytes*, are usually found in small numbers in normal blood, and may appear in large quantities when the number of circulating erythrocytes is depressed. They can be distinguished from mature erythrocytes under proper staining conditions by the presence of a fine reticular structure throughout their cytoplasm.

White blood cells (leukocytes) can be divided into two types — the nongranular lymphocytes and monocytes, and the granulocytes. Lymphocytes are spherical, and contain a large basophilic nucleus and little cytoplasm. The larger monocytes have oval or kidney-shaped nuclei and more cytoplasm than lymphocytes. Granulocytes have a lobulated nucleus (2 to 5 lobes), and contain specific granules in the cytoplasm. On the basis of the differential staining of the granules, these cells can be classed as acidophils (coarse granules staining with acidophilic dyes), basophils (finer granules staining with basophilic dyes), and neutrophils (fine granules staining with neutral dyes). Leukocytes are one of the chief defenses against infection. Granulocytes migrate to a site of inflammation and phagocytose (engulf) and destroy foreign particulate matter such as bacteria. Lymphocytes are involved in certain inflammatory and immunological processes.

Blood platelets are tiny colorless pinched-off pieces of cytoplasm from the megakaryocytes of the bone marrow. They are important in blood coagulation.

Blood cells are formed in the hematopoietic (blood-forming) tissues. Lymphocytes and monocytes are produced in lymphatic tissue — lymph nodes, thymus, the "white pulp" of the spleen, and some scattered small areas in the body. Erythrocytes, the granular leukocytes, and platelets are formed in the myeloid tissue, which consists primarily of the red, or active, bone marrow.

All hematopoietic tissue has the same general structure. The stroma, or framework, of the tissue is a fibrous meshwork composed of reticular fibers and cells, and contains fixed macrophages and primitive reticular cells. The former cells are nonmotile, but are active phagocytes in their fixed location. The primitive reticular cells are not active phagocytes, but are capable of differentiating into a wide variety of blood and connective tissue elements. Within the meshwork of the stroma, hematopoiesis (formation of blood cells) occurs.

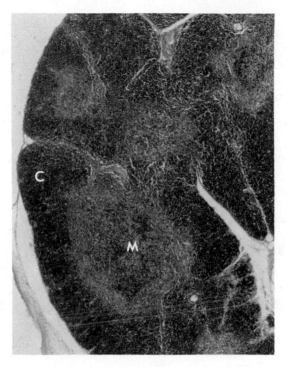

(A)

Fig. 9.1. Histological sections of thymus from young adult rats. (A) Normal, showing the cortex (C) and medulla (M). The cortex contains lymphocytes in nodules. (B) One day after total-body exposure to 750 R, illustrating absence of lymphocytes in cortical region.

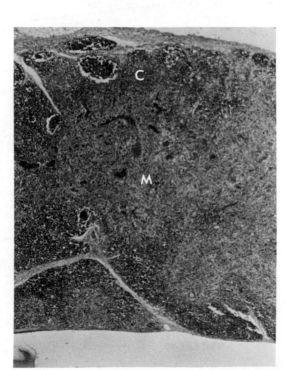

(B)

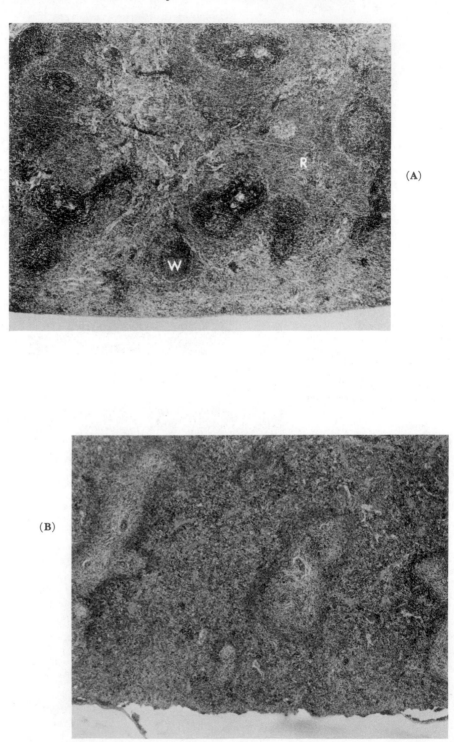

(C)

Fig. 9.2. Histological sections of spleens from young adult rats. (A) Normal, illustrating red (R) and white (W) pulp areas. The white pulp contains lymphocytes in nodules. (B) One day after total-body exposure to 750 R, illustrating absence of lymphocytes in nodules of white pulp. (C) Three days after exposure to 1500 R, illustrating marked shrinkage of organ and absence of lymphocytes.

Embedded in the stroma of lymph nodes are lymph nodules, the site of lymphocyte production. The central portion of the nodule, or germinal center, contains large and medium sized lymphocytes. Surrounding this central portion is a peripheral zone comprised primarily of small lymphocytes. These small lymphocytes migrate into the sinuses (passage ways) of the stroma and are subsequently carried into the lymph stream and the circulating blood.

The thymus is divided into lobes and, like the lymph nodes, has germinal areas of lymphocyte production (see Figure 9.1A). This organ is most active in the young animal. It atrophies (shrinks and becomes nonfunctional) as the animal grows older, and is believed to be the source of some humeral factor which, in young animals, will initiate the formation of antibody-producing lymphocytes in the other lymphatic tissues. Small round structures of concentrically arranged cells, called *Hassall's corpuscles*, are also contained in the thymus. Their function is unknown at present.

The spleen (Figure 9.2A) can be subdivided into the white and red pulp areas. The white pulp contains germinal centers for the production of lymphocytes. The red pulp is composed of sinuses which contain red blood cells. These sinuses serve as a "filter" for the blood. In certain species such as the rat, erythrocytes, platelets, and granulocytes can be formed in the red pulp of the spleen. This is termed *extramedullary hematopoiesis*.

The active, red bone marrow is the primary site of production of erythrocytes, platelets, and granulocytes. The bone marrow stroma is looser than that in lymphatic tissue, the meshes are larger, and the cellular organization is more diffuse. Four major types of cells can be recognized. They are (1) hemocytoblasts or free stem cells which are precursors to all the other types [these are nongranular, ameboid (capable of movement) and look rather like large lymphocytes]; (2) erythroblasts, which are nucleated precursors to the erythrocytes (under proper conditions many forms of erythroblasts can be distinguished in various stages of transition into erythrocytes); (3) myelocytes, which are precursors to granulocytes; (4) megakaryocytes, from which platelets are formed. These are giant cells with an irregular surface. Portions of the cytoplasm are pinched off to form blood platelets. As they are formed, the mature cells penetrate the walls of the stromal sinuses and enter the circulating blood.

RADIATION EFFECTS ON BLOOD AND HEMATOPOIETIC TISSUE

The radiosensitivity of the blood-forming tissues was recognized early in the history of radiobiology. It has since become apparent that damage to the hematopoietic system is a major factor in the mortality following acute radiation exposure.

Whole-body irradiation in the moderate dose range (500 to 1000 rads) will lead to a decreased concentration of all of the cellular elements of the blood. This can be due to (1) direct destruction of the mature circulating cells, (2) loss of cells from the circulation by hemorrhage or leakage through capillary walls, and (3) loss of production of cells. Since the blood cells all have a finite life span, they must be constantly renewed by cell production in the hematopoietic tissues. If such cell production is inhibited, replacement of circulating cells will not occur.

The adult, nondividing small lymphocytes are extremely sensitive to radiation. Mature granulocytes and erythrocytes are radioresistant. Loss of all types of cells from the circulation by hemorrhage and leakage certainly occurs following moderate doses of radiation. Since this is hard to quantitate, the contribution of this mechanism to cell depletion is not clear. Depressed hematopoiesis (production of blood cells) is a major factor in decreasing the numbers of circulating cells. Accordingly, the response of these tissues to radiation will be considered next, followed by a discussion of the change in numbers of blood cells.

Lymph nodes. The stroma of lymph nodes is relatively radioresistant. Lymphocytes of all sizes, however, are very sensitive to even very small amounts of radiation. A decrease in the number of small lymphocytes in lymph nodes can be detected following total-body exposures as low as 10 R. This appears to be primarily a direct effect of the radiation on the lymphocytes, although indirect effects from irradiation of other tissues may be involved. Such "abscopal" effects are known to occur with large exposures.

Within 15 minutes after a moderate dose of total-body radiation, there is a very marked reduction in cell division among lymphocytes. Many show evidence of necrotic changes. This continues for 6 to 7 hours after exposure (see Table 9.1), at which time the macrophages begin the process

Table 9.1. SUMMARY OF CHANGES IN IRRADIATED LYMPHATIC ORGANS OF RATS FOLLOWING MODERATE RADIATION DOSES

Phase	Lymph nodes	Thymus	Spleen
Destructive	1–7 hours	1–8 hours	1–8 hours
Debris removal	7–24 hours	8 hours to 2 days	8–24 hours
Inactivity	1–2 days	2–9 days	1–14 days
Regeneration	2–25 days	10–28 days	14–49 days

of phagocytosis (engulfment of the dead cells). Most of the debris is removed by 24 hours after irradiation, and the lymph nodes appear almost devoid of lymphocytes. The lack of lymphocytes and collapse of lymph nodes produce an apparent increase in the number of reticular (stromal) cells. The stromal architecture, ordinarily obscured by lymphoid cells, becomes obvious in histologic preparations. With such moderate doses, regeneration may start 2 to 5 days after exposure but will not be complete for at least 3 weeks.

Following exposure to small doses of radiation, regeneration may start before all of the dead cells have been removed from the node. With very low doses, the increase in rate of mitosis in lymphocytes may overcompensate for the loss of cells and the lymph nodes may contain more than normal numbers of lymphocytes. Large radiation doses may permanently inhibit regeneration, and fibrosis of the node may occur after a considerable period of time.

Thymus. The stroma and Hassall's corpuscles of the thymus are morphologically resistant to radiation. The lymphocytes show changes similar to those described for lymph node lymphocytes. For example, 1 day after a moderate dose, most of the lymphocytes have disappeared from the thymus; this is illustrated in Figure 9.1B. Regeneration occurs later in the thymus, and is usually not completed until at least 4 weeks after exposure.

Spleen. The white pulp of the spleen, the area of lymphocyte production, is rapidly depleted of lymphocytes after a moderate radiation exposure. Most of the cellular debris is removed by 24 hours after irradiation, and the spleen is frequently less than half normal size at that time, due to a marked decrease in lymphocytes (see Figure 9.2B). Large doses of radiation will cause a very severe depletion of cells and marked shrinkage of the spleen, as illustrated in Figure 9.2C. Regeneration of white pulp is slow in the spleen. Lymphocytes can be observed about 2 weeks after exposure, but regeneration is not usually complete for almost 7 weeks.

If erythropoiesis (formation of red blood cells) is occurring in the spleen of an irradiated animal, erythroblasts are damaged by moderate doses. Recovery starts in 3 or 4 days, and by 5 or 6 weeks there may be a hyperplasia (greater than normal number of cells) of the red pulp due to this "extra-medullary hematopoiesis."

Bone marrow. The fat cells, macrophages, reticular cells, and connective tissue of the bone marrow are relatively resistant to radiation. Most of the cells which are precursors to the red blood cells, granulocytes, and platelets are radiosensitive.

Erythroblasts (red blood cell precursors) are the most sensitive of these cells and show a marked decrease within a few hours after exposure (see Table 9.2). They reach minimum numbers between 4 and 5 days after

Table 9.2. SUMMARY OF CHANGES IN HEMATO-POIETIC CELLS OF RAT BONE MARROW FOLLOWING A MODERATE DOSE OF RADIATION

Phase	Erythroblasts	Myelocytes	Megakaryocytes
Detectable decrease in numbers	3 hours	2 days	3 days
Minimum numbers	4–5 days	4–5 days	4–13 days
Recovery	7–8 days	12–40 days	14–41 days

irradiation, but with moderate radiation doses are about normal in numbers by 7 to 8 days. There are some indications that there may be a disturbance of red cell maturation with large doses of radiation.

Myelocytes, precursor cells to the granulocytes, are somewhat less sensitive than erythroblasts. The radiosensitivities differ somewhat for myelocytes in different stages of maturation. Mitosis is inhibited in young dividing myelocytes. Maturing myelocytes are not damaged to any great extent. Their reduction in numbers after several days reflects the time of their maturation and release from the marrow. Moderate doses of radiation have little, if any, effect on their maturation or release. Recovery of myelocytes is slow. It begins at about day 12, and is complete by about 40 days after exposure.

Megakaryocytes, the giant cells from which platelets are formed, are decreased in number from about 3 days postexposure. Minimum numbers of these cells are present between days 4 and 13, and recovery proceeds slowly from about 2 weeks to 6 or 7 weeks after irradiation.

Small radiation doses result in a smaller depression in cell counts and more rapid recovery. Large doses produce a greater decrease of cells and a longer time for recovery. Hemorrhage characteristically occurs in the marrow and may contribute to the overall damage. It is also possible that rapid cell loss in the marrow cavity may lead to vascular distention and mechanical breakdown of the vessels.

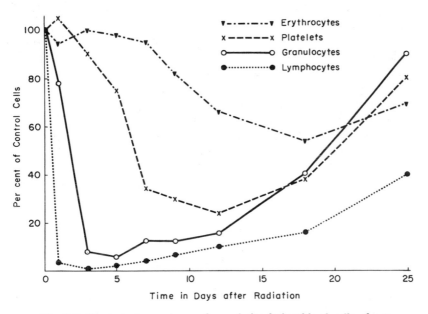

Fig. 9.3. Typical change in numbers of circulating blood cells of rats at various times after exposure to a moderate dose of radiation.

Blood cells. Within 15 minutes after exposure to moderate doses of total-body irradiation there is a detectable decrease in the number of circulating lymphocytes. Within hours, very few lymphocytes can be found in the circulation (see Figure 9.3). Doses as low as 5 to 25 rads can produce a drop in the number of these cells! Lymphocytes begin to reappear in a few days but the recovery is very slow. Normal numbers of cells are not measured until at least 3 or 4 weeks after irradiation. Direct damage is probably a major factor in the early lymphopenia (decrease in lymphocytes). Disturbed lymphopoiesis is probably responsible for the slow replacement of circulating lymphocytes.

Lymphocytes are so extremely radiosensitive that they have been used to quantitate radiation exposures. Within certain dosage ranges, the decrease in number of lymphocytes can be correlated with the radiation dosage.

Furthermore, certain "bilobed"[1] lymphocytes, and lymphocytes containing extranuclear chromatin material, have been identified in the blood of irradiated mammals and are claimed to be specifically radiation-induced. Correlation of these cells with radiation dosage is being attempted in order to quantitate exposure.

Mature granulocytes are radioresistant but have a relatively short life span in the circulation (probably less than a day). A rather sharp granulocyte decline following moderate doses of radiation begins after the first day in mice and rats, after the second day in rabbits and after the third day in dogs and monkeys. It has been suggested by Patt and Quastler (1963) that the differences in time of onset of the sharp drop may reflect, in part, differences in the numbers of maturing, nonmitotic myelocytes. When the source of replacement cells is depleted due to mitotic inhibition of the young myelocytes, the number of circulating cells falls rapidly as a result of their normally short life span. Lowest levels are reached by 3 to 5 days. Recovery begins at about the ninth day and is about complete by day 25. As indicated in Figure 9.3, the cell decrease is less rapid and slightly less severe than with lymphocytes, and the recovery is more rapid. Doses of 50 rads will produce a detectable depression in numbers of granulocytes. Since the majority of granulocytes are neutrophils, these results reflect primarily changes in neutrophils. However, basophils and eosinophils probably exhibit a similar sensitivity to radiation.

Blood platelets have not been studied as extensively as granulocytes, but the general pattern seems to be similar. The platelets themselves are radioresistant. They are supplied to the circulation as long as megakaryocytes are present in the bone marrow (about 3 to 5 days). Platelets then decrease in the circulation at a rate which is consistent with a 4 to 5 day life span.[2] Minimum numbers of these cells are present in the second and third week. This contributes to the hemorrhage and to the difficulty in clotting which characteristically occur at this time. Recovery of platelets is usually somewhat slower than recovery of granulocytes. Radiation doses larger than 50 rads are needed to produce a detectable drop in platelets. Very high doses of radiation may result in a temporary increase in number of platelets, possibly due to a mobilization of cells from the marrow.

Mature erythrocytes are radioresistant and have a normal life span of about 4 months. Although there is a dramatic decrease in the production of erythrocytes in the bone marrow after moderate radiation doses, the number of circulating cells does not reflect this drop. This is due to the slow turnover rate in the circulation. Starting 5 to 7 days after a moderate radiation dose,

[1]"Bilobed" lymphocytes are actually binucleate cells containing two discrete nuclei within a single cell. Certain authors claim to have found these cells in nonirradiated animals, and, therefore, deny that they are specifically produced by radiation.

[2]The numbers given are for several species of animals. The rate of decrease in circulating platelets after irradiation is slower in man. This may be related to a longer platelet life span (8 to 9 days) and a slower turnover of megakaryocytes.

a depression in erythrocytes can be measured. Severe anemia may develop during the first week or two after exposure. This is partly due to the decreased production of cells and partly to the hemorrhage and leakage of blood through capillary walls. Minimum levels of erythrocytes are reached at about 3 weeks after exposure. Recovery starts at about the fourth week and is usually complete by the fifth or sixth week.

Reticulocytes normally constitute 1 to 2% of total circulating red blood cells. During the third and fourth weeks after radiation exposure, the number of reticulocytes often increases to 500 to 700% of normal, indicating accelerated erythropoiesis. (The bone marrow is pushing out immature cells.) Moreover, although 300 rads total-body dose is required to produce a detectable depression in total red blood cells, a decrease in reticulocytes can be detected several days after a dose of 25 rads. This depression reflects the sensitivity of erythrocytic precursors. It may also be due, in part, to a delayed release of maturing reticulocytes from the bone marrow.

Plasma. Direct radiation effects on the constituents of blood plasma are relatively slight. The concentration of total plasma proteins is reduced slightly at first, but remains relatively constant. There is some variation in the specific protein fractions. The albumin fraction has been reported to decrease by as much as 50%, and the α_1, α_2, and β globulin fractions increase. This results in an increase in the globulin/albumin ratio.

There are a number of changes in plasma which are secondary to changes elsewhere in the body. An increase in amino acid nitrogen, nonprotein nitrogen (NPN), and urea reflects cell destruction. Increase in serum iron and in the concentration of bile pigments probably follows an increased destruction of red blood cells. A decrease in sodium and potassium may reflect alterations in absorption through the gastrointestinal tract.

DIGESTIVE SYSTEM

The various portions of the digestive tract (mouth, pharynx, esophagus, stomach, small intestine, and large intestine) form a continuous tube most of which is lined with a mucous membrane. Layers of submucosal connective tissue and muscle surround the mucous membrane in certain portions of the tract. The mouth, pharynx, and esophagus serve primarily as pathways for the passage of food. Salivary glands such as the parotid, submaxillary, and sublingual glands empty their secretions into the mouth, thereby aiding in the digestion of food.

The mucous membrane of the stomach forms many invaginations or "gastric pits." At the bottom of these are simple branched tubes, or glands. Enzymes which help in digestion and mucus for lubrication of the membrane surface are elaborated by the glandular cells. Food which enters the stomach is moistened, softened, and partially dissolved by the chemical secretions from the glands. It is mixed mechanically by the contractions of the muscle layers surrounding the mucous membrane and moved into the small intestine by contraction of other muscle layers. Surface cells of the stomach are continually dying, are shed, and are replaced by new cells which are formed in the gastric pits.

The epithelial surface area of the membrane of the small intestine is greatly increased by the formation of circular mucosal folds and finger-like outgrowths of mucosal membrane called *villi* (see Figure 9.4A). Between the bases of the villi are the glands, called *crypts of Lieberkühn*, where new cells are formed and secretions are elaborated. As new cells are formed in the crypts, they migrate up to the tips of the villi. They are then sloughed off into the lumen. The formation, migration, and sloughing ordinarily take place in 1 to 3 days. Absorption of amino acids (from protein), glycerol and fatty acids (from neutral fat), and glucose (from carbohydrates) takes place through the villi of the small intestine. The largest glands of the small intestine are the liver and pancreas, which secrete bile and pancreatic juice, respectively. Their products assist in the digestive process.

The large intestine has no villi and few folds in the mucosa, but has well-developed muscle layers. Scattered lymphatic nodules are usually found in the submucosa along the large intestine.

RADIATION EFFECTS ON THE DIGESTIVE SYSTEM

For years it has been recognized that many of the syndromes which characterize the radiation responses of an organism are associated with changes in the digestive system. Histologic damage can be observed soon after exposure, and functional alteration is apparent. Many gross symptoms such as anorexia (loss of appetite), nausea, vomiting, and diarrhea are characteristically present.

Histologic changes. The oral mucosa (mouth) frequently becomes ulcerated or eroded following large doses of radiation. The ulcers are a possible site for entry of bacteria or other sources of infection. The pharynx and esophagus are relatively radioresistant, although massive doses may cause necrosis of the epithelium and underlying tissues.

Several types of histologic change can be seen following irradiation of the stomach. Degenerative changes are apparent within 30 minutes after exposure to moderate doses, especially in the cells of the glandular epithelium. The chief cells, which elaborate pepsinogen, are most sensitive. They show morphologic evidence of degeneration following small radiation doses. The parietal cells, which secrete hydrochloric acid, seem to be resistant as judged by histologic criteria. However, a decrease in their hydrochloric acid secretion indicates functional impairment. Histologic recovery of the stomach is rapid and is nearly completed by about 5 days after exposure. Gastric ulcers have been reported to occur several weeks after irradiation. Hemorrhaging (bleeding) in the stomach is also common at that time.

The small intestine is the most radiosensitive portion of the digestive tract. The duodenum (portion nearest the stomach) is the most sensitive segment and the epithelium in the crypts is more sensitive than that of the surface. Thirty minutes after a moderate exposure, about half of the crypt cells of the duodenum show nuclear fragmentation, swelling, or other evidence of cellular disintegration. Debris accumulates in the lumen of the crypts. Within the first few hours most mitotic activity ceases. The cells which do undergo division have bridges or acentric fragments. Within a day, the epithelial surface contains only a thin layer of cells and the villi are shortened (see Figure 9.4B).

Recovery of the small intestine is rapid. There is evidence of mitotic activity 1 day after exposure but many of the new cells are abnormal. By the third day, the rate of mitosis in the crypts is greater than normal although the villi are covered by only a thin layer of stretched cells (see Figure 9.4C). By the end of a week, new cells have covered the villi and the intestinal epithelium appears about normal. Removal of cellular debris is slow. Regeneration commonly occurs in the presence of degenerative changes.

The initial damage is more severe following large doses and recovery of the epithelium may not occur. The villi appear to be shortened and only partially covered by degenerating cells (see Figure 9.4D). Death of the animal usually occurs between the third and fifth day without regeneration of the epithelium of the small intestine.

The large intestine is less radiosensitive than the small. It is about as sensitive as the stomach; however, it is particularly prone to ulceration. This characteristically occurs either 1 to 2 weeks or 1 to 2 months after irradiation. In addition to ulceration, an increase in the mucous secretion of the large intestine has been reported.

Histologically, the early changes in blood vessels of the gut wall are not prominent. Petechial ("pin point") hemorrhages are often seen in the stomach or intestine about 10 days after exposure to moderate doses, or in the mouth following large doses. Capillary leakage and localized edema in the connective tissue layers can also be seen.

After massive radiation exposures, some slight changes are apparent in the muscular layers of the gastrointestinal tract. These include vacuolization of muscle fibers, nuclear irregularities, and hyaline changes in the contractile material of the fibers.

The intestinal epithelium can develop an apparent resistance to radiation under certain conditions. For example, when radiation doses of 80 rads or less are given repeatedly, there is, at first, a decrease in mitotic figures and evidence of minimal destructive changes. As the radiation continues, how-

(A)

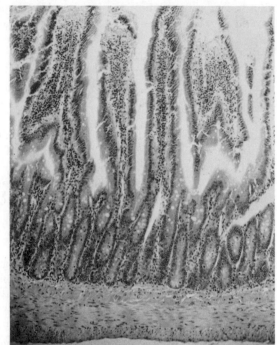

(B)

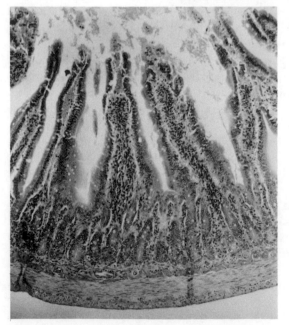

Fig. 9.4. Histological sections of small intestine (duodenum) of adult rats. (A) Normal, showing finger-like villi which project into the lumen. Mitosis occurs in the crypts at the base of the villi. (B) One day after total-body exposure to 750 R, illustrating shortening of villi and decrease in mitotic activity in crypts. (C) Three days after exposure to 750 R, illustrating marked depletion of epithelial cells along villi and beginning regeneration in crypts. (D) Three days after exposure to 1500 R, illustrating vacuolated, necrotic cells on shortened villi and absence of regeneration.

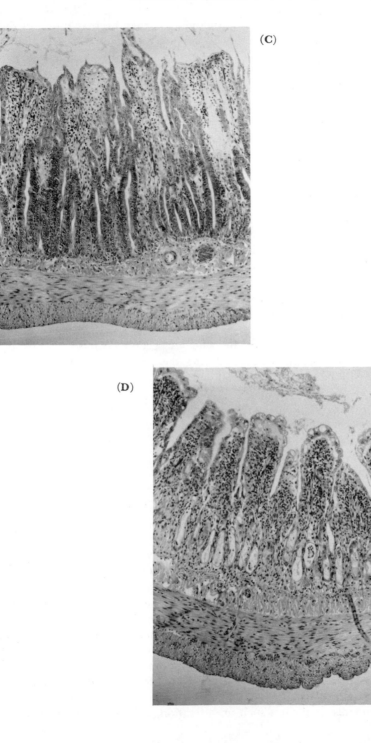

(C)

(D)

ever, these changes become less apparent. It may be that the most radio-resistant cells have survived and there has thus been a selection of the "fittest" cells.

Functional changes. Anorexia (decrease in appetite) characteristically follows exposure to radiation. There is a measurable decrease in the food intake of rats exposed to 50 rads. Food consumption decreases to about 10% of normal in rats irradiated with a moderate dose. The mechanism for this depression of the appetite center is unknown.

The stomach emptying time is greatly prolonged in irradiated animals. Food will remain in the stomach of rats for 3 to 4 days following exposure to moderate doses. Since relatively little absorption occurs through the stomach wall, this means that nutrient material is not entering the body by intestinal absorption during this period.

The absorption of nutrients through the irradiated small intestine has been studied both in isolated, exteriorized segments and in intact sections. Absorption of thiamine, glucose (active transport), and xylose (passive transport — not against a concentration gradient) is maximally inhibited 3 days after moderate doses of radiation. There seems to be no detectable effect on lipid absorption. The 3 day period coincides with the appearance of new, abnormal, epithelial cells in the villi. A second decrease in metabolite transport and absorption occurs during the second week, which seems to correlate with a swelling and loss of mitochondria in the intestinal epithelial cells.

Alterations in both segmented (exerting churning action on food) or propulsive (moving food along) contractions have been measured after irradiation. In the stomach, segmented contractions decrease in amplitude, starting within 10 minutes after the beginning of exposure. The frequency of contraction is unchanged.

Segmented contractions in the small intestine increase in amplitude with doses as low as 100 rads. This is apparently a local effect as it has been demonstrated in exteriorized gut loops. It is probably due to parasympathetic stimulation at the level of the enteric ganglia.

Propulsive motility has been measured by the rate of movement of a charcoal gum acacia meal. Within an hour, following a moderate dose of radiation, the degree of motility is about 60% greater than in nonirradiated animals. A marked depression in movement then occurs and persists for about 4 days. Sometimes antiperistalsis (reverse movement) occurs during the first few hours. This delivers the contents of the small intestine into the stomach where it may remain for several days.

Two or three days after irradiation, the small intestine appears translucent, dilated, and filled with a thin, yellow, foul-smelling, gaseous liquid. This suggests an altered bacterial content. Diarrhea is common during this period and also indicates such an alteration, since bacteria normally contribute to the controlled development of formed feces. The diarrhea results in loss of fluids and electrolytes from the body and is probably one of the major contributing factors in the death of animals receiving doses above 1000 rads.

Guinea pigs and hamsters experience less severe dehydration and electrolyte imbalance than do mice and rats. It has been suggested that the presence of a longer colon (portion of large intestine) in the former animals permits a greater reabsorption of the fluids and electrolytes, and, therefore, accounts for less diarrhea.

Radiation effects on accessory organs to digestive system. The liver appears to be a very radioresistant organ. This is due, at least in part, to its large regenerative capacity. In mice, even very large doses to the liver produce few necrotic changes. Such doses do, however, result in reversible cytologic effects such as a decrease in hepatic glycogen, hyperglycemia, and fatty infiltration of the liver. These changes are probably due in large part to stress on the remainder of the body and to the nonspecific adaptation syndrome. Similarly, an increase in liver mass 6 hours after small doses is correlated with a rise in water and lipid content and can probably be attributed to increased adrenal cortical activity.

As a result of damage to the vascular system, there may be degeneration of some liver cells 1 to 2 weeks after exposure. Also, in response to the presence of degenerative changes in other tissues there is a swelling and activation of Küpfer cells (phagocytes of the liver). This is part of the mobilization of the reticuloendothelial system (RE system) which follows irradiation.

The pancreas is markedly radioresistant. Following moderate doses, only very slight degenerative changes have been noted. However, with such exposures, the volume of pancreatic secretion decreases temporarily in dogs and the quantity of several enzymes (amylase, lipase, and trypsin) is depressed for 1 to 2 days.

Exposure of the heads of dogs to large doses results in atrophy and fibrosis of salivary glands and some alteration in the morphologic details of the cells. Dryness of the mouth is common following head irradiation. This is probably due to the occlusion of the glands by mucous plugs. Regeneration of irradiated salivary glands is rapid and is complete unless the radiation dose has been extremely high.

VASCULAR SYSTEM

The vascular system is composed of the heart and the tubelike vessels: the arteries, veins, and capillaries. The arteries which carry blood from the heart to the capillaries are lined with endothelial cells and contain varying amounts of muscle and elastic tissue in their walls. The veins which carry blood from the capillaries back to the heart have thinner, softer, less elastic walls, and are larger in diameter than the corresponding arteries. Arteries and veins vary greatly in size. They are largest near the heart and decrease in diameter while increasing in number toward the capillaries.

Capillaries are very narrow tubes with a wall that normally consists of a single layer of flattened endothelial cells. The cells are held together by a mucopolysaccharide substance which is often referred to as *intercellular cement*. Transfer of fluids and gases normally takes place through the endothelial cells.

In certain organs, sinusoids connect the arteries and veins. These are passageways with irregular diameters lined with a discontinuous layer of phagocytic and nonphagocytic cells.

RADIATION EFFECT ON THE VASCULAR SYSTEM

Damage to the vascular system contributes indirectly to many of the changes in other tissues. Some aspects of this interaction are described in an earlier section of this chapter.

The heart and the large arteries and veins are radioresistant.[3] The endothelium of capillaries is radiosensitive. Occlusion of capillaries and small arteries is produced by moderate doses of radiation (see Figure 9.5). This can be caused by several different mechanisms.

1. Endothelial cells may swell as a result of the irradiation and may block the narrow capillary lumen.
2. Certain low doses of radiation may damage a few endothelial cells. The presence of the injured cells may stimulate the remaining cells to divide. If more new cells are produced than were initially destroyed by the radiation, there will be a piling up of cells which may occlude the capillaries. This is termed *overcompensatory hyperplasia*.
3. If radiation destroys cells, there will be a tendency for clot formation. Cells will stick to capillary walls and block the lumen.
4. Inflammation in the vicinity of the capillary may exert external pressure and cause it to collapse.
5. In larger vessels, if the endothelium is injured, fluid and cells may leak through the vessel walls. This may lead to a progressive fibrotic reaction with a replacement of normal muscle with connective tissue and a narrowing of the vessel.

Capillary occlusion will block the blood supply not only to the tissue in the immediate vicinity of the occluded area but to all tissue further along the capillary. Moreover, even if new "collateral" capillaries are formed to re-

---

[3]The endothelium of arteries and veins is only slightly less sensitive than that of capillaries to the changes which are described. However, because of the large diameter of these vessels, even if endothelial proliferation and swelling does occur, the vessels will not be occluded.

Fig. 9.5. Arteriole in renal cortex of rat sacrificed 160 days after a single injection of polonium-210, an alpha emitter. Note almost complete occlusion of the lumen by enlarged and proliferating endothelial cells. (Courtesy G. Casarett [1964] and Academic Press, Inc.)

place the damaged portion, there may be tissue cells which are no longer in contact with a capillary.[4]

An increase in permeability of capillaries has been demonstrated after moderate doses of radiation. It has been postulated that this is due to a depolymerization of the mucopolysaccharide "cement substance."

For example, Willoughby (1960) has shown that 1 day after abdominal irradiation of rats, there is an increase in the permeability of vessels to Trypan blue (a protein-bound dye). This effect is maximal by the third or fourth day. Pretreatment of the animals with antihistamine drugs decreases the response in the first day but has no subsequent effect; pretreatment with antiesterase drugs decreases the response after the second day. This suggests that the initial effect is due to a release of histamine and that this is maintained by some enzyme system involving a protease or esterase.

BONE

A variety of types and shapes of bone serve as structural support for the body and as a body store of calcium. Bone is not simply a framework. It is a dynamic living tissue.

In typical long bones (such as the femur) the diaphysis or shaft consists of compact bone surrounding the bone marrow cavity. The epiphysis, at the

---

[4]This has been noted particularly in the kidney. When a capillary entering a glomerulus is blocked, the new vessel often goes around the glomerulus instead of in, if revascularization occurs. This new vessel is *not* replacing the function of the old.

end of the shaft, consists of spongy bone covered with a layer of compact bone. In immature animals, growth in length occurs in a cartilage plate between the epiphysis and diaphysis.

Figure 9.6A shows the histologic structure of this cartilage plate. At the epiphyseal end is a zone of resting, primitive cartilage cells. Next is a zone of cell multiplication, composed of dividing cartilage cells, followed by a region of growing cells. A zone of full-grown cells and a zone of degenerating cells are at the diaphyseal end of the plate. The vascular connective tissue of the marrow extends between the columns of degenerating cells. As these latter cells disappear, a calcified matrix is left between the channels of invading marrow. Osteoid material is deposited on this matrix by osteoblasts and then is resorbed by other bone cells (osteoclasts) to form the shaft of the bone. Growth occurs in a longitudinal direction.

RADIATION EFFECTS ON BONE

Nongrowing portions of bone are relatively radioresistant, but the epiphyseal-diaphyseal plate in growing long bones can be damaged by moderate radiation doses. For example, if radiation is given to the knee region of 30 day old rats, mitosis ceases within 2 or 3 days, many of the cells of the cartilage plate disappear or are swollen and degenerate, and the columnar organization is disrupted. The invading blood vessels from the marrow are engorged with blood or hemorrhaged. Figure 9.6B illustrates some of these changes. Within a week, many of the remaining cells are dead or degenerate. Long projections of unresorbed cartilage extend into the shaft, and marrow invasion of the cartilage plate is incomplete. Fat replaces many of the marrow cells and the number of blood vessels in the marrow is decreased. By 2 weeks after irradiation there may be a complete separation of the cartilage plate from the marrow. Regeneration may start 2 to 6 weeks after exposure. Although recovery of the epiphyseal-diaphyseal plate may appear to be complete several months after irradiation, there will usually be a residual shortening of the bone, as a result of the period of growth inhibition.

If large doses of radiation are given to a growing bone, most cartilage cells degenerate, osteoblasts are destroyed, and an acellular, avascular (without cells or blood vessels) bonelike substance may be formed from the cartilage. This will gradually be resorbed and replaced by bone to form a solid plate which is incapable of growth. The result will be a shortened bone, as illustrated in Figure 9.7. If some of the resting cartilage cells survive the radiation, they may migrate distally to form a new functioning epiphyseal line.

Irradiation of formed bone may result in a derangement of the synchronization of resorption and new bone deposition. This may produce either

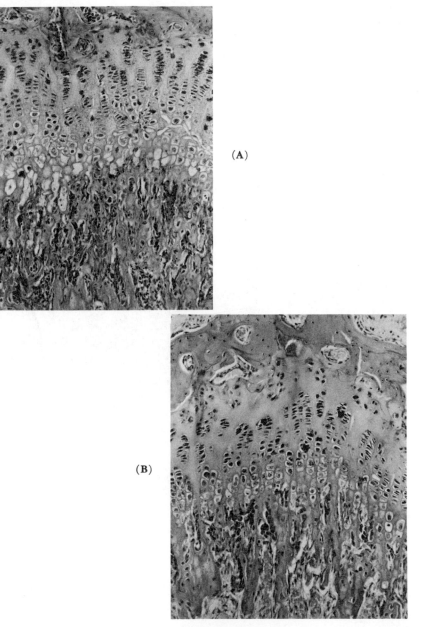

Fig. 9.6. Histological sections of epiphysial-diaphysial cartilage plates from femurs of young adult rats. (A) Normal. Note longitudinal columns of developing cartilage cells in the plate. The shaft of the bone is at the bottom; the head of the femur is at the top; bone marrow cells are visible in both regions. (B) Three days after total-body exposure to 750 R, illustrating degeneration of cartilage cells and narrowing of the plate. Note also the decrease in cellularity of the bone marrow, especially at the head (upper) end.

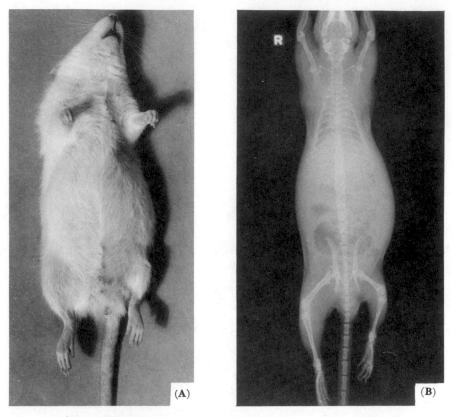

Fig. 9.7. (A) Gross appearance of 3 month old rat whose left leg was exposed to 1000 R when the animal was 21 days old. (B) X-ray of same rat showing shortened bones in left leg.

excessive absorption or an overgrowth of bone. In irradiated bone, fractures are common due either to structural damage to the bone matrix or to defective bone mineralization.

The effects of radiation on bone are of particular concern when dealing with radioactive isotopes of elements such as plutonium, radium, and strontium which localize in bone. In addition, localized radiation therapy frequently results in massive doses to bones.

TEETH

The radiation effect on teeth is similar to that on bone. Small to moderate doses produce a characteristic lesion — a chalky zone extending the full width of the tooth. This appears in the incisors of rats at 1 to 2 months after exposure. It usually disappears from the continually growing teeth by the end of the fourth month by attrition at the occlusal edge. Large doses to growing and developing teeth will produce stunting and malformation.

SKIN AND HAIR

The skin consists of a surface epithelium, called the *epidermis;* a connective tissue layer, called the *dermis;* and a subcutaneous layer of loose, fatty connective tissue. The deepest zone of the epidermis, the germinal layer, contains cells in division. More superficial layers contain nondividing, more differentiated, and cornified cells which are periodically sloughed and replaced from the germinal layer. In certain portions of the body, the dermis may be ridged, forming typical patterns, as seen in fingerprints.

Blood vessels are frequent in the subcutaneous layer, with branches extending up into the dermis. Hair follicles, which are tubular invaginations of the epidermis, extend from the skin surface into the subcutaneous layer.

RADIATION EFFECTS ON SKIN AND HAIR

The germinal layer of the epidermis is the most radiosensitive portion of the skin. Radiation exposures as low as 35 R will decrease the rate of cell division in this zone! Somewhat higher doses will cause cell death and may also increase the rate of cellular differentiation at the expense of cell division; that is, more than normal numbers of cells will differentiate into cells characteristic of the superficial layers of the epidermis and fewer cells will remain in the germinal layer. The differentiated cells in the more superficial epidermal layers are radioresistant. These cells, however, will disappear by normal sloughing and will have to be replaced from the germinal zone.

Moderate radiation doses are required to produce changes in the dermis. With such doses, the tight-ridged structure may be changed to a looser tissue organization. Characteristic fingerprints may then disappear, as a result of the loss of dermal ridges.

When large radiation doses destroy the cells in the germinal layer, the superficial layers will be sloughed, leading to a condition called *dry desquamation.* Massive doses may lead to a similar condition called *wet desquamation* where fluid exudate is present in the area where the epidermis has been removed. Ulceration, or holes extending down through the dermis, may follow increasingly larger doses and acute necrosis may follow exposure to 5000 R. In the latter case, the degenerating area may extend as far as the bone structure.

If the area irradiated is small or if the dose is low so only epidermal cells are damaged, regeneration usually occurs. If necrotic changes extend into the dermis or underlying connective tissue, any regeneration that occurs is usually abnormal and scar tissue is formed.

In studying the effects of radiation on humans, it was noted that the skin of some individuals is considerably more radioresistant than that of other people. For example, radiation produces less effect in the dark skin of Negroes than in the lighter skin of Caucasians.

Cattle which were exposed to fallout from the July 1945 nuclear weapons test (local beta irradiation of the skin) showed certain characteristic lesions of the skin. Along the central area of the back of many of these cattle, as late as 13 years after exposure, there were areas of greying hair and hornlike projections of keratin and cutaneous tissue.

A reddening of the skin which resembled sunburn was among the first of the changes that were noted in humans who were exposed to radiation. This "radiation erythema" was found to be due not to a change in the cells of the skin, but rather to a dilation of the capillaries that supply the skin. The increase in blood gives an appearance of redness. Several phases of erythema may occur following irradiation. (1) From several hours to 1 or 2 days after exposure, erythema may occur. This is apparently due to a release of histamine. (2) Reddening commonly occurs between 10 or 15 days to about 30 days after irradiation. Occlusion of arterioles (small arteries) may occur during this period. It has been postulated that the decreased blood flow resulting from occluded arterioles produces a condition of ischemia (decrease in oxygenation) in the tissue. This condition, through cardiovascular compensation, leads to capillary dilation and an appearance of reddening in the skin. (3) Erythema sometimes occurs between 4 and 6 weeks after irradiation. Collapse of venules (small veins) apparently leads to an increase in size of the capillaries which lead into the venules, since the capillaries are unable to drain properly. This is called *prestenotic capillary dilation*.

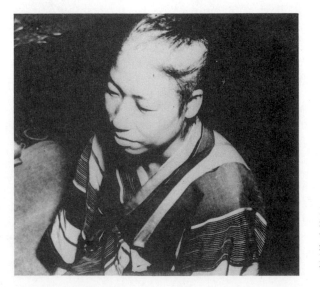

Fig. 9.8. Example of epilation in the Japanese who were exposed to radiation from the atomic bomb at Hiroshima. (Courtesy J. W. Howland.)

Fig. 9.9. Depigmentation of hair 11 months after polonium administration. The rat on the right is normal. (From Sproul *et al.* [1964], courtesy R. Baxter and Academic Press, Inc.)

Hair follicles are radiosensitive. Growth of hair can be inhibited by small doses of radiation. Epilation (loss of hair) frequently occurs 1 to 3 weeks after exposure to moderate doses (Figure 9.8). Regrowth of hair will usually occur about 1 month after irradiation if the exposure has not been massive. After temporary epilation, the new hair may be coarse if sebaceous glands have been destroyed, and it may be white or grey if melanoblasts (cells which produce pigment) have been damaged.

Although most laboratory animals have shown a change from dark to white or grey hair following irradiation (Figure 9.9), the hair from certain of the older Japanese survivors has reportedly changed from white to grey following irradiation!

RESPIRATORY SYSTEM

The respiratory system is composed of the lungs, in which gaseous exchange takes place, and of the passages through which air is conducted to and from the lungs. These include parts of the nose, pharynx, larynx, trachea, and bronchi.

All portions of the system are usually considered to be relatively radioresistant. However, an inflammatory reaction (radiation pneumonitis) frequently occurs in the lungs following exposure of the chest to large doses of radiation. This is characterized by edema (fluid accumulation), followed by hyalinization and fibrosis (thickening of connective tissue) of the alveolar walls. Resulting impairment of ventilatory and diffusion capacities of the lung may be significant in terms of the long-term effects on the animal. Hemorrhage in the lung may occur, also, as a result of changes in the blood vessels. Inhaled radioactive materials deposit in the lung and may produce carcinogenic or other changes.

URINARY SYSTEM

The urinary system is composed of the kidney, in which urine is formed, and the bladder and excretory ducts, through which the urine is removed from the body. The latter are radioresistant structures.

The parenchymal cells of the kidney are radioresistant as far as short-term changes are concerned. Months or years after exposure, however, secondary kidney effects may occur which are due, at least in part, to changes in the blood vessels. For example, a decrease in blood supply (ischemia) will produce hypertension, an increase in blood pressure. This causes a thickening of the walls of renal vessels and, therefore, a decrease in the size of the capillary lumen. The ischemia which follows results in a further increase in blood pressure, and so on. A diminished excretory function, retention of nitrogen, proteinuria (protein in the urine), hematuria (blood in the urine), and diuresis have all been observed in dogs several months after kidney irradiation with moderate to large doses.

MUSCLE AND CONNECTIVE TISSUE

Muscle is among the most radioresistant tissues of the body. Only slight muscular atrophy in rats is produced by localized irradiation with massive doses. Total necrosis may appear if the dose is as high as 72,000 rads. Intermediate doses produce a range of changes from atrophy to necrosis. Greater than 50,000 rads results in early fatigue and decrease in the amplitude of contraction in frog muscles, according to some authors.

Connective tissue has been described as morphologically radioresistant. However, low doses have apparently increased the permeability of dermal connective tissue, as indicated by an increased rate of absorption of intracutaneous saline, glucose, or dyes. It has been suggested that this effect is due to a depolymerization of the mucopolysaccharides which form the ground substance of the connective tissue.

NERVOUS SYSTEM

The nervous system is composed of the central nervous system (CNS), which is the brain and spinal cord, and the peripheral nervous system, which includes the nerves that carry impulses between the CNS and other portions of the body. Information on the external and internal environment of the body is channeled into the CNS by sensory nerves. This information is either stored in the memory or it is acted upon by sending out appropriate impulses by way of the motor nerves.

Nerve cells, or neurons, are postmitotic, highly-differentiated cells. Many of them contain long cytoplasmic extensions called *axons* or *nerve fibers*. Conduction of nerve impulses along these axons is accompanied by a change in the electrical potential of the nerve.

Among the nerve cells and their processes in the brain are several types of small cells called *neuroglia* which compose the supporting tissue of the brain. Numerous blood vessels penetrate the various regions of the CNS.

RADIATION EFFECTS ON THE NERVOUS SYSTEM

The nervous system of adult animals is usually considered to be extremely radioresistant, in terms of morphologic changes. However, a variety of physiologic responses have been reported following relatively low doses of radiation.

Morphologic effects. Small to moderate radiation exposures rarely are reported to produce any detectable morphologic changes in nervous tissue. Doses of several thousand rads result in a "delayed radionecrosis" of the CNS which occurs after a latent period of months or years. This is apparently due indirectly to alterations in the interstitial glial cells and to altered permeability of the capillaries and impaired capillary circulation, rather than to direct effects on the neurons themselves. Some authors consider these glial cell changes equivalent to the "reproductive death" which occurs with a much shorter latent period in other tissues (see Chapter 5).

Above about 3000 rads (total-body or locally defined irradiation) some interstitial or vascular cells appear swollen and pyknotic within hours after the radiation exposure. There is an increased permeability of the capillaries and some evidence of hemorrhage. Some neurons may also be damaged directly by very high doses. With increasing amounts of radiation, more cells are destroyed immediately. Doses of 10,000 rads or more destroy virtually all of the glial cells, the endothelial cells of the capillaries and many of the neurons located in the path of the radiation beam. With such doses, delayed radionecrosis can no longer occur, as no cells remain to initiate the process.

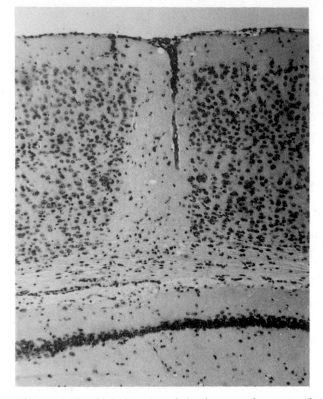

Fig. 9.10. Histological section of visual cortex of a mouse 45 days after irradiation. A circular collimated 10 Mev proton beam of 0.25 mm was directed on the visual cortex at the site of a penetrating artery. A dose of 36,000 rads was given in 6 seconds. The nerve cells in the beam path are irretrievably lost, but the glia and the interstitial elements are present and the blood vessels appear reasonably normal. This rather strange effect is due to the repopulation of the irradiated tissue by interstitial elements invading from the peripheral nonirradiated brain tissue, and also from the fact that isolated interstitial elements have not been fatally injured. (Technical data: paraffin sections 6 μ stained with PAS Gallocyanin, 80:1.) (Courtesy W. Zeman [1963] and G. Thieme.)

Very high doses of finely collimated beams of radiation produce discrete lesions in the central nervous system as illustrated in Figure 9.10.

Burros have shown an unusual nervous system sensitivity. Under specific dose-rate conditions, exposure of the heads of burros to only moderate doses of radiation is lethal to many of the animals within a few days. Death is preceded by symptoms which are suggestive of changes in the central nervous system, such as lethargy and ear-droop. The deaths are tentatively attributed to a peculiar sensitivity of the vasculature of the nervous system of the burro.

Physiologic effects. A variety of physiologic responses have been reported to follow exposure to radiation. The amount of radiation required to evoke these responses depends greatly on the specific effect which is being measured. For example, high-energy x-rays delivered at a rate of 6.5 rads/sec will produce a change in the electroretinogram of a frog after a total accumulated dose of only 0.25 rad, that is, after a latent period of less than 40 msec. In contrast, a dose of at least 6000 rads to the cat cerebellum is needed to depress the action potential which can be evoked by stimulation of the sciatic nerve. Moreover, it has been reported that 125,000 rads will produce no change in the conductivity of an isolated giant axon of a squid.

The presence of x-rays can be detected when the radiation is viewed with a dark-adapted retina. Individuals have "seen" x-rays after receiving a total dose of only 0.5 mrad, delivered at a rate of 2.0 mrads/sec. This and many other experiments suggest that nerve cell bodies, especially those engaged in receptor activity, have a high radiosensitivity as measured by physiologic changes. Nerve fibers appear to have a low sensitivity.

Much of the Soviet literature on radiation effects has centered around observations of conditional reflexes.[5] A postirradiation alteration either in the response to a previously established stimulus or in the establishment of a new reflex has been taken as an indication of functional changes in the nervous system. In a few experiments, extremely small doses of radiation (10 to 15 rads) have been reported to produce changes in conditional reflex activity. However, moderate, and even massive, doses have frequently been needed to demonstrate such changes.

It is interesting to note parenthetically that many Soviet papers stress the role of the nervous system in producing changes in other organs after irradiation. For example, in 1956, Livshits[6] reported an experiment in which he attempted to establish a conditional reflex that would lead to the same changes in blood cells as are produced by irradiation. After a series of irradiation sessions, the animals were left undisturbed for periods of 40 days to 10 months, sufficient time for a return to relatively normal blood values. The animals were then subjected to sham irradiation, accompanied by a variety of stimuli to which they had been "conditioned" by their association with irradiation. There was a decrease in lymphocytes, but not in erythrocytes or platelets in these animals. The experiments have been interpreted as indicating the presence of a reflex mechanism which produces some of the blood changes following irradiation.

[5]These have frequently been referred to as *conditioned reflexes*, but it now appears that the correct translation of the Russian word is "conditional." Perhaps the best known examples of the conditional reflex are the experiments of Pavlov in which dogs salivated when presented with a stimulus (bell, buzzer, etc.) as a result of a previously established association between the stimulus and food.
[6]Reported in Van Cleave (1963).

Behavioral effects. A variety of different experiments have indicated that animals can detect radiation in some way which is apparently not related to visual reception. For example, a conditioned avoidance of food and water has been demonstrated when x-rays are used as a conditioning stimulus. Rats with an established preference for saccharin water and who are given saccharin water to drink during irradiation exhibit a strong decrease in their previous preference for the saccharin in postirradiation tests. This conditioned aversion is still apparent 30 days after exposure to doses as low as 30 rads.

Spatial avoidance behavior has also been demonstrated. For example, rats were irradiated in one compartment of an enclosed alley and kept in another compartment for a similar period of time without irradiation. Later, given a choice between the two compartments, the rats showed a decreased preference for the one in which they were irradiated.

Radiation doses required to demonstrate conditioned avoidance differ markedly in various experiments. One report indicates that 7.5 or 15 rads of neutrons will produce aversion behavior in rats. Other experiments suggest that there may be a threshold of several hundred roentgens for this effect.

The receptor mechanism, the endocrine state, or even the anatomical region which is required for the perception of the stimulus has not been established.[7] It is not clear whether it is a direct effect of radiation on some portion of the nervous system or whether it is the result of neural mediation of radiation effects in other systems.

The ability to learn is not a radiosensitive process. Exposure to small radiation doses does not impair, but actually improves, the learning capacity of monkeys! Animals exposed to large doses of total-body radiation maintain their ability to solve new complicated learning problems until they are so weak that response is no longer possible. This enhanced performance in some irradiated animals has been tentatively attributed to decreased general activity and distractibility.

In rats, small doses of total-body radiation apparently have slight, if any, effect on the retention of a task, such as maze running, that was learned before irradiation. However, such doses have been reported to slightly decrease the ability of the rats to reorganize a preirradiation response pattern into a new, postirradiation pattern, such as performance in a different maze.

EYE

The most prominent radiation effect on the eye is development of cataracts, which are opacities of the lens of the eye. Head irradiation with low doses of

---

[7]Radiation effects on behavior have tentatively been grouped into those resulting from abdominal irradiation and those from irradiation of the head. Recent experiments have indicated that, in the latter case, the olfactory bulb may be involved in the perception of radiation.

x-rays produces minimal lens change in mice months or even years after exposure; severe alterations within 6 months to a year are produced by moderate to large doses. Neutrons are particularly effective in producing cataracts.

At low doses, the chief changes in the lens appear to be due to damage to individual cells of the lens epithelium and involve a decrease in mitotic activity of the germinative cells, abortive attempts to elaborate into normal lens fibers, and a resultant accumulation of abnormal cells and debris at the lens poles. With higher doses, all the structures within the capsule may eventually be involved and the lens may become completely opaque.

The immediate response of the retina (photoreceptor organ) to very low doses of radiation has already been mentioned. An electroretinographic response is elicited by an x-ray flash of as low as 7 mR. The distinction must be maintained, however, between this physiologic response from which the tissue recovers completely and the morphologic damage produced in the lens by much higher exposures.

Extremely high radiation exposures may produce permanent damage in other portions of the eye. Only minimal, reversible damage to the cornea has been reported after doses as high as 18,000 rads to the eyes of adult rabbits. The undifferentiated retina of immature mice, however, shows changes with several thousand rads.

EAR

The ear is relatively radioresistant. Moderate doses have no apparent effect on hearing in rats. A temporary hearing loss and tinnitus (ringing in the ears), however, have been reported in some patients following large doses to the head area. These changes have been attributed to a transient inflammation of the vessels of the inner ear. The inner ear has also been implicated as probably responsible for the acute alteration in space orientation which has been reported following large doses of x-rays to the brain of hamsters.

MALE REPRODUCTIVE ORGANS

The male reproductive organs include the paired testes in which sperm are produced, the epididymides in which sperm are stored, and the ducts through which sperm pass prior to ejaculation. The seminal vesicles and prostate elaborate secretions which constitute a large part of the liquid portion of the semen.

The testes contain connective tissue, blood vessels, Leydig cells, which elaborate the hormone testosterone, and many tubules in which the sperm are produced. Within these "seminiferous tubules" are Sertoli cells, which

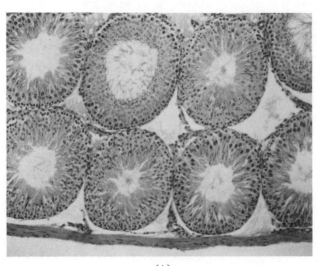

(A)

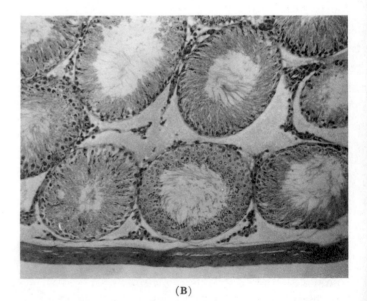

(B)

Fig. 9.11. Cross sections of seminiferous tubules from testes of young adult rats. (A) Normal, showing several types of cells within each tubule. (B) Twenty-five days after total-body exposure to 750 R, showing maturation depletion. Some maturing sperm are present in most tubules and regenerating spermatogonia are present in several. Some tubules also contain a few spermatocytes and spermatids.

provide nutrient material to developing sperm, and the germinal epithelium, which contains the varied cells of the spermatogenic series. The most primitive cells of the germinal epithelium are the spermatogonia. These cells proliferate and differentiate to form spermatocytes which undergo meiotic division to form small, haploid spermatids. Spermatids do not divide, but undergo a long series of complex transformations to become sperm. Following a maturation period, during part of which the sperm are stored in the epididymides, the highly differentiated sperm are ready for ejaculation and fertilization.

Cross sections of normal seminiferous tubules contain a number of combinations of germinal cells in different phases of spermatogenesis (see Figure 9.11A). Usually four cell types are present and each type is in a different stage of maturation. Recognition of the sequence of cell development has made it possible to determine the timing[8] of the normal spermatogenic cycle and has permitted an accurate estimate of the relative radiosensitivities of the various cells in the sequence.

RADIATION EFFECTS ON
MALE REPRODUCTIVE ORGANS

The male accessory organs (prostate and seminal vesicles) and the epididymides and vas deferens are radioresistant. Within the testes, Leydig cells, Sertoli cells, and the interstitial connective tissue and vascular elements are also relatively radioresistant. Many of the cells of the spermatogenic series, however, undergo necrotic changes, chromosome abnormalities, and inhibition of division following small to moderate radiation exposures. Genetic changes in these cells occur with very small radiation doses but such effects will not be discussed in this chapter.

Spermatogonia are the most radiosensitive testicular elements. The most primitive of the spermatogonia (Type A) of rodents show mitotic inhibition following doses of 50 rads. The slightly more differentiated Type B spermatogonia are killed by several hundred rads.[9] However, some Type A spermatogonia can survive massive radiation exposures, and, after a considerable period of mitotic inhibition, will repopulate the testes.

Moderate radiation doses are required to damage the more mature spermatocytes. Cell death and certain abnormalities such as chromosome bridges occur, but this damage usually is not obvious until the cell enters meiotic division.

[8]There is a considerable species variation in the spermatogenic cycle time. It appears, for example, that rats have about a 48 day cycle, and mice a $34\frac{1}{2}$ day cycle from the initial spermatogonial division to completion of the testicular portion of sperm maturation.

[9]In certain species, such as the mouse, 20 to 25 rads is sufficient to kill a large proportion of the Type B spermatogonia.

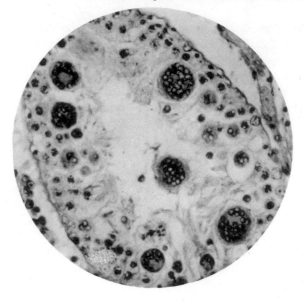

Fig. 9.12. Multinucleate giant cells (coalescent degenerate spermatids) from section of seminiferous tubule of testis of rat given 5 μCi/kg of polonium 517 days before death. (Courtesy G. Casarett [1964] and Academic Press, Inc.)

Postmitotic spermatids and sperm are altered morphologically with large doses. Clumping of spermatids results in multinucleate giant cells (see Figure 9.12), and structural malformations are visible in sperm.

If a series of testes from rats which have been exposed to small radiation doses are examined histologically, a characteristic pattern, called *maturation depletion*, is observed. Only the division of Type A spermatogonia is affected directly by such radiation doses. One day after irradiation the testes appear normal with the exception of a marked depression in mitotic activity of spermatogonia. One week after irradiation there are almost no spermatogonia and only a few early spermatocytes. Other cells are present, since they were beyond the spermatogonial stage at the time of irradiation and have continued to mature. By 3 weeks postexposure, sperm, spermatids, and a few Type A spermatogonia are present; by 4 weeks only sperm and Type A spermatogonia can be seen. The testes have thus been depleted of spermatogenic cells by the normal maturation of cells which are not replaced. Recovery occurs with division of the surviving Type A spermatogonia and may start before the process of maturation depletion is completed (see Figure 9.11B).

Massive radiation doses will produce degeneration and a rapid depletion of all germ cells, including Type A spermatogonia. Fibrosis of the testis may occur with no renewal of spermatogenesis.

The changes produced in the testis by small or moderate doses of radiation are direct effects; that is, the same changes will be produced by the same total dose of radiation delivered to the whole body, or to the testis alone. Likewise, if a testis is shielded and other tissues irradiated, there will be little or no change in the shielded testis.

The effect of chronic or protracted irradiation of the testis depends primarily on the rate at which the radiation is delivered. For example, more than 300 R delivered to dogs at a rate of 0.12 R/day (5 days/week for over 10 years) resulted in no effect on spermatogenesis that could be attributed to the radiation. When similar radiation was given to litter mates of these dogs at a rate of 0.6 R/day, however, there were marked changes in spermatogenesis in most of the dogs within 1 year with an accumulated exposure of only about 150 R.

Radiation effects on sperm. Mature sperm have been shown to be morphologically resistant to radiation. Even massive doses produce no effect on the motility of sperm or on their morphology. Many sperm irradiated with very high doses, however, are unable to fertilize ova. If fertilization does occur, the ova are not implanted. In addition, genetic changes occur even with very low exposures (see Chapter 6).

Rat sperm which are formed from cells irradiated as developing spermatids or immature sperm are morphologically normal if small doses are used. Sperm which have developed from cells irradiated as postmitotic spermatogonia or spermatocytes are less motile and are structurally weak (for example, the sperm heads may be separated from the sperm tails). Sperm which are produced in the initial period of regeneration (irradiated as spermatogonia) may have structural defects such as degenerate heads or abnormal sheaths on the sperm tails.

Radiation effects on male fertility. A period of fertility follows radiation exposures of small or moderate doses. This is due to the presence of sperm which were mature at the time of exposure and to sperm which developed from cells which were spermatocytes, spermatids, or immature spermatozoa at irradiation. Litter sizes during this period may be slightly smaller than normal. The reduction is likely due to the induction of chromosome aberrations which do not interfere with fertilization but which cause death of the embryo *in utero*.[10] In addition, although many of the sperm can fertilize an ovum and produce a viable embryo, genetic changes may have occurred which will result in nonviable or abnormal offspring.

The length of the initial fertile period depends on many factors such as the dose of radiation and the lifetime of mature sperm in the species. A sterile period follows the initial fertile period. No sperm are produced during this period as a result of inhibition of spermatogonial division and maturation depletion. If the dose of radiation is very small, the mitotic inhibition may be incomplete and some sperm will be formed. In such an instance, a situation of "semisterility," rather than complete sterility, will exist. The length

[10]Sperm which are irradiated as spermatids or spermatozoa do not undergo division prior to fertilization. Accordingly, there is no opportunity for those containing gross abnormalities to be weeded out of the population by the stress of division.

of the sterile period depends largely on the extent of inhibition of spermato-genesis and, therefore, on the dose of radiation.

If recovery of spermatogenesis occurs, a period of fertility follows the sterile period. Subnormal numbers of sperm may be produced if the radiation dose was sufficient to kill some of the primitive spermatogonia. In general, reduction in litter size is less during this second fertile period than during the initial fertile period. Some spermatogonia which contain chromosomal defects are eliminated, perhaps as a result of the series of divisions which these cells must undergo.

FEMALE REPRODUCTIVE ORGANS

The female reproductive organs consist primarily of the ovaries, where ova develop and mature, and the fallopian tubes, uterus, and vagina. The ovaries of a sexually mature female contain follicles of all sizes and stages of maturation in the thick peripheral cortical layer. A central medulla is composed of connective tissue and many large blood vessels. Many small, spherical, primary follicles at the periphery of the cortex each contain a single large ovum surrounded by a layer of small granulosa cells. Larger, more centrally located, intermediate follicles contain many mitotically active granulosa cells and an irregular fluid-filled space (see Figure 9.13A). Mature follicles occupy the width of the cortex and even form bulges on the free surface of the ovary. At ovulation, mature follicles rupture and the ova are released from the ovary. Meiotic division of an ovum starts just before ovulation, and the cell remains in the second meiotic metaphase until fertilization occurs. After ovulation, the ruptured follicle becomes a corpus luteum.

RADIATION EFFECTS ON
FEMALE REPRODUCTIVE ORGANS

The excretory ducts of the female are morphologically radioresistant. Corpora lutea of the ovary are also resistant, but ova and granulosa cells are radiosensitive.

Numerous pyknotic granulosa cells can be seen in developing follicles within several hours after total-body exposure to small or moderate doses (see Figure 9.13B). The ovum in most of these follicles also shows degenerative changes. In most species, intermediate follicles are the most sensitive, mature follicles somewhat more resistant, and primary follicles most resistant. In rats, for example, small doses will damage the intermediate and mature follicles, but primary follicles will survive and regeneration will occur within about a month.

The primary follicles of mice are peculiarly sensitive to radiation and will be completely destroyed by less than 100 rads. Doses as low as 50 rads

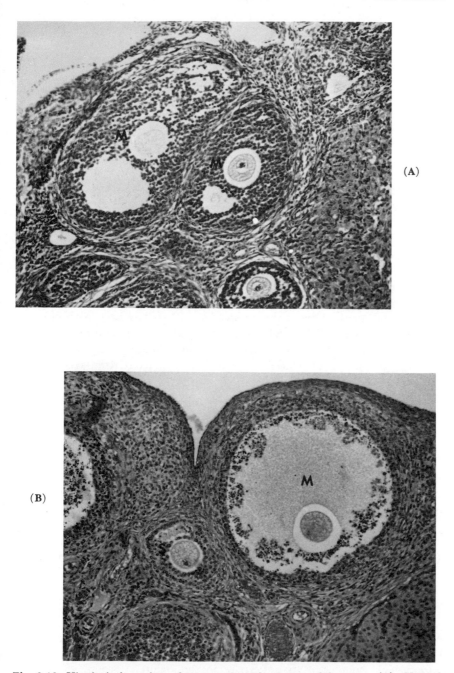

(A)

(B)

Fig. 9.13. Histological sections from ovaries of young adult rats. (A) Normal, showing position of ova and surrounding granulosa cells in two medium follicles (M). The section of the slightly larger follicle does not contain an ovum. (B) One day after total-body exposure to 750 R, illustrating degeneration of granulosa cells in the medium follicle (M). Some granulosa cells in the smaller follicle (S) are also degenerating.

have been reported to produce permanent sterility in mice. In addition, tumors are common in the ovaries of these mice.

Radiation effects on female fertility. In most species, an initial fertile period occurs immediately after small or moderate radiation exposures. This results from the presence of the relative radioresistant mature follicles. A sterile or semisterile period follows the initial fertile period, corresponding to the time at which the radiosensitive, medium sized follicles would normally have matured. The resistant small follicles survive moderate doses and eventually mature, resulting in renewed fertility. As with the male, genetic or chromosomal changes may have occurred in these cells which, during the periods of fertility, would lead to a decrease in litter size, or to abnormal offspring.

In humans, small doses will produce temporary sterility (12 to 36 months) in most individuals. Moderate or massive exposures are required to produce permanent sterility, artificial menopause, and psychic symptoms. A number of years ago it was a common practice for physicians to irradiate the ovaries of women who were sterile in order to induce ovulation. Pregnancies occurred after many of these exposures. This has been attributed to rupture of mature follicles and ovulation resulting from radiation-induced congestion and hyperemia. Irradiation for sterility was stopped, however, with the recognition of genetic changes resulting from irradiation of the ovaries.

ENDOCRINE SYSTEMS

The endocrine glands are specialized tissues which secrete specific substances (called *hormones*) into the circulation. They are subject to control by the central nervous system and by each other, in a complex series of interrelationships.

The thyroid, parathyroid, pituitary, adrenals, and endocrine portions of the reproductive organs are all rather radioresistant to direct damaging effects of radiation. However, changes occur in these organs which are secondary to effects in other tissues of the body.

The parathyroid glands are apparently extremely resistant to radiation. Massive doses are required to induce morphologic changes in the thyroid glands of the adults of most species of mammals. In one series of experiments, hyperactivity (increase in function) of the gland in rats followed small doses; hypoactivity (decrease in activity) resulted from moderate doses. Other authors have concluded that there is no significant alteration of thyroid activity after moderate doses of external radiation.

The radioactive isotopes of iodine concentrate in the thyroid. If the concentration of radioiodine is very high, the thyroid gland may be destroyed. Use is made of this selective tissue destruction in radiation therapy.

Local irradiation of the pituitary gland (with, for example, a well-columnated beam of high-energy deuteron particles) will produce immediate destruction of the gland, if the dose is 20,000 rads or more. After exposures of about 1000 R, there may be morphologic changes in the anterior lobe of the pituitary. These are not usually apparent, however, for many months. Degranulation and disappearance of acidophils occur first; basophils seem to be less severely damaged. Massive doses to the pituitary will also result in degenerative changes in other endocrine organs such as testes and thyroid. These changes suggests an alteration in hormone secretion by the pituitary. The growth rate of animals is often depressed following pituitary irradiation. This, also, probably indicates a change in secretory activity.

Adrenal glands are morphologically radioresistant. Massive doses produce few degenerative changes in the adrenal gland of rats. In the moderate dose range, irradiation usually results in hypersecretion of adrenal cortical hormones especially the 17-ketosteroids. This is a typical stress reaction and probably results indirectly from irradiation of other tissues of the body, rather than directly from irradiation of adrenal tissue. The adrenal medulla, a source of adrenergic hormones, is extremely radioresistant and apparently has little influence on the radiation response of an animal.

EFFECTS ON STRESS REACTION

Certain endocrine glands of higher animals respond to almost any form of stress in a characteristic manner. The pattern of response to generalized stress has been described in detail by Selye and has been termed the *general adaptation syndrome*. One of the most important of the stress changes is an increase in the secretion of adrenal cortical steroids.[11]

Numerous authors have suggested that radiation may act on the body as a nonspecific stress; that is, it may evoke the general adaptation syndrome. Many of the characteristics of the syndrome would be masked by other radiation-induced changes, but certain of them have been observed after irradiation.

Almost immediately after moderate doses of radiation, there is a decrease in the ascorbic acid and cholesterol content of the adrenal. This is generally considered to be an indication of increased adrenal secretion and is typical of the response to a great variety of stresses. A second phase of apparent adrenal hyperactivity and hypersecretion begins about 2 days after irradiation and continues for several days, or until the death of the animal.

[11]It is believed that the stress "stimulates" the hypothalamic nerve centers either directly or through adrenaline as a mediator. The hypothalamic control over the anterior pituitary results in an increase in ACTH secretion. This causes a hypersecretion from the adrenal cortex.

It is likely that certain phases of the stress-induced adrenal hypersecretion may be slightly deleterious to the animal. The middle (fasciculata) zone of the adrenal is usually considered to respond to stress by the elaboration of the glucocorticoids, such as cortisone. These steroids are known to depress the rate of regeneration of bone marrow, produce involution of lymphatics, and disturb the physiology of connective tissue, permitting the spread of infection. All of these effects oppose the repair processes needed after radiation.

Some authors have concluded that there is also a rise in mineralocorticoid secretion (e.g., desoxycorticosterone) of the outer (glomerulosa) zone of the adrenal during the first phase of the reaction. These hormones help to conserve sodium and, thus, to counteract the initial salt loss associated with damage to the digestive tract. Some adrenal activity has been shown to be essential for survival after moderate exposures (see Figure 11.13).

EFFECTS ON IMMUNE MECHANISMS

Several different tissues or portions of tissues contribute to the immune mechanisms of the body; that is, the reactions of the body which resist and overcome invasion by foreign materials or infectious agents. Some immune mechanisms are natural, others are acquired.

NATURAL IMMUNITY

Certain "natural" or innate mechanisms are present, such as the mechanical barriers of the skin and mucous membranes, the macrophage system with its ability to phagocytose (engulf foreign material), and certain serum proteins which are thought to have bactericidal activity.

There are few definitive studies on the radiation effects on natural immunity. However, the following concepts are well substantiated. Radiation disrupts the normal mechanical barriers such as skin and mucous membranes of the digestive tract. This permits the entry of infectious agents. After irradiation macrophages, although relatively radioresistant, are not as able to kill bacteria which they have ingested. The mechanism of this effect is not known. Normal bactericidal activity of rabbit serum is decreased by radiation. Similarly, properdin levels in dogs and rats are reduced significantly. This serum protein is thought to have bactericidal and viricidal properties.

ACQUIRED IMMUNITY

Acquired immune mechanisms largely involve antibodies which arise as a result of the introduction into the body of foreign proteins called *antigens*. A distinction can be made between the primary antibody response, which occurs following the initial introduction of the antigen into the body, and a secondary or "anamnestic" response, which occurs after a second injection of the same antigen.

There is a delay of some days following the initial injection of an antigen (latent period), then antibody appears in the serum, increases in concentration to peak or maximum titer, and eventually decreases to a very low level. The secondary response involves a similar sequence of antibody levels but it usually occurs faster; that is, the latent period is shorter. Also, the peak titer is often higher in the secondary response. The primary and secondary responses to an antigen are illustrated in Figure 9.14.

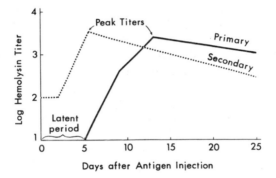

Fig. 9.14. Primary and secondary antibody responses in rabbits following intravenous injection of antigen (sheep red blood cells) and a similar reinjection. This experiment measures a "hemolysin" response, since the antibodies produced by the rabbit in response to the sheep red blood cells will "lyse" or break up red blood cells. (Modified from Taliaferro *et al.* [1963],courtesy Academic Press, Inc.)

There are apparently several steps involved in the primary antibody response. Immediately after antigen injection, there is a short period of a few hours or less during which some event occurs which is essential for initiating the development of the antibody-synthesizing mechanism. The suggestion has been made that, during this time, certain of the hemocytoblasts (stem cells for all blood cells) or lymphocytes are in some way "directed"[12] to produce antibody-containing cells. This is called the *induction process.* Changes in this process seem to correlate with the amount of antibody formed, or peak titer.

Following this short induction period, there is a time during which the synthetic mechanism is developed. During this period large numbers of

[12]Several mechanisms have been proposed. For example, there may be some "instructive" process involving an antigen-induced change in the DNA of the cells, or the genetic information may already be available in cells to produce antibody to all foreign antigens to which the body can respond.

antibody-producing cells[13] are formed by division and differentiation. The latent period is considered to extend until the presence of antibody can be detected in the serum. Continued rapid production of the antibody-producing cells and their synthesis of antibody results in a sharp rise in antibody titer to peak titer levels. A prolonged latent period and a decreased rate of rise of the antibody level are usually considered to reflect a change in the proliferation and maturation of antibody-forming cells. The normal decrease in antibody level following peak titer can be attributed to a decreased production of cells and a decrease in antibody synthesis.

Radiation alters the immune response in a number of ways. If antigen is given *before* irradiation, the peak antibody titer is not reduced, indicating no effect on induction processes. Indeed, the peak titer is often higher than normal when antigen precedes irradiation.[14] Production of antibody is delayed, however, as indicated by a lengthening of the latent period and a less rapid rise in antibody titer. These changes are illustrated in two of the curves in Figure 9.15 which show the hemolysin response of irradiated rabbits following antigen injected at different intervals before and after irradiation.

When antigen is injected *after* irradiation, all of the parameters of the immune response are found to be depressed. Decreased peak titers indicate interference with induction, and prolonged latent periods with slow rise of

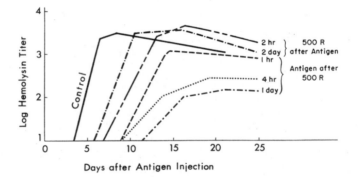

Fig. 9.15. Primary hemolysin response in five irradiated rabbits (dashed lines) following one intravenous injection of $10^{8.3}$ sheep red cells per kg 2 days before to 1 day after 500 R and in a control nonirradiated rabbit (solid line). (Modified from Taliaferro *et al.* [1963], courtesy Academic Press, Inc.)

[13]It is generally agreed that the antibody-producing cell is the plasma cell. Recently, the suggestion has been renewed that lymphocytes may also be a source of antibody. Some authors have suggested that plasma cells are formed from lymphocytes possibly by transformation into morphologically primitive "blastlike" cells which further differentiate and divide into plasma cells.

[14]This increase has been tentatively attributed to the breakdown of cells by radiation, and thus to the presence of partially depolymerized nucleic acids to serve as "building blocks" for DNA synthesis and production of plasma cells. In nonirradiated animals, antibody production can be increased by the injection of certain breakdown products of nucleic acid.

antibody titers suggests a decreased rate of production of antibody. Indeed, if antigen is injected as late as 28 days after irradiation, the peak titer is subnormal, although there is almost no delay in the appearance of the antibody in serum.

The secondary antibody response is less radiosensitive than the primary response. Considerably larger doses are required to prolong the latent period and to decrease the peak titer. For example, one experiment[15] measured the effect of different doses of radiation on the depression of the normal antibody response to fluid tetanus toxoid given to mice 1 hour after irradiation. A 50% depression in the primary response was produced by 153 rads; a 50% depression in the secondary response required 437 rads.

It has been repeatedly pointed out, however, that a valid comparison between the responses is difficult. This is due, in part, to the marked differences in peak titer values, time of appearance of antibody, and even the physiochemical properties of the two responses.

It is possible to draw some general conclusions about the immune response from experimental results. The induction process apparently requires an intact, nonirradiated animal in order to proceed with maximum effectiveness. A rabbit that has been irradiated as much as a month prior to antigen injection is unable to produce a normal peak hemolysin titer. In contrast, a rabbit given antigen only 2 hours before radiation can produce antibody in normal, or even greater than normal, amounts. The fact that radiation has so much less effect on the secondary response has been interpreted by some authors as an indication that induction, as it occurs in the primary response, does not occur in the secondary reaction.

An animal which is irradiated during the time when antibody synthesis is occurring, or prior to the initiation of the synthesis, will have a prolonged latent period and a slower rise in antibody titer. This suggests a slower production of antibody-producing cells and, therefore, a decreased rate of antibody synthesis. This is to be expected, since a decrease in mitosis of hematopoietic cells is commonly observed in irradiated animals. Indeed, it is surprising that a dose of 500 rads, which will inhibit division of most cell types and directly destroy most lymphocytes, will permit as much antibody production as does occur!

GENERAL REFERENCES

Bacq, Z. M., and Alexander, P. *Fundamentals of Radiobiology*, Chapter 18, 2d ed. Pergamon Press, New York (1961).
Bloom, W., and Faucett, D. W. *A Textbook of Histology*, 8th ed. W. B. Saunders Co., Philadelphia (1962).

[15]Stoney and Hale in Leone (1962).

Errera, M., and Forssberg, A. *Mechanisms in Radiobiology*, Vol. II, Chapters 1 & 2. Academic Press, Inc., New York (1960).

Hollaender, A. *Radiation Biology*, Vol. I (Pt. 2), Chapters 14, 15, 16, 17. McGraw-Hill Book Co., Inc., New York (1954).

Krayevskii, N. A. *Studies in the Pathology of Radiation Disease*, A. Lieberman (trans.). Pergamon Press, New York (1965).

Patt, H. M., and Quastler, H. Radiation Effects on Cell Renewal and Related Systems. *Physiol. Rev.*, **43:** 357–396 (1963).

ADDITIONAL REFERENCES CITED

Casarett, G. W. Pathology of Single Intravenous Doses of Polonium. *Radiation Res.*, Suppl. 5: 246–321 (1964).

Evans, T. C., Richards, R. D., and Riley, E. F. Histologic Studies of Neutron- and X-irradiated Mouse Lenses. *Radiation Res.*, **13:** 737–750 (1960).

Harris, R. J. C. (ed.). *Cellular Basis and Aetiology of Late Somatic Effects of Ionizing Radiation*. Academic Press, Inc., New York (1963).

Leone, C. A. (ed.). *Effects of Ionizing Radiations and Immune Processes*. Gordon & Breach, New York (1962).

Sproul, J. A., Baxter, R. C., and Tuttle, L. W. Some Late Physiological Changes in Rats after Polonium-210 Alpha-Particle Irradiation. *Radiation Res.*, Suppl. 5: 372–388 (1964).

Taliaferro, W. H., Taliaferro, L. G., and Jaroslow, B. N. *Radiation and Immune Mechanisms*. Academic Press, Inc., New York (1964).

Tessmer, C. F. Radioactive Fallout Effects on Skin. *Arch. Pathol.* **72:** 175–190 (1961).

Van Cleave, C. D. *Irradiation and The Nervous System*. Rowman and Littlefield, Inc., New York (1963).

Willoughby, D. A. Pharmacological Aspects of the Vascular Permeability Changes in the Rat's Intestine Following Abdominal Radiation. *Brit. J. Radiol.*, **33:** 515–519 (1960).

Zeman, W. Radiosensitivities of Nervous Tissues, pp. 176–199 in *Fundamental Aspects of Radiosensitivity*. Brookhaven Symposia in Biology No. 14, Brookhaven National Laboratory, Upton, New York (1961).

Zeman, W. Metabolic and Histochemical Studies on Direct Radiation Induced Nerve Cell Necrosis. *Proc. Intern. Congr. Neuropathol.*, *4th*, **1** (1962). Vol. I Histochemistry and Biochemistry, Georg Thieme, Stuttgart.

ACUTE
RADIATION EFFECTS IN
WHOLE ANIMALS

10

When an animal is exposed to a sufficient amount of radiation, there will be changes in many organs of the body. As a result of either the effects in one particular organ or the interaction of effects in several organs, the animal as a whole will show characteristic syndromes.[1] Some syndromes result inevitably in death. Others may or may not be lethal, depending on the extent of the tissue damage. The time of appearance of the syndromes, their duration, and the survival of the organism depend on many factors such as the radiation dose, the number of exposures, and the species of animal.

This chapter will deal primarily with the syndromes which occur acutely after irradiation — within the first month or two. The long-term radiation effects which are not apparent for months or years will be described in Chapter 12.

LETHALITY

Radiation in sufficient quantity can produce death in any organism. The minimum lethal dose differs considerably, however, when such widely differing organisms as mammals, fish, insects, or protozoa are considered.

It has become customary, in considering acute radiation lethality in higher organisms, to refer to the $LD_{50(30)}$ value. This is the dose of radiation which will kill about 50% of the exposed organisms within 30 days after

---

[1] A syndrome may be defined as the collection of signs and symptoms characterizing the response of an organism to a particular stimulus, in this case radiation.

exposure (literally, "lethal dose to 50% in 30 days," sometimes also written LD_{50}^{30}). Thirty days is a convenient interval for determining the radiosensitivities of various species[2] since many studies with laboratory animals have indicated that the prominent acute radiation syndromes and the acute radiation death usually occur within the first month after exposure.

In order to estimate the $LD_{50(30)}$ of a certain type of radiation for a given colony of animals, the following procedure is usually followed. Several groups of animals are exposed to different amounts of radiation. Hopefully, most of the doses used will be lethal to at least some of the individuals in each group, and not more than one dose will be so high that it will kill all the organisms exposed to it. It is most desirable to have doses that will result in lethalities between 10 and 90%. The number of animals in each group that

Table **10.1.** TYPICAL EXPERIMENTAL DATA TO
DETERMINE THE $LD_{50(30)}$ FOR A COLONY OF RATS
EXPOSED TO TOTAL-BODY X-IRRADIATION

Dose(rads)	Number of rats exposed	Dead at 30 days(%)
0	15	0
650	9	11
675	13	23
750	23	48
825	12	83
900	7	100

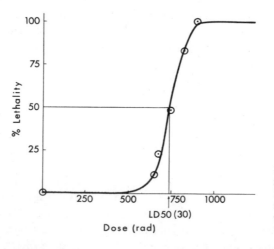

Fig. **10.1** Typical graph for determination of $LD_{50(30)}$ for rats exposed to total-body x-irradiation. (Data from Table 10.1.)

[2]Organisms which have a short life span are usually compared on the basis of a dose which is lethal in some shorter period which is convenient for the individual experiment. For example, as indicated in Chapter 7, loss of reproductive ability of bacteria (usually called lethality) is measured by the decrease in colony formation several days after irradiation. The lethal dose for such organisms is often extremely high.

die within 30 days is then tabulated, as illustrated in Table 10.1. A non-irradiated, control group is included in this experiment to provide an indication of the general condition of the colony.

The survival data are plotted as illustrated in Figure 10.1. Note the characteristic sigmoid (S-shaped) curve. Below a certain threshold dose, no animals die. With increasing radiation doses, more of the animals die, and above a certain exposure, none survive. This curve differs markedly in shape from the exponential survival curve from "single target" microorganisms, which was discussed in Chapter 7. The $LD_{50(30)}$ value of 740 rads can be estimated directly from the graph as indicated in Figure 10.1.

Figure 10.2 illustrates another way to plot survival data. The logarithmic probability paper essentially transforms the sigmoid survival curve into a straight line.[3] This is advantageous since it is simpler to fit a straight line to data than a curve. Moreover, the slope and variance of a straight-line function can be determined, so that two different types of radiation exposure can be compared in a rigorous manner. Values for 0 and 100% mortality can be included in this graph only by a complex manipulation.[4]

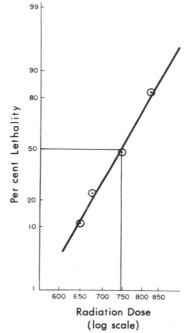

Fig. 10.2. Data from Table 10.1 plotted on logarithmic-probability paper, illustrating a method of determination of the $LD_{50(30)}$ for rats exposed to total-body x-irradiation.

Table 10.2 indicates some typical $LD_{50(30)}$ values for several species of animals. These values are only estimates since they are dependent on many factors such as dose rate, type of radiation, strain of animal, and even the time of year in which the experiments are done.

From the values in Table 10.2, one may conclude that among the vertebrates, mammals are generally more radiosensitive than birds, fishes, amphibians, or reptiles. Most invertebrates have been shown to be more resistant than vertebrates.[5] For example, adult *Drosophila* (fruit flies) are reported to survive radiation exposures of 64,000 R, and *Habrobracon* (a parasitic wasp) has been reported to have an increase in length of life after 100,000 R given

---

[3]Implicit in the use of this method of analysis is the assumption that susceptibility to a toxic agent, such as radiation, occurs in a population with a normal distribution of sensitivity in regard to some function of the dose.

[4]See, for example, Litchfield and Wilcoxon (1949).

[5]See Bacq and Alexander (1961) for details.

Table 10.2. Typical LD$_{50(30)}$ Values for Total-body Exposure of Animals to x- or Gamma Radiation

Organism	LD$_{50(30)}$ (rads)
Dog	350
Guinea pig	400
Man	250–450[a]
Mouse	550
Monkey	600
Rat	750
Rabbit	800
Chicken	600
Song sparrow	800
Goldfish	2300
Frog	700
Tortoise	1500

[a]Many attempts have been made to estimate the LD$_{50(30)}$ for man. Most of these suggest that the value is between 250 and 450 rads. However, a precise value cannot be assigned until more data are available from accidental overexposures or from radiation therapy.

under starvation conditions.[6] Adult *Artemia salina* (a primitive crustacean living in salt water) are killed by doses in the 200,000 to 300,000 rads range. Most coelenterates are extremely radioresistant, although *Hydra fusca* will die within about 20 to 40 days after doses of 4000 to 6000 rads.

Unicellular organisms are generally extremely radioresistant. About 100,000 rads will usually kill amoebae; paramecia may survive doses of 300,000 rads. Microorganisms vary in their sensitivity but are generally radioresistant (see Chapter 7).

Many suggestions have been made to explain the wide differences in radiosensitivity among animal species. Contributing factors may be the complexity of organism, mitotic index, oxygen tension, the presence or absence of sensitive enzyme systems, and the number or size of chromosomes.

ACUTE RADIATION SYNDROME IN MAMMALS

Within 30 days following irradiation, a variety of specific tissue changes or syndromes occur in most mammals. Some are grossly visible (see Figure 10.3), some are visible on a microscopic level. Some of these changes are

[6]The sterilization dose for such insects is in the range of 5000 to 10,000 rads. Since this is considerably below the lethal dose it has been possible to eliminate certain undesireable insect populations by sterilization of the males.

Fig. 10.3. Rat 3 days after exposure to 1200R of x-rays showing gross characteristics of the acute radiation syndrome.

lethal, some are not. The time of appearance is characteristic of the specific syndrome; the appearance at those times is dependent on the amount of radiation received by the system. Figure 10.4 indicates a typical relationship between the radiation exposures and the times of death.

Doses of 100,000 rads or more usually result in death during irradiation or immediately after. This is called *molecular death* since such massive doses presumably cause an inactivation of many substances which are needed for the basic metabolic processes of the cells and tissues.

Doses in excess of about 10,000 rads produce death within a day or two after exposure. The terminal symptoms of hyperexcitability, incoordination, respiratory distress, and intermittent stupor suggest damage to the nervous system as the cause of death. Therefore, this is called the *central nervous system (CNS) syndrome*. Some experiments have indicated another type of death within a few hours after radiation doses in excess of 15,000 rads. This type of death is apparently associated with lesions of the lung. Since there have not been extensive histologic or biochemical studies of the animals which die from these massive exposures, the actual mechanisms of the deaths are not clearly defined.

In the wide dose range between about 900 and 10,000 rads most animals die between 3 and 5 days after exposure. During this period severe morphologic changes are seen in the gastrointestinal tract and lethality is probably related to these changes.[7] Accordingly, deaths during the 3 to 5 day period are attributable to a gastrointestinal (GI) syndrome.

Depending on the radiosensitivity of the species and of the individual animals, deaths may or may not occur following radiation exposures of 300 to 900 R (see Table 10.2 for sensitivities of various species). Death, if it occurs, is usually between 10 and 15 days after exposure. Since most irradiated individuals show characteristic alterations in the blood cells and blood-

---

[7]The fluid and sodium loss at this time is considerable. It has been suggested that the level of sodium in an irradiated animal dying during this period is insufficient to maintain body function, and that this is the direct cause of death in these animals.

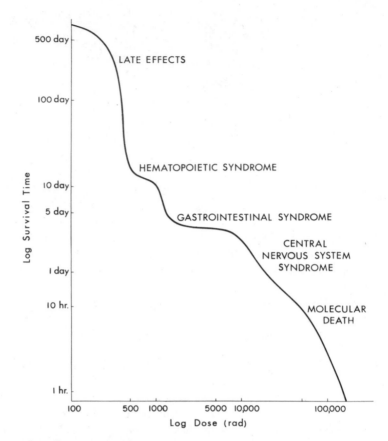

Fig. 10.4. Relationship between dose and survival time for adult rats, following a single total-body exposure of x-rays. The syndromes are indicated which are associated with the various times of death.

forming organs, deaths are said to be associated with the hematopoietic syndrome. This does not necessarily imply that the animals die due to a decrease in the number of their blood cells. Death is probably related, in many cases, to hemorrhage or infection, which may result secondarily from changes in the hematopoietic system.

Below 300 rads many organ systems will be damaged but few, if any, deaths will occur within the 30 day period after exposure.

INTERRELATIONSHIP OF ORGAN SYSTEMS IN THE ACUTE RADIATION SYNDROME

It must not be assumed from the preceding discussion that a dose of radiation produces effects in only one organ system. A very large dose will

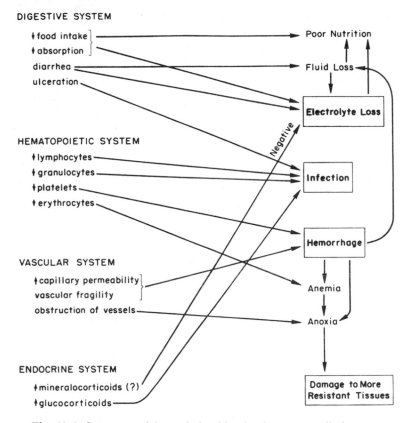

Fig. 10.5. Summary of interrelationships in the acute radiation syndrome (see text).

certainly cause tremendous damage to the hematopoietic system and to the gastrointestinal system. If the animal dies within the first day from the CNS syndrome, the hematopoietic and GI changes will probably not contribute greatly to the death; indeed, changes in these organs may not even be apparent by that time. With lower doses, however, when irradiated animals survive for at least a week or two, the effects on several systems will contribute to the overall radiation injury. The composite effect on the animal from the several systems is often termed the *acute radiation syndrome*.

Figure 10.5 summarizes some of the characteristics of the acute radiation syndrome. The contributions of various systems to the organism and the probable causes of death are indicated. The inhibition of mitosis and direct destruction of cells in the GI tract (especially small intestine) lead to a sloughing of the surface epithelium. Without an intact surface, there is a decrease in the absorption of nutrient elements. The animal also has a decreased appetite and has lost a considerable amount of fluids and electrolytes as a result of diarrhea. All of these factors contribute to a poor nutritional state and a further loss of electrolytes. It has already been indicated that electrolyte loss may be among the causes of death of irradiated animals.

Ulceration in the digestive tract removes one of the "natural barriers" of the body and permits the entry of foreign material which may result in infection. Concomitant with the removal of this barrier is a decrease in granulocytes, the white blood cells which remove foreign material by phagocytosis. In addition, the dramatic reduction in lymphocytes is probably related to a decrease in antibody production following irradiation. This lowers the ability of the body to resist infection.

As indicated in Chapter 9, the increased secretion of glucocorticoids following irradiation also encourages the spread of infection by disrupting the normal physiology of the connective tissue and by inhibiting regeneration of hematopoietic tissue. Adrenal mineralocorticoids may be increased immediately after irradiation and may be beneficial to the animal by diminishing the loss of body sodium.

Increase in capillary permeability and in vascular fragility permits blood to leave the vascular system; decrease in platelets means that the escaped blood may not clot. As an indication of this, hemorrhage is commonly observed in many organs following irradiation. The resulting anemia and anoxia can lead to damage to radioresistant tissues such as brain or kidney. As indicated in Figure 10.5, the decrease in red blood cells and obstruction of vessels also contribute to anoxia of tissues.

ACUTE RADIATION INJURY IN MAN

The effects which radiation produces in animals can, with many modifications and uncertainties, be extrapolated to predict the effects in man. In addition, there are four direct sources of information on acute exposures in man: (1) Exposure of the Japanese at Hiroshima and Nagasaki; (2) exposure of the Japanese fishermen and Marshallese to fallout radiation during the atomic tests in 1954; (3) reactor and radiation accidents in the United States, USSR, and Yugoslavia; and (4) exposure of patients to therapeutic radiation. The radiation exposures from the first three categories are seldom known with accuracy. However, it is usually possible to associate radiation effects with different exposure ranges. In therapeutic irradiation, the exposure is usually known quite precisely, but the radiation is frequently limited to a portion of the body. The syndrome which results will depend on the tissue which was irradiated. Also, the condition being treated may influence the radiation sensitivity of the individual. A rough picture has emerged from the available data.

Exposure of humans to doses of total-body radiation greater than 50 rads produces certain characteristic symptoms which collectively have been called *radiation sickness*. The onset, duration, and severity of the symptoms depend largely on the dose of radiation. The symptoms may include headache, dizziness, malaise, abnormal sensations of taste or smell, nausea, vomiting, diarrhea, decrease in blood pressure, decrease in white blood cells and blood platelets, increased irritability, and insomnia.

As in animals, three organ systems in man seem to be most important in the acute radiation syndrome. The central nervous system is most involved with exposures of several thousand rads or more. Between 500 and 2000 rads, the gastrointestinal system is of major importance. Exposures of less than 500 rads produce changes which are primarily associated with the hematopoietic system.

Central nervous system syndrome. Only a few instances of accidental human exposure have involved doses high enough to produce a central nervous system syndrome. The symptoms in those cases have been consistent with those observed in animals following comparable exposures. In 1958, for example, an accident caused massive overexposure of one man during a routine plutonium salvage operation at Los Alamos Scientific Laboratory.[8] The averge total-body dose was estimated to be greater than 3000 rads. The dose to the front of the head and chest was about 10,000 rads. Within minutes, the individual was in a state of shock and only semiconscious. Lymphocytes virtually disappeared from the circulating blood within 6 hours. Several hours after exposure, the shock condition lessened and he appeared more comfortable. More than 30 hours later, however, he became worse, developed severe abdominal cramps, became cyanotic, and lapsed into coma. Death, 34 hours after exposure, was associated with cardiac failure.

Another fatal accident occurred in a uranium-235 recovery plant in Rhode Island in 1964.[9] The individual received a total-body exposure of mixed neutron and gamma radiation, estimated at about 8800 rads. It is interesting to note that, although initially stunned, he did not lose consciousness and was able to run from the accident area. Almost at once he vomited and complained of abdominal cramps and headache. Periodic diarrhea, difficulty in enunciating words, edema, decreased blood pressure, badly diminished vision, and disorientation preceded his death which occurred 49 hours after the accident. Death was associated with circulatory failure.

Other individuals who have been exposed to massive radiation doses (greater than 5000 rads) characteristically show initial listlessness and prostration, followed frequently by convulsions and ataxia prior to death. Total-body radiation in doses sufficient to cause the central nervous system syndrome is inevitably fatal.

Gastrointestinal syndrome. Total-body doses between 500 and 2000 rads usually produce symptoms which relate to the gastrointestinal syndrome.

[8]Shipman *et al.* (1961) describes this accident.
[9]Described in Karas and Stanbury (1965).

Individuals lose their appetites and feel lethargic and depressed. Nausea and vomiting are so characteristic that the approximate dose received can be estimated from the time of onset, duration, and severity of the vomiting. Diarrhea for 2 to 3 days usually accompanies doses greater than about 1000 rads. This is sometimes referred to as the *N-V-D syndrome*.

Normal food and fluid intake is depressed; fluid loss also results from the persistant diarrhea and decreased absorption from the gastrointestinal tract. The abdomen is distended and peristalsis is absent; there is dehydration and hemoconcentration. Death occurs in about a week, due usually to circulatory collapse.

Hematopoietic syndrome. Nausea, vomiting, and some diarrhea also follow doses of less than 500 rads to the whole body. These are transitory and within a day or two individuals appear quite recovered. The hematopoietic organs, however, are atrophying. Two or 3 weeks after exposure there is an onset of chills, fatigue, petechial (pinpoint) hemorrhages in the skin, and ulceration of the mouth and pharynx.[10] Epilation occurs at this time and intestinal ulceration. Susceptibility to infection is markedly increased by the decrease in white blood cells, the impairment of immune mechanisms, and the hemorrhagic ulceration permitting entry of bacteria. Death, if it occurs, is usually between the third and sixth week. Some authors indicate that infection is probably not a major cause of death in humans, where antibiotic therapy can be provided on an individual basis as needed.

Survivors of the hematopoietic syndrome may show other changes at later periods. If the individual is young, there may be a decrease in bone growth. Males may have a decrease in the rate of sperm production; females may have interference with their menstrual cycle. Hair loss may be permanent in some areas; regrown hair may differ in color from the original. Months or years after exposure, irradiated individuals may show symptoms characteristic of the late effects of radiation. These will be discussed in Chapter 12.

RADIATION EFFECTS ON
PRENATAL DEVELOPMENT

EFFECTS OF RADIATION ON MAMMALIAN EMBRYOS

Ionizing radiations have been shown, repeatedly, to produce very marked effects on embryos in terms of lethality and in the production of abnormali-

---

[10]Oropharyngial ulceration was reported in the Japanese in Nagaski and Hiroshima; this effect has not been reported in Caucasians.

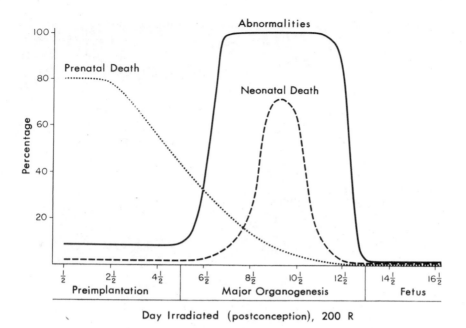

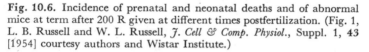

Day Irradiated (postconception), 200 R

Fig. 10.6. Incidence of prenatal and neonatal deaths and of abnormal mice at term after 200 R given at different times postfertilization. (Fig. 1, L. B. Russell and W. L. Russell, *J. Cell & Comp. Physiol.*, Suppl. 1, 43 [1954] courtesy authors and Wistar Institute.)

ties. Russell and Russell (1954) have summarized a series of comprehensive studies of the effects of 200 R on mouse embryos as shown in Figure 10.6. These authors have found it convenient to divide the total embryological period into three parts: (1) The preimplantation period, between fertilization and the attachment of the embryo to the uterine walls; (2) the period of major organogenesis during which the organs are being formed; and (3) the fetal period when growth is occurring. In the mouse these correspond approximately to days 0-5, 6-13, and 14-20.

Preimplantation irradiation results in a high percentage of prenatal deaths (those which occur before birth). This shows up primarily as a decrease in the size of individual litters, rather than as a decrease in the number of litters. There are few morphologic abnormalities, and most authors have indicated an absence of a particular pattern in the types of malformation which are found following preimplantation irradiation. The few abnormal animals resemble those found with a high frequency after irradiation at certain later postimplantation stages. It has been suggested that the abnormalities which are produced in the preimplantation period may not be caused entirely by direct irradiation effects on the embryo. They may result from a change in the intrauterine environment occurring when considerable numbers of embryos in a litter are killed. Rugh (1962) has reported that small

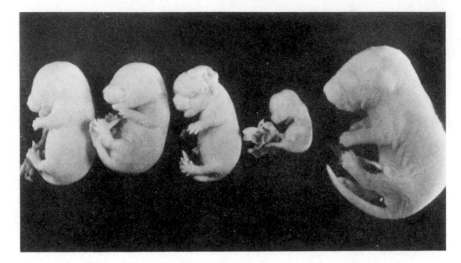

Fig. 10.7. Four mouse fetuses on left from a litter exposed to 50 R at day 3.5. One has brain herniation (exencephaly); one is a dead fetus; two appear grossly normal but are obviously stunted when compared with the unirradiated control on the right. (Courtesy R. Rugh [1959] and the Association of Military Surgeons of the U.S.)

irradiation doses in the preimplantation period produce at least one specific abnormality in his mice. This is exencephaly, the protrusion of the brain through the top of the skull. Figure 10.7 illustrates this abnormality, as well as the stunting of growth which Rugh has also reported. Details of exencephaly are also shown in Figure 10.8.

A series of experiments by Rugh (1962) with extremely low radiation doses have demonstrated the remarkable radiosensitivity of very young mouse embryos. Pregnant mice were exposed to x-rays 0.5 day after conception,[11] before the first cleavage of the fertilized egg. The dose received by the ovum was estimated to be 5 rads. At 6 and at 24 hours after irradiation the oviducts and uteri of the mice were examined and the embryos were studied. Even with such a small dose of radiation, the first cleavage was delayed in most embryos and there was a significant increase at 24 hours in the number of abnormal embryos. Other embryos were similarly irradiated, but not examined until 18.5 days. Analysis of implantation sites showed an increase in resorptions over the controls.

Russell *et al.* (1966) have further characterized the radiation sensitivity of the earliest portion of the preimplantation stage. The effects studied were embryonic mortality and sex-chromosome loss in survivors. They indicate

[11]Many of the irradiations of mouse and rat embryos are done at the "half-day" periods such as 0.5 day, 9.5 days or 13.5 days. This has occurred because of the convenience of irradiating the animals during the day, although most mice and rats mate between 10 p.m. and 3 a.m.

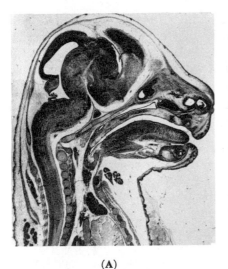

(A)

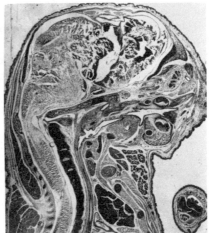

(B)

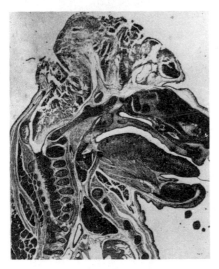

(C)

Fig. 10.8. Sagittal sections through head of mouse fetuses at 18.5 days. (A) Normal. (B) Mouse fetus exposed to 200 R x-rays at 8.5 days, showing almost complete degeneration of brain and spinal cord and some damage to skeleton. The soft tissues, viscera, and heart appear normal. (C) Similarly exposed mouse fetus, showing exencephaly. (Courtesy R. Rugh and Columbia Presbyterian Medical Center.)

that sensitivity is extremely high shortly after sperm entry (completion of the second meiotic division) and in the early pronuclear stage. In the later pronuclear stage (probably post-DNA synthesis) the sensitivity becomes low. On day 1, the sensitivity is relatively low in the resting 2 cell stage, but becomes very high at the beginning of the second cleavage. Their results suggest that death occurs at about the time of implantation, probably as the result of aneuploidy (fewer than normal numbers of chromosomes).

Irradiation during the period of major organogenesis causes fewer prenatal deaths, but a high incidence of many types of morphologic abnormalities.

For example, micropthalmia (small eyes), limb abnormalities, or snout abnormalities may be produced. The period for the production of a specific defect is short. It is usually related to the first morphologic evidence of differentiation of the organ or portion of the organ involved. For example, abnormalities in the long bones of mice occur with greatest frequency following irradiation on day 10. Reduction of eye size occurs from irradiation on day 8, but narrowing of the iris is produced by irradiation after day 10. The liver of mice is particularly sensitive to irradiation on day 8. Raising the radiation dose widens the period of sensitivity of most systems.

The developing nervous system of embryos is a particularly radiosensitive tissue. The experiments of Rugh (1959) have shown that the incidence of exencephaly is especially high following irradiation of 8.5 day embryos. Active development of the nervous system is occurring at that time. Rugh has reported a litter in which seven of nine embryos were exencephalic following 200 rads at 8.5 days. Histologic examination of embryos irradiated at that time has also revealed that many which are grossly normal in appearance have severely damaged brains and spinal cords (Figure 10.8).

The incidence of neonatal deaths (dead at birth) is high when irradiation is given during the period of organogenesis. This may be related to a high incidence of abnormalities of types which are lethal as the fetuses approach term. Many of the deaths may occur at the time of separation of the placenta and during the delivery of the young.

Relatively few obvious abnormalities result from irradiation during the fetal period. Higher doses are required to kill embryos during this period than during earlier periods of development. Moreover, death, when it does occur, is usually in the first 2 weeks after birth and resembles the acute radiation syndrome produced in adult animals. Gastrointestinal and hematopoietic changes are prominent. Fetal irradiation may also produce effects which do not show up until later in the life of the animal. For example, eye opacities, skin defects, changes in the reproductive organs and tumors have been reported as late effects of fetal irradiation.

Implantation. The failure of certain irradiated embryos to implant in the uterus is apparently due primarily to defects in the embryo rather than to an alteration in the uterus itself, although, as indicated in the previous section, the presence of dead embryos may influence the intrauterine environment to some slight extent. The uterus of a castrated rat can be stimulated to thicken and undergo the changes which normally precede implantation. This can be accomplished by mechanical trauma, such as running the point of a needle along the inner surface of the uterus. Moderate doses of radiation interfere only slightly with the usual structural and biochemical changes in the uterus. In contrast, such radiation doses cause extensive cellular damage in a preimplantation embryo.

Direct vs. indirect effects on embryos. A series of experiments have been described by Brent (1960) which indicate that direct irradiation of the rat embryo is primarily responsible for the production of abnormalities by x-irradiation. In one study, the embryonic site was shielded, while the remainder of the animal was exposed to 400 R (or 1000 or 1400 R partial-body) on the ninth day of gestation. No external malformations were observed with this treatment, although only 25 to 100 R is required to produce fetal malformations in a 9 day, total-body irradiated embryo. The fetal mortality was higher in the irradiated groups. In another experiment, the placental tissue of 12 day embryos was exposed to 400 R without exposure of the fetus. There was no change from controls in fetal growth or mortality.

RADIATION EFFECTS ON NONMAMMALIAN EMBRYOS

Malformations and decreased survival have been reported in chick and fish embryos following irradiation. Most reports indicate that 2 day chick embryos show a relative radioresistance, with sensitivity increasing sharply between 2 and 9 days, decreasing up until 15 days and increasing again in the 17 day embryos. Death occurring during the first day after irradiation is characterized by hemorrhage and general circulatory breakdown. Embryos of all ages that die between 2 and 5 days after irradiation show a decrease in growth and severe edema. Animals surviving to hatching frequently have malformations of limbs and are smaller than chicks from nonirradiated eggs.

Higher radiation doses are apparently required to damage fish embryos. Rugh et al. (1959) have reported that 500 R at the 2 cell stage of the fish *Fundulus heteroclitus* prevents neural development, while an exposure of as much as 2000 R to a much later stage results in stunted but otherwise normal embryos.

RADIATION EFFECTS ON HUMAN EMBRYOS

Rugh[12] has said, "Every irradiation-produced anomaly in the human fetus has been experimentally produced in the mouse or rat embryo by x-irradiation at a comparable stage of development so that it is justifiable to extrapolate from effects on the mouse or rat to the human fetus in regard to specific organ susceptibilities with the proviso that any statement is prefaced with the word 'probable.' "

Implantation in the human occurs at about 11 days after fertilization; the period of major organogenesis extends to about day 38. It is not surprising that irradiation, even with very low doses, during the first 2 weeks of pregnancy results in a high number of spontaneous abortions and in gross

[12]Rugh in Errera & Forssberg, Vol. II (1960).

abnormalities. Between the third and sixth week of pregnancy, when the major organ systems are being formed, gross abnormalities may be produced by small radiation exposures. Beyond day 40, the embryo is more radioresistant, although certain radiosensitive neuroblasts and germ cells may be damaged with resulting damage to nervous system and reproductive organs.

Repeated examinations have been made of children who were exposed *in utero* at Nagasaki and Hiroshima.[13] In one study, 7 of 11 women who were within 1200 meters of the hypocenter at Hiroshima had children with microcephaly (small brain) and mental retardation; those beyond 1200 meters had no such effects. Congenital dislocation of hips, mongolism, congenital heart disease, and other abnormalities were reported among other exposed children, presumably with a higher frequency than in the population as a whole.

Among 30 women exposed during pregnancy at Nagasaki, almost half of the pregnancies resulted in fetal, neonatal, or infant deaths. Among the 16 surviving children, four were mentally retarded. These women were within 2000 meters of the hypocenter which has been estimated to have given a total dose in excess of 20 rads.

There are many other reports of fetal death and congenital malformations which may have resulted from radiation exposures. It is apparent that diagnostic x-ray examinations of pregnant women should be avoided whenever possible, especially during the first trimester of pregnancy. Indeed, in certain countries, because of the high probability of abnormalities without spontaneous abortions, therapeutic abortions have been suggested following radiation exposure to as little as 10 R.

RADIATION EFFECTS
ON REGENERATION

Radiation has been found to influence the process of regeneration, that is, the replacement of parts of an organ or organism that have been removed. A majority of the early studies on regeneration were done with amphibians, since these animals possess the greatest ability among vertebrates to regenerate lost organs. More recently, certain mammalian organs such as liver have been studied. Not only have these experiments provided information on the action of radiation but many of them have also contributed to our understanding of the process of regeneration.

Brunst (1950) has reviewed and summarized his own and other work on amphibian regeneration. A number of interesting conclusions can be drawn on the subject. An animal such as a newt, salamander, or axolotl is capable

---

[13]It is difficult to determine the total dose of gamma and neutron radiation that was received by the survivors who were at different distances from the hypocenter. However, estimates suggest that at 1200 meters from the hypocenter at Hiroshima approximately 500 rads of gamma and 200 rads of neutron radiation were received.

of replacing certain portions of its body. When a limb is amputated, there is a short period during which the wounded area heals. Then a new limb is formed. It appears that local cellular elements are responsible for re-generation. Radiation studies such as the following have helped to elucidate this process.

If an entire limb is irradiated (massive doses) and then amputated, re-generation will not occur. If the distal (away from the body) portion of the limb is irradiated and the proximal (toward body) part is cut, the entire limb will regenerate, presumably from cells in the nonirradiated proximal part. If, instead, the proximal part is irradiated and part of the shielded distal part is amputated, a regenerated distal area will be produced, despite the fact that it is connected to the irradiated proximal part. With very high radiation doses to the proximal section, the regenerated distal area may be connected to the body by only a thin, soft stalk. It appears from these results that radiation must produce its major effect directly on the cells which will regenerate and that these are located close to the site of amputation.

If a regenerating limb is irradiated *after* amputation, several types of effects may be seen. (1) There may be stunting of the limb, if regeneration has started but is not complete. Normal differentiation will occur, but over-all size will be decreased. (2) If the limb has already regenerated, it may re-main alive and morphologically normal but be unable to regenerate if sub-sequently amputated. (3) There may be a "dedifferentiation" process in which the newly formed tissues are actually destroyed (probably by activity of macrophages). This can occur even after regeneration appears morpholog-ically to be complete, although it is not found in adult organs or in mamma-lian systems. The result is a shortening of the limb, sometimes to the bud stage. While the doses that are required to produce these effects are extreme-ly high in terms of mammalian effects (5000 to 10,000 rads), amphibians are generally very resistant to radiation. Total-body doses of about 10,000 rads are required to kill these animals.

Recent experiments with protozoa, reported by Giese and Lusignan (1961), have provided additional information on regeneration. If the an-terior half of a protozoan is cut off, it will be replaced. Since the anterior portion of many protozoa contains the mouth parts, the appearance of food vacuoles within the cell is a good end point to indicate regeneration of the front portion. If the posterior part of a bisected cell is irradiated, there is a delay in the appearance of food vacuoles, and, presumably, in overall regeneration. Under ordinary conditions, doses between 20,000 and 80,000 rads are required for this effect, with the larger doses resulting in the greatest delay. Above 100,000 rads the cell may be "killed," and regeneration per-manently inhibited. Division of the cell (after regeneration has occurred) is also delayed by the same doses of radiation. Again, the largest doses produce the greatest effect.

A number of investigators have demonstrated that radiation influences the rate of regeneration of certain mammalian organ systems. The liver has been a convenient system for regeneration studies since removal of a large portion of the liver from an animal such as a rat causes the remaining hepatic cells to divide to replace the liver mass which has been removed.

Radiation interferes with this regeneration in several ways. If radiation (small to moderate doses) is given just prior to partial hepatectomy, the onset of both DNA synthesis and mitosis is delayed. Moreover, if partial hepatectomy is delayed for days or even weeks after irradiation, both DNA synthesis and mitosis are still delayed. That is, the effect of the radiation remains in the normally nondividing interphase hepatic cells. In addition, cytologic studies of irradiated, regenerating livers have indicated the presence of many abnormal cells containing chromosome bridges, polyploid nuclei, and other types of abnormalities. This is obviously not "normal" regeneration.

It appears from these experiments that radiation can influence regeneration in a variety of ways, depending primarily on the relative amounts of cell division and differentiation that are required. Within a single cell (protozoan) where differentiation processes are involved, massive doses are required to inhibit regeneration. In animals such as amphibia, where both differentiation and cell division processes are required in the reforming of entire limbs, cell division is more radiosensitive than differentiation, although both can be inhibited by very large exposures. Regeneration of mammalian liver following partial hepatectomy takes place primarily by cell division. Abnormalities in the mitotic process are produced by small doses of radiation.

GENERAL REFERENCES

Bacq, Z. M., and Alexander, P. *Fundamentals of Radiobiology*, Chapters 16 & 21, 2d ed. Pergamon Press, New York (1961).

Bond, V. P., Fliedner, T. M., and Archambeau, J. O. *Mammalian Radiation Lethality: A Disturbance in Cellular Kinetics.* Academic Press, Inc., New York (1965).

Errera, M., and Forssberg, A. *Mechanisms in Radiobiology*, Vol. II, Chapters 1 & 2. Academic Press, Inc., New York (1960).

Hollaender, A. *Radiation Biology*, Vol. I (Pt. 2), Chapter 13. McGraw-Hill Book Co., Inc., New York (1954).

United Nations Scientific Committee on the Effects of Atomic Radiation. Seventeenth Session Suppl. #16 (A/5216) (1962).

ADDITIONAL REFERENCES CITED

Brent, R. L. The Indirect Effect of Irradiation on Embryonic Development. II. Irradiation of the Placenta. *Am. J. Diseases Children*, **100**: 103–108 (1960).

Brent, R. L., and McLaughlin, M. M. The Indirect Effect of Irradiation on Embryonic Development. I. Irradiation of the Mother While Shielding the Embryonic Site. *Am. J. Diseases Children*, **100**: 94–102 (1960).

Brunst, V. V. Influence of X-Rays on Limb Regeneration in Urodele Amphibians. *Quart. Rev. Biol.*, **25:** 1–29 (1950).

Giese, A. C., and Lusignan, M. W. Regeneration and Division of *Blepharisma* Following X-Irradiation. *Exptl. Cell Res.*, **23:** 238–250 (1961).

Karas, J. S., and Stanbury, J. B. Fatal Radiation Syndrome from an Accidental Nuclear Excursion. *New Engl. J. Med.*, **272:** 755–761 (1965).

Litchfield, J. T., Jr., and Wilcoxon, F. A Simplified Method of Evaluating Dose-Effect Experiments. *J. Pharmacol. Exptl. Therap.*, **96:** 99–113 (1949).

Rugh, R. Ionizing Radiations: Their Possible Relation to the Etiology of Some Congenital Anomalies and Human Disorders. *Military Med.*, **124:** 401–416 (1959).

Rugh, R. Low Levels of X-Irradiation and the Early Mammalian Embryo. *Am. J. Roentgenol., Radium Therapy Nucl. Med.*, **87:** 559–566 (1962).

Rugh, R., and Grupp., E. Ionizing Radiations and Congenital Anomalies in Vertebrate Embryos. *Acta Embryol. Morphol. Exptl.*, **2:** 257–268 (1959).

Russell, L. B., and Montgomery, C. S. Radiation-Sensitivity Differences within Cell-Division Cycles during Mouse Cleavage. *Intern. J. Radiation Biol.*, **10:** 151–164 (1966).

Russell, L. B., and Russell, W. L. An Analysis of the Changing Radiation Response of the Developing Mouse Embryo. *J. Cellular Comp. Physiol.*, Suppl. 1, **43:** 103–149 (1954).

Shipman, T. L., Lushbaugh, C. C., Petersen, D. F., Langham, W. H., Harris, P. S., and Lawrence, J. N. P. Acute Radiation Death Resulting from an Accidental Nuclear Critical Excursion. *J. Occupational Med.*, Suppl. 3: 145–192 (1961).

11 MODIFICATION OF RADIATION INJURY

The sequence of events which characterize the response of an organism to radiation has been described in previous chapters. Briefly, energy is transferred from the ionizing radiation to the organism. The molecules which absorb the energy may be altered and directly or indirectly may produce biochemical lesions in the cells of the organism. Depending on the extent, character, and distribution of altered cells, physiologic and anatomic lesions may result in the organism. If the lesions are sufficiently extensive, the organism will die. This sequence of events is neither invariable nor inevitable. It can be modified by changing the conditions of the experiment. For example, the biological response of an organism is dependent on the characteristics of the radiation itself — the type of radiation and the rate at which the dose is given. Also, organisms differ in their response to radiation depending on such factors as age, metabolic rate, extent of oxygenation, and nutritional status. Numerous chemical agents have been shown to modify the development of an acute radiation syndrome, especially when these agents are given prior to irradiation. Even after the syndrome is developing, certain treatments have been found which help to support the organism and increase the probability of survival.

A number of the physical and biological factors and chemical protective agents will be considered in this chapter. Some of the current concepts of the treatment of radiation injury will also be discussed.

PHYSICAL MODIFICATION OF RADIATION EXPOSURE

The *spatial* distribution of the radiation dose may be grossly changed by shielding either portions of the body or certain organ systems. Likewise, radiations with different rates of energy loss will give a different spatial distribution of energy on a micro scale. A change in the dose rate of a single exposure or the fractionation of the dose into several separate exposures alters the *temporal* distribution of energy transfer. Such changes in the physical characteristics of the radiation exposure can make great differences in the biological response to the radiation.

PARTIAL-BODY RADIATION

When only a portion of the body is irradiated, the response of the organism is determined primarily by the nature of the irradiated tissue and by the radiation dose.[1] Consider, for example, an experiment in which rats are exposed to radiation over only the lower portion of the body (below the xiphoid process at the lower end of the breast bone), or the upper portion of the body (above the xiphoid process). The total-body $LD_{50(30)}$ for these rats is about 750 rads and the acute radiation syndrome for total-body irradiation is as described in Chapter 10.

The $LD_{50(30)}$ for rats for lower-body irradiation is about 1100 rads. Following such a dose, animals show a characteristic gastrointestinal syndrome with death, if it occurs, usually between 3 and 5 days after exposure. There are changes in the irradiated hematopoietic organs and in the leukocyte content of the blood, but these are much less severe than in an animal exposed to total-body irradiation in the total-body $LD_{50(30)}$ range. Furthermore, animals which survive beyond the 5 day period seldom die during the second week in the period characteristic of the hematopoietic syndrome death.

In contrast, rats in which the upper body is irradiated do not undergo a characteristic gastrointestinal syndrome, although the mucous membranes of the oropharyngeal and esophageal regions may be severely damaged. The most dramatic change is atrophy of hematopoietic organs in the upper part of the body. Death, if it occurs, is usually in the second or third week after irradiation and may be associated with pneumonia, weight loss, infection, or anemia. The $LD_{50(30)}$ for upper-body irradiation is about 1750 rads.

Irradiation of other portions of the body will produce changes which are, for the most part, characteristic of the particular tissues that are irradiated. For example, irradiation of either the head of a rat or mouse or of the part of the head that contains the anterior part of the mouth may result in a characteristic "oropharyngeal syndrome." Death in the second week after such an exposure has been attributed by some authors to starvation and dehydration

---

[1]It must be remembered that radiation dose is expressed in rads; that is, the energy absorbed per gram of (in this case) tissue. The *total* energy absorbed by the organism is the dose in rads multiplied by the amount of tissue exposed (in grams). This is the gram-rad dose (nearly equivalent to the earlier gram-roentgen dose). A radiation dose of a number of rads may be given to the entire body, or to a part of the body, and although the exposure is the same, the total energy absorbed may differ greatly.

since survival is markedly increased if nutrition and hydration are maintained by feeding through in-dwelling gastronomy tubes.

Radiation therapy usually involves partial-body irradiation in which the exposure field is limited as much as possible to the tissue being treated. There may be some interaction between irradiated and nonirradiated tissues, but the major effects are in the exposed areas. As another example of partial-body irradiation, most internal emitters are not uniformly distributed throughout the body. The response to irradiation depends, as with localized external irradiation, on the specific tissues exposed and the dose which they receive.

Small volumes of tissue which are shielded from external radiation can greatly modify the acute effects of otherwise total-body irradiation. For example, the protective influence of shielded spleen tissues was demonstrated by Jacobson and co-workers in 1949, and has been repeatedly confirmed. In a series of experiments with CF-1 mice, Jacobson showed that shielding the exteriorized spleen increased the $LD_{50(28)}$ from about 550 R to 975 R. In further experiments, Jacobson found that the shielded spleen gave no protection if it was removed within 5 minutes after exposure. If, however, the shielded spleen was not removed until an hour or more after irradiation, some protection was demonstrated. Other experiments have shown that protection to irradiated mice is provided by the injection of spleen homogenates (containing intact cells) after irradiation. Injection of spleen extracts (containing no intact cells) does not afford protection.

The results of these and many other experiments suggest that cells from nonirradiated or shielded spleens may act as a source of stem cells to "seed" or repopulate irradiated hematopoietic tissue. The hematopoietic syndrome is, therefore, less severe, and the probability of survival is increased when such cells are present.

Shielding of other tissues can also protect irradiated animals. For example, shielding the thigh while the remainder of the animal is irradiated provides some protection. However, if the bone is removed from the thigh prior to shielding and irradiation, shielding the thigh is not effective. This effect suggests that the stem cells from the marrow of the shielded bone may repopulate irradiated hematopoietic tissue.

The preceding examples have been given to show the importance of the distribution of radiation to exposed and shielded tissue. It is usually possible to predict the overall response of a partially-irradiated animal by considering the specific response of the irradiated tissues, the relation of this response to the total physiology of the body, and the influence that nonirradiated tissue can have in altering the response of the exposed portions.

RBE AND THE INFLUENCE OF LET

The many different types of ionizing radiations have generally been found to produce effects in biological systems which are qualitatively similar or indistinguishable. There are often, however, important quantitative differences; that is, some types of radiations are more effective, or efficient, than others in producing certain changes.

As indicated in Chapter 3, the expression Relative Biological Effectiveness (RBE) is commonly used in radiobiology to compare the effectiveness of two radiations in producing a given change.

$$\text{RBE} = \frac{\text{dose of baseline radiation needed to produce a given magnitude of a certain effect}}{\text{dose of another radiation needed to produce the same magnitude of the same effect}}$$

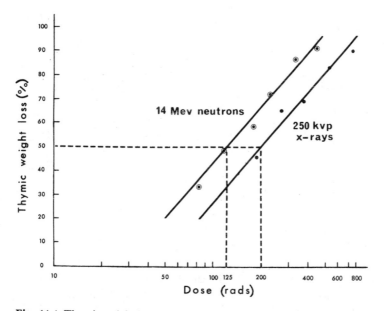

Fig. 11.1 Thymic weight loss in mice as a function of dose, 5 days after exposure to 14 Mev neutrons and 250 kvp x-rays. The RBE for 50% reduction in thymic weight is about 1.6. (Modified from Storer *et al.* [1957], courtesy Academic Press, Inc.)

The RBE is determined by comparing the amount of radiation to produce a particular effect; it is *not* determined by comparing the relative effect of equal doses of two different radiations. For example, in Figure 11.1, the thymic weight loss in mice is plotted as a function of dose of fission neutrons and of x-rays. The RBE of fission neutrons, relative to x-rays, *for this effect* may be estimated by comparing the dose of each which will produce a 50% reduction in thymic weight. Approximately 125 rads from neutrons or 200 rads from x-rays give this effect. Therefore, the RBE for fission neutrons for the production of 50% thymic weight loss in mice can be estimated[2] from these data as approximately $\frac{200}{125}$, or 1.6.

To a considerable extent, the RBE of various radiations depends on the rate of energy loss (LET)[3] along the paths of the individual ionizing particles or photons. In order to understand how the LET of a radiation can influence the biological response, one must again consider the spatial distribution of the ionizations. As illustrated in Figure 2.16, radiations with low LET values produce diffuse ionization throughout the medium. In contrast, the individual tracks of particles with high LET will be short and dense so that some volumes of the irradiated medium will receive very little. Figure 11.2 illustrates the relative distribution of ionization from low and high LET radiations in a series of hypothetical sensitive volumes or "targets."

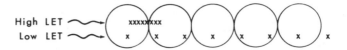

Fig. 11.2. Diagrammatic illustration of the relative ionization density per "target" from a single track each of high and low LET radiation.

In some chemical or simple biological systems, the effect being measured appears to be produced by a single ionization; that is, the kinetics of the simple target theory are applicable (see Chapter 7). In Figure 11.2 all of the targets are "hit" at least once by the low LET radiation. Each target will be changed if only one ionization is required to produce the effect. Two of the targets are "hit" twice, but the second ionization is wasted, since one will produce as much effect as two within the volume.

The high-energy LET is represented in Figure 11.2 as producing the same total number of ionizations. However, most of these have occurred within one target volume. Only the first two "targets" will be changed; many of the ionizations have been wasted. Thus, when only one ionization is needed in a target, the low LET radiation is more effective (higher RBE) since all five targets are hit as compared with only two from the high LET radiation.

If, however, two or more hits are required within a target, the low LET radiation may not be the most effective. In Figure 11.2 two targets have each received two ionizations from the low LET source; two targets have received two or more ionizations from the high LET radiation. In each case, if two hits are required to produce the effect, two of the targets will be affected. The single hits from the low LET radiation will be wasted; the hits in

[2]A more sophisticated determination of the RBE made by comparing the regression equations for the lines gives an RBE of 1.71.
[3]See Table 2.1 for LET values of some commonly used radiations.

excess of two in the first target from the high LET radiation will be wasted. If three or four hits are required, only the high LET radiation will produce an effect in any of the targets in the scheme presented.

As the total dose of radiation is increased, the number of tracks will increase. Then, more than one "track" of a low LET radiation may go through a target. The effects which require multiple hits are thus produced by individual ionizations from several particles or photons. Figure 11.3

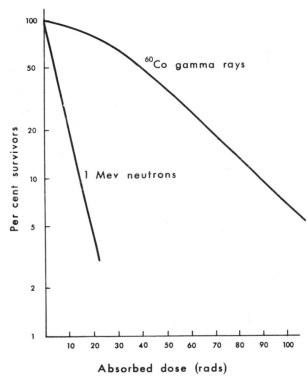

Fig. 11.3. Survival curves for late type A and type B spermatogonia irradiated with 1 Mev neutrons and cobalt-60 gamma rays. (Modified from Rossi [1962], courtesy Radiobiological Society of North America.)

illustrates such an effect. The survival curve for spermatogonia irradiated with neutrons is exponential, but the curve for gamma rays has a shoulder (see Chapter 7 for a discussion of the shapes of curves for single- and multiple-event interactions). It appears that several ionizations are required to produce death of a spermatogonium. The LET associated with the neutrons is high so that the passage of a single track will put enough ionizations into a traversed cell to produce death. With increasing doses of radiation, more tracks will be formed, and more cells will be killed according to the kinetics of the simple target theory. Gamma radiation produces diffuse ionization. Therefore, "hits" must be accumulated in a cell by the passage of several photons before the cell dies. The curve which results is characteristic of multiple-event kinetics. Note that the neutrons are much more efficient at

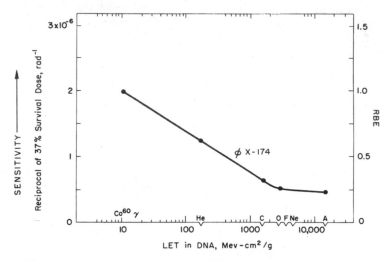

Fig. 11.4. Sensitivity (reciprocal of 37% survival dose) of ØX-174 for cobalt-60 gamma rays and heavy ions. An RBE scale is constructed on the right-hand ordinate. (Modified from Schambra & Hutchinson [1964], courtesy Academic Press, Inc.)

producing death of the spermatogonia (RBE for 1 Mev neutrons for the reduction of numbers of cells to 37% of the nonirradiated value is about 6.5).

There is considerable difference among various test systems in the relative effectiveness of radiations with different LET values. In general, chemical systems and simple biological systems act as if they require only one ionization in each sensitive volume (simple target theory). In those systems, the radiations with low LET values are most effective.

Such a relationship is illustrated in Figure 11.4. The survival of phage ØX-174 is measured. This is a relatively simple bacteriophage which is

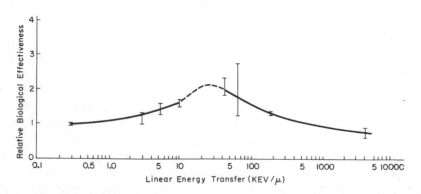

Fig. 11.5. Relation of RBE to LET for a variety of mammalian systems (see text). (Modified from Storer *et al.* [1957], courtesy Academic Press, Inc.)

known to contain single-stranded nucleic acid. The effectiveness of cobalt-60 gamma radiation is compared with that of charged ions from the Yale heavy-ion accelerator. The LET is plotted in units of energy loss in DNA (Mev-cm²/g) instead of in units of energy loss per micron of tissue, but this does not change the analysis of RBE. The gamma radiation from cobalt-60 has an LET of about 10 Mev-cm²/g. An RBE scale is indicated on the right, using the cobalt-60 gamma effect as a baseline of 1. The decreased effectiveness with high LET radiations is presumably due to wasted ionization, as multiple ionizations occur within a target which can be inactivated by only one ion pair.

In contrast, many of the effects (such as lethality) which are measured in complex biological systems apparently require many ionizations. Then, radiation with a nonrandom ionization pattern is most effective. Figure 11.5 shows the relationship of RBE to LET for a series of mammalian systems. This graph presents the results of a long series of experiments done at Los Alamos. Radiations used included: 250 kvp x-rays; 1.1 to 1.3 Mev gamma rays from cobalt-60; 4 Mev gamma rays from graphite capture of thermal neutrons; 6 kev beta particles from tritium; thermal neutrons from a reactor; 14 Mev neutrons from the deuterium-tritium reaction in an accelerator; fission neutrons; alpha particles and lithium recoils from thermal neutron capture by boron; and fission fragments from the *in vivo* fission of plutonium by fission neutrons. These radiations cover a range of average LET from about 0.3 kev/μ to about 4000 kev/μ.

The biological effects measured in mice and/or rats included lethality, atrophy of several organ systems, depression of iron-59 uptake by red blood cells, and duration of depression of mitotic activity of the skin. When RBE is plotted against LET (expressed in terms of kev/μ) as determined for each of these biological end points, the curve in Figure 11.5 is obtained. The verticle lines on the curve indicate the range of RBE values that were observed for each of the radiations. The RBE values of the various radiations are expressed relative to cobalt-60 gamma radiation which has an LET value of 0.3 kev/μ.

The RBE values increase with LET to about 30 kev/μ, and then decrease as LET gets larger. The maximum effectiveness presumably represents the distribution of ionization which allows optimal utilization of ionization with minimal loss due either to insufficient ions within some "target" or to an excess above those needed to produce the effect. The term *target* as used here means any volume or volumes within which ionizations must be produced in order to produce the effect. It may be part of a cell, an entire cell, or a portion of a tissue in which a block of cells must be affected in order to produce the effect which is being measured.

Certain radiations have a high RBE for the production of specific effects. For example, in many experiments, values of 10 to 15 have been found for the production of eye opacities by fast neutrons, relative to x-rays. In contrast, values of only 1 to 2 are observed for the production of some long-term effects by neutrons of the same energy.

DOSE RATE EFFECT

Radiation may be delivered to a test system at a very rapid rate, with the total exposure lasting only a short time, or it may be given slowly. There may be one exposure of the system or the total dose may be split into a number of treatments. When the radiation is given intermittently — that is, the total exposure is split into a number of parts with nonirradiation periods between — it is usually called *pulsed* irradiation if the total exposure time is short, or *fractionated* if the overall time of irradiation from the first to last exposure is days or weeks. When the exposure is either continuous or fractionated, but is given over a long period of time (weeks, months, or years), it is called *chronic* irradiation.

The term *dose rate*, as used in this book and in most radiobiological literature, refers to the rate at which radiation is delivered to the test system during actual exposure. For example, an exposure might be given at a dose rate of 5 rads for each minute that the radiation is actually on. The term *average dose rate* is sometimes used for pulsed exposures, and is the total dose divided by the total time from beginning to end of the irradiation (including the time during and between irradiations).

The majority of experiments which have compared the effectiveness of various dose rates have used $LD_{50(30)}$ as the criterion of effect. Table 11.1 illustrates for mice the type of response which is usually obtained in these experiments. The $LD_{50(30)}$ increases with decreasing dose rate; that is, radiation delivered at a rapid rate appears to be more efficient than when given slowly at a low dose rate. Recent experiments have indicated that this apparent dose-rate effect is actually an effect of altering the total exposure time. This has been demonstrated by using a high dose rate, but pulsing the expo-

Table 11.1. $LD_{50(30)}$ OF MICE EXPOSED TO X-RAYS AT DIFFERENT DOSE RATES[a]

Dose rate (rads/min)	Duration of exposure	$LD_{50(30)}$ (rads)
706	1 min	788 ± 24
68	11–13.5 min	850 ± 12
8	2 hr	948 ± 8
$4\frac{1}{2}$	4 hr	1030 ± 10
3	6 hr	1040 ± 12
$2\frac{1}{2}$	8 hr	1097 ± 31

[a]From Neal, 1960.

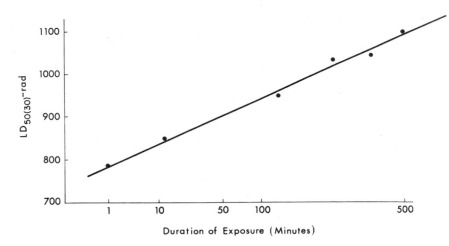

Fig.11.6. Relationship between duration of exposure and $LD_{50(30)}$ for mice exposed to x-rays at different dose rates. Data is given on Table 11.1. The abscissa is scaled to give a straight line relationship between the $LD_{50(30)}$ and the cube root of the duration of exposure.

sure over a number of hours. The lethality pattern is similar to that from a continuous exposure at a low dose rate in which the total time from beginning to end of radiation is the same as in the pulsed experiment.

The relationship between duration of a single radiation exposure and $LD_{50(30)}$ is plotted in Figure 11.6 for the data in Table 11.1. The use of cube root of exposure time as abscissa has no identifiable biological significance, but results in the best linear "fit." If pulsed radiation exposures are given, the relationship shown in Figure 11.6 [$LD_{50(30)}$ ∽ (exposure time)$^{1/3}$] is approximately true, as long as the total exposure time does not exceed 6 hours. If the total time of the pulsed exposures is greater than 6 hours, there is no systematic relationship between total time and lethal dose. The number of fractions and their spacing become important. It is generally true, however, that the total dose required to produce death is greater when the periods between irradiations are extended.

The greater lethality of short-time exposures can be explained by postulating that there is recovery from radiation damage. If a certain amount of injury must be accumulated in order to produce death and if recovery starts as soon as the injury is produced, some of the initial damage will usually be repaired before the lethal amount has been produced. Thus, by prolonging the exposure time there will be less effective damage existing *at any one time*, although the *total* energy transfer will be the same. As the exposure time is increased further, there will be more repair of early damage before the conclusion of the irradiation and, thus, less effective damage.

On a cellular or subcellular basis, these relationships are suggestive of multiple-hit kinetics. If several ionizations are required within a target to produce an effect and if part of the change produced by a single ionization is reversible or repairable, then a high dose rate will permit accumulation of the necessary number of changes in the target before much significant recovery of the earlier changes has occurred.

Both linear and exponential kinetics have been postulated for the time course of repair in the recovery process. Short-term (hours) experiments often suggest linear repair, while experiments involving fractionation over several days usually predict exponential kinetics. It may be that the linear repair only represents the more rapid, early phase of exponential kinetics.

The dose-rate effect can be seen quite clearly in mammalian cells in tissue culture. In a typical experiment, cell cultures are given a continuous exposure to gamma radiation at differing dose rates. The highest dose rate (shortest exposure time) produces the highest amount of killing and is, therefore, most effective. Similar cultures are irradiated, all at the highest dose rate of the continuous exposures but with pulsed radiation, so that the total exposure times correspond to the exposure times of the continuous irradiations. The shortest total exposure time is again the most effective, despite the fact that the dose rate is the same.

The dose rate effect has also been demonstrated with many other experimental systems. As was indicated in Chapter 5, "two-hit" chromosome aberrations are usually produced more efficiently by x- or gamma irradiation when the total exposure time is decreased. Mutation production has recently been shown to be dose-rate dependent (see Chapter 6).

Ionizing radiations with a high LET do not usually show a dose-rate effect. For example, Figure 11.7 shows the results of an experiment on the growth rate of *Vicia faba* roots exposed to alpha, gamma, or x-irradiation. Except for anomalous 2 hour values, the effect of an exposure to alpha radia-

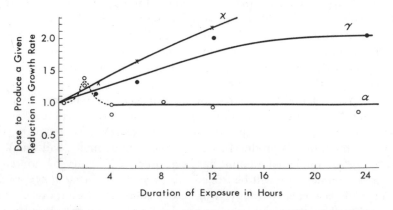

Fig. 11.7. Relationship between duration of exposure and the dose which produces a chosen reduction in growth rate of *Vicia faba* root tips. The dose of the various radiations used is not equivalent and is expressed as a relative scale. (Redrawn from Gray & Scholes [1951], courtesy British Institute of Radiology.)

tion is independent of the duration of treatment for at least as long as 24 hours. In contrast, the dose of x- or gamma radiation which is necessary to produce a chosen reduction in growth rate increases considerably when the exposure time is increased. This result suggests that multiple, coincident events are required for this type of injury. Such multiple events are produced almost instantaneously by a single track of high LET radiation within the target, therefore there is no time for repair between events.

CHRONIC IRRADIATION

Very often the response of a system to chronic radiation is characterized by a plateau in the development of damage. During the initial exposures, some measurable effect may be produced, such as a decrease of certain cells in a particular organ. With increasing exposures the cell populations decrease further. Repair or replacement of the damaged cells is also taking place, however, and, at some point, the rate of recovery is equal to the rate of damage from the radiation. If irradiation is discontinued, the cell population usually increases. The level of the equilibrium plateau depends on the size of the individual radiation doses, not on the total dose. With very small doses the plateau may be near the control level; with high individual doses it may represent a nearly maximum effect.

This plateau effect is illustrated in Figure 11.8. The number of early spermatocytes in the testes of chronically irradiated rats is plotted as a func-

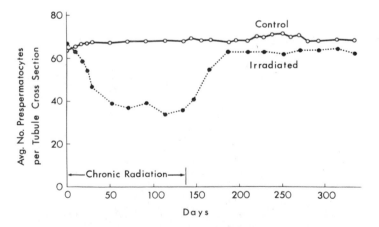

Fig. 11.8. Average number of prespermatocytes/tubule in testes of control rats and rats irradiated with 3 R/day, 5 days/week for 20 weeks. This illustrates the "plateau" which is frequently observed in chronic irradiation experiments. Notice that the rate of decrease of cells is approximately the same as the rate of increase during the post-irradiation period.

tion of time. The number of cells decreases during the first 50 days, remains about constant until the end of the radiation period, and then increases. Notice that the rate of decrease of cells (day 0 to day 50) is approximately equal to the rate of increase during the recovery period (day 140 to day 190).

When animals are given chronic irradiation until death, the time of death (and total dose) depends on the daily dose. A comprehensive study of survival following continuous irradiation at a variety of daily doses has been done by Sacher and Grahn (1964). Some of their results are shown in Figure 11.9, where the mean aftersurvival (days survived after the start of irradiation) is plotted against the mean accumulated dose (MAD).

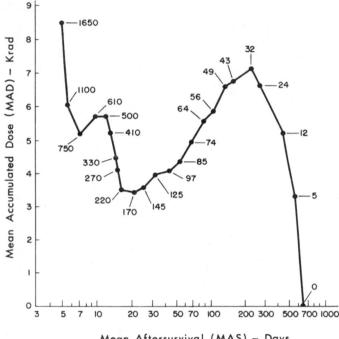

Fig. 11.9. Relation between mean accumulated dose (MAD) on a linear scale to mean after-survival time (MAS) on a logarithmic scale for female mice given daily exposures ranging from 5 R to 1650 R. The daily dose is recorded at each point. (Modified from Sacher & Grahn [1964], courtesy National Cancer Institute.)

At extremely high exposures (60,000 to 200,000 R/day), not shown in Figure 11.9, the animals survive less than a day. Death is associated with a hyperexcitability syndrome, presumably because of damage to the central nervous system. At exposures greater than 1000 R/day, as shown in Figure 11.9, the animals survive only a few days, but the MAD is high. Much of this high dose represents "wasted" radiation, since the exposure from the first day or two would be lethal, even if the remaining treatments were not given.

Mice exposed to radiation in the range of hundreds of roentgens per day live longer and have lower MAD values than those receiving higher daily doses. These intermediate dose rates represent the most "efficient" killing of the animals since the mean accumulated dose is least. With lower daily doses, recovery occurs between exposures so that more total radiation is needed to produce lethality. Below 32 R/day the MAD decreases and the total life span approaches that of the control mice.

BIOLOGICAL FACTORS WHICH MODIFY THE RADIATION RESPONSE

AGE OF ANIMAL

The age of animals at irradiation is an important factor in their radiosensitivity. The marked sensitivity of embryos has already been indicated in Chapter 10. In general, immature animals are radiosensitive, young adults are most resistant, and adult animals increase in sensitivity with increasing age. Figure 11.10 illustrates the relationship between age and $LD_{50(30)}$ for rats. The maximum sensitivity ($LD_{50(30)}$ = 225 rads) occurs at 1 day of age, the earliest period tested in these experiments.

Some species variation in age-sensitivity has been demonstrated. In contrast to the high sensitivity of newborn rats, newborn mice have been shown to be relatively insensitive or radioresistant. This resistance rapidly disappears, however, and the sensitivity passes through a maximum value at 3 to 4 weeks of age.

A comparison of the acute pathologic changes in rats irradiated at weaning (21 days) with rats irradiated as young adults indicates the same general pattern of effects in these two groups. This is suggestive of similar mechanisms of death, but a greater radiosensitivity of the tissues of the younger animals. The increased sensitivity may be related to the higher mitotic rate of most tissues in young, rapidly growing rats compared with adults. The extreme sensitivity of the very young rat may also be related to the poor ability of such animals to respond to stress prior to the establishment of the hypothalamic-pituitary-adrenal system.

Blair has suggested a way to correlate the decrease in $LD_{50(30)}$ with increasing age of adult animals. As a part of his theory of life span shortening (see Chapter 12), he postulates that the lethal threshold of injury decreases directly as the life expectancy. Life expectancy may be regarded as decreas-

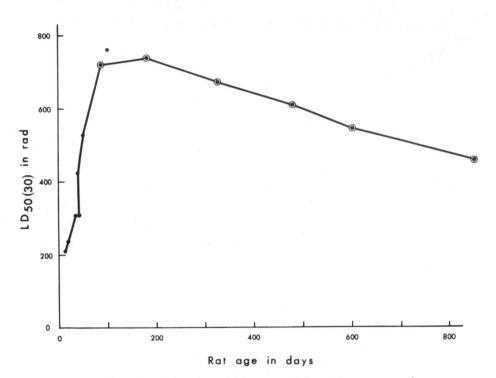

Fig. 11.10. Relationship between age and $LD_{50(30)}$ for rats exposed to a single dose of x-rays. (Data for adult rats courtesy of J. B. Hursh; young rats, unpublished data of author.)

ing approximately linearly with age. Figure 11.10 and data from other experiments have indicated a decrease in $LD_{50(30)}$ with age. The decline, however, is not as rapid as would be expected from Blair's postulate.

It does not appear that a simple linear relationship exists between age and radiosensitivity. Rather, the radiosensitivity of aging animals is probably correlated with their impaired ability to repair or replace cells and tissues which have been damaged by radiation. Since many different tissues and cells are involved in such a response, it is not surprising that a strict linear relationship does not appear.

GENETIC CONSTITUTION (STRAIN DIFFERENCES)

The wide range of radiosensitivities of different species has been described in Chapter 9. Even within species, the radiosensitivity may vary considerably. This has been demonstrated in bacteria, where *E. coli* strain B/r is considerably less radiosensitive than strain B. These two strains differ by mutation of not more than a few genes. In mammals, a comparative study of lethality and death rate has been made using six strains of mice. Kohn and Kallman

Table 11.2. Effect of Strain on Acute Lethality of Mice[a]

Strain	$LD_{50(30)}(R)$
BALB/c	544
C57BL	618
A/He	632
C3H	665
CAF_1	656
ACF_1	649

[a]From Kohn and Kallman (1965).

found the $LD_{50(30)}$ to vary from 544 R for BALB/c mice to 665 R for C3H mice (see Table 11.2). These values are the results of samples pooled from a number of individual experiments among which the variation was high.

SEX OF ANIMAL

In most species females are slightly less radiosensitive than males. The difference is small, however, and is not always apparent. It is most likely related to hormonal differences, since testosterone (male hormone) has been shown to enhance mortality in x-irradiated mice and estradiol benzoate (female hormone) administered to mice 9 to 10 days before irradiation decreases the mortality slightly.

HEALTH OF ANIMALS

Animals in poor health are usually more radiosensitive than healthy animals. For example, Figure 11.11 shows the radiation mortality curves for normal

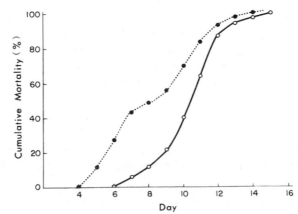

Fig. 11.11. Radiation mortality curves for normal (solid line) and pseudomonas-infected (dotted line) mice given a single lethal (900 R) exposure of x-rays. (From Hollaender [1960], courtesy Pergamon Press.)

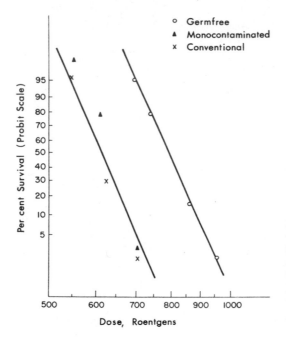

Fig. 11.12 Thirty day survival of irradiated germfree, conventional, and *E. coli* monocontaminated mice. There is no statistical difference between the latter groups. (From McLaughlin *et al.* [1964], courtesy Academic Press, Inc.)

mice and mice carrying the organism *Pseudomonas* in their intestinal flora. Under ordinary circumstances this organism would not affect them; however, the infected mice die sooner after a single lethal radiation exposure than do noninfected mice given the same dose.

Germ-free mice have been used to evaluate the influence of bacterial contamination on survival. Figure 11.12 illustrates the results of a typical experiment with these mice. The experimenters found that the $LD_{50(30)}$ for germ-free mice was 840 ± 9 R compared with 610 ± 15 R for conventional animals and 640 ± 12 R for mice which had been contaminated with *E. coli* but were free of other organisms. The germ-free mice survived longer. Following massive radiation exposures (40,000 R) they failed to show the convulsions[4] and bloody diarrhea which characterized the response of conventional mice. The authors indicated that it was not clear whether the altered radioresponse of the germ-free mice was due to the absence of bacteria or bacterial products or to basic physiological differences in germ-free animals as a result of the bacterial-free state.

DIET

Animals which are maintained on good diets have the best survival following irradiation. Rats on a protein-free diet, for example, are more susceptible

---

[4]The germ-free animals did have tremors after massive-dose irradiation and occasionally had convulsions as a terminal event, but not consistently following the exposure.

to whole-body radiation than animals given protein. Rats fed diets containing a 2 or 10% content of either margarine fat or cottonseed oil live longer following irradiation than comparable rats fed either 20 to 30% fat or fat-free diets. Irradiated rats fed basal diets supplemented with high vitamin levels or raw vegetables appear to have an increased survival time, although the data are not conclusive.

ENDOCRINE STATUS

A functional adrenal-pituitary system is necessary for optimal radioresistance. This has been demonstrated in an experiment in which rats were exposed to 600 rads of x-rays when they contained known volumes of adrenal tissue (as determined by the time of regeneration after surgical adrenal enucleation).

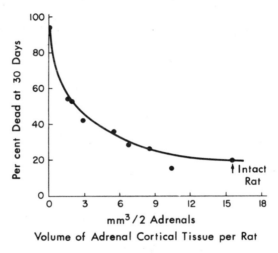

Fig. 11.13. Relationship between volume of adrenal tissue and radiation lethality of rats after 600 R x-rays. (Modified from Casarett & Brayer [1961], courtesy Academic Press, Inc.)

As indicated in Figure 11.13, most of the animals with little or no adrenal tissue died from this exposure; most animals with half or more of the control adrenal cortical volume survived. Furthermore, deaths occurred 3 to 4 days after irradiation which suggests mechanisms similar to the gastrointestinal (G.I.) syndrome; that is, imbalance of both sodium and fluid levels. These are two parameters that are normally under adrenal control. Similarly, hypophysectomized rats are more radiosensitive than intact animals.

OXYGEN CONCENTRATION

Probably the most effective biological means of modifying the radiation syndrome is by the production of hypoxia; that is, by a decrease in the oxygen concentration in the organism during irradiation. This oxygen effect has

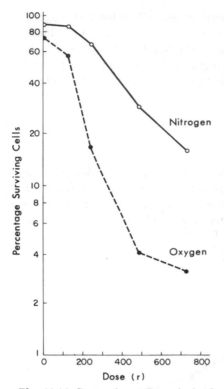

Fig. 11.14 Comparison of survival of HeLa cells irradiated in oxygenated and nitrogenated media, expressed as a percentage of nonirradiated controls. (From Weiss [1960], courtesy Taylor & Francis.)

been demonstrated repeatedly in a wide variety of biological systems. Hypoxia during irradiation will decrease the mutation frequency in *Drosophila*. Hypoxic tumor cells are about 2.5 times as resistant as well-oxygenated cells when telophase abnormalities are scored. Hypoxia increases the radio-resistance of mammalian cells in tissue culture by a factor of 2 to 3. Figure 11.14 illustrates this effect. The survival of cells irradiated in an oxygen environment is compared with cells irradiated in nitrogen.

In intact animals, hypoxia provides a high degree of radioprotection for 4 day old chick embryos (factor of about 2). Rats kept in a 5% O_2-95% N_2 environment during irradiation have an $LD_{50(30)}$ of between 1200 and 1400 rads, instead of about 700 rads. Similarly, the LD $_{50(30)}$ of mice is approximately doubled by irradiation in an oxygen deficient atmosphere. Many other experiments have also indicated that the radioresistance of most systems increases by a factor of about 2 when the oxygen tension is markedly decreased.

A few systems have been found which do not show this oxygen effect. Response of T_2 phage is not dependent on oxygen concentration, but when the system irradiated is a T_2 phage-*E. coli* monocomplex there is an oxygen effect. Also, with radiations of high LET, there is either no oxygen effect or it is much smaller than with x-radiation.

Several theories have been proposed to explain the relationship between oxygen tension and sensitivity. It has been suggested, for example, that fewer peroxyl and hydroperoxyl radicals are formed at low oxygen tension. The importance of these radicals in producing biochemical changes was indicated in Chapter 4. Furthermore, a larger proportion of the chemical changes which do occur are apparently reversible if oxygen is excluded from the system.

If a given effect appears to require the production of several ionizations (that is, multiple-hit theory) and if these several ionizations are produced within the required volume by a high LET radiation, the presence of oxygen does little to increase the effect. If, however, only a few ionizations are produced in the volume, oxygen may make the ionizations more "efficient" in terms of molecular alterations. It is also possible that the presence of oxygen may inhibit the repair process, so that altered molecules from a second ionization can interact with those from the first.

Changes in the radiosensitivity of tissues in whole animals cannot be reasonably correlated with alteration in the oxygen turnover or basal metabolic rate. Rather, they correlate with changes in the level of tissue oxygen. This will be apparent in the discussion of radiosensitivity of hibernating animals in a later section.

Increasing the oxygen content of the atmosphere does not greatly influence the radiosensitivity of intact animals unless the oxygen is delivered at pressures of several atmospheres. Under this condition the oxygen itself is toxic, and animals die from a combined effect of oxygen and radiation toxicity. There is speculation and some evidence that the mechanisms of action of high oxygen pressures and of ionizing radiation have features in common. One of these is probably the production of free radicals.

When the percentage of oxygen in inspired air is increased at the normal atmospheric pressure of sea level, the amount carried by hemoglobin increases only slightly. At the usual alveolar oxygen tension of 100 mm, hemoglobin is already 95% saturated. With increasing inspired oxygen there is an increase in the amount of oxygen dissolved in the blood. This contributes very little, however, to the level of oxygen in tissues. Thus, under most circumstances, no substantial increase in tissue oxygenation accompanies the increase in atmospheric oxygen. There is, then, no basis to expect a greater radiosensitivity.

An increase in the radiosensitivity of certain tumors can be produced by increasing oxygen in inspired air. Tumors which have a relatively limited blood supply are predominantly anaerobic. A small increase in oxygen concentration from the slight additional oxygen in the blood has a much more marked effect on these tumors than on normal tissues which have an adequate blood supply.

TEMPERATURE

For many years it has been recognized that chilling mammals during irradiation will protect them against radiation injury. For example, an experiment by Storer and Hempelmann (1952) has shown that the mortality of infant mice which are chilled to 5°C during irradiation is about half that of litter mates exposed to the same dose of total-body x-irradiation at room temperature.

Cooling adult animals will also provide protection. The $LD_{50(30)}$ is increased and the cellular depletion of spleen and bone marrow is less marked in animals which are irradiated at low temperature. Jamieson and van den Brenk (1963) have demonstrated that protection by hypothermia is not due to the depression in temperature *per se*, but rather to the resultant reduction in partial pressure of oxygen in the tissues and to the oxygen effect which was discussed in the previous section.

Increasing the temperature of rats above normal during irradiation also provides slight protection. Jamieson and van den Brenk have measured the oxygen tension in various tissues during hyperthermia and have demonstrated a decrease in oxygenation in the spleen. The authors suggest that the respiratory and circulatory stimulation may not keep up with the heightened metabolic demand. Splenic hypoxia could result. Alternately, arterio-venus shunts may open to bypass the spleen. Again, the result would be splenic hypoxia.

The ultimate radiation mortality of cold-blooded animals such as frogs is not influenced by a decrease in the body temperature during irradiation. The survival time of these animals, however, may be greatly prolonged if they are kept in the cold. Since the tissues of a cold-blooded animal are well oxygenated at low temperatures, the lack of a "temperature effect" on mortality is presumably due to the presence of oxygen in the tissues.

HIBERNATION

Hibernation delays the radiation response of marmots and ground squirrels but does not seem to decrease the lethal dose. When normal temperature is regained, the animals show the same sequence of radiation syndromes as do nonhibernating animals.

Circulatory and respiratory systems are functional during hibernation and tissues are well oxygenated. The metabolic rate is decreased. The latter apparently helps to delay the onset of obvious damage, but does not modify its magnitude.

CHEMICAL PROTECTION

Many hundreds of chemical compounds have been tested for radio-protective action. A number of them have been shown to provide some protection against radiation damage in isolated chemical systems, in independent cell systems, and in intact animals. Only a few of the most promising compounds will be discussed in this chapter. The mechanisms of the protective action of most of these agents are not fully established, but a number of theories are presented to explain the observed effects.

Many of these agents must be present in concentrations which are close to lethal (or at least toxic) and usually must be present within the system at the time of irradiation. The latter observation suggests that the agents are influencing the initial events in the production of the damage rather than in facilitating repair processes.

There is general agreement that most chemical protective agents act, in effect, to reduce the *effective* radiation dose received by the system. For example, rats given the maximum tolerated amount of cysteine (a radioprotective agent) prior to irradiation appear as if they had received about 40% less radiation than they actually did; for example, if they received 1000 rads, they would show the same response as animals receiving 600 rads without cysteine; if they received 500 rads, they would resemble untreated animals after 300 rads, etc.

The effectiveness of most of the radioprotective chemical agents is evaluated by the determination of a "Dose Reduction Factor" (DRF). This is sometimes expressed as the ratio of the $LD_{50(30)}$ for protected animals to that for unprotected animals.

THIOLS

The most effective class of chemical protectors are the aminothiols. These are, specifically: The naturally occurring amino acid cysteine; 2-mercaptoethylamine (cysteamine) (MEA); S,2-aminoethylisothiourea dihydrobromide (AET), which rearranges to 2-mercaptoethylguanidine (MEG); and related structures. The formulas for several of these are given in Figure 11.15. This class of compounds is characterized by -SH and $-NH_2$ groups separated by two carbon atoms. They all have DRF's of about 1.7 under optimal conditions and probably have the same mechanisms of protective action. Differences in

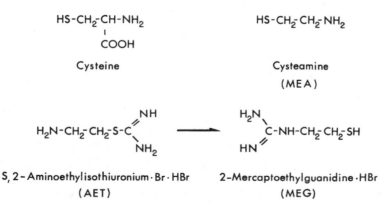

Fig. 11.15 Several amino thiols. According to convention, the AET is diagrammed without the associated HBr.

the action of these compounds may be related to variations in penetration, distribution, and detoxification.

Cysteine, for maximum DRF, must be given in almost toxic amounts. The maximum safe dose given intravenously to rats is usually considered to be 1200 mg/kg; the lethal dose is about 1500 mg/kg. About 1200 mg/kg is required to demonstrate protection. Cysteine is more effective when given by the intravenous rather than by the intraperitoneal route; oral administration is almost useless. A DRF of 1.7 applies for a variety of criteria such as acute mortality, decrease in spleen weight, and loss of lymphocytes or granulocytes. Cysteine also protects against delayed effects such as the production of lens opacities and hair loss.

Cysteamine is similar to cysteine in the scope of its action. It is sometimes referred to as more "efficient" than cysteine since a smaller quantity (150 mg/kg) will give the same DRF. This may be related to the slower oxidation of cysteamine. From experiments in rodents, AET has been judged to be slightly more effective than cysteine. It is more stable, has a more effective tissue distribution, and a decreased rate of detoxification. It is effective given orally to mice, but not to rats. In humans, AET results in nausea, vomiting, and circulatory disturbances.

A number of explanations have been offered to account for the radioprotective action of these aminothiol compounds. In general, the proposed mechanisms are concerned with: (1) Competitive removal of free radicals, (2) repair by donation of a hydrogen atom to the target molecule, (3) interaction with cellular components, and (4) production of tissue hypoxia. At present, no one of these mechanisms can explain all of the observations, nor can the protective effects brought about by one single compound under different conditions always be explained by the same mechanisms.

(1) Competitive removal of free radicals. The inactivation of free radicals or "radical scavenger" hypothesis assumes that the indirect action of radiation is of primary importance. It suggests that certain aminothiol compounds (such as cysteamine) are oxidized by free radicals (such as OH°) and form resonance-stabilized free radicals which are incapable of reacting with cell components. By this mechanism, the free radicals are prevented from interacting with vital cell constituents. The spatial relationship of the –NH$_2$ and –SH groups is favorable to the formation of such a structure. Thus, as illustrated in Figure 11.16, cysteamine (in the Zwitterion form) is oxidized by a hydroxyl radical (gives an electron to the hydroxyl) to form a free radical. The new radical can undergo any one of a variety of reactions such as reaction with another oxidizing free radical to form an acid, regeneration of the original compound, or reaction with a similar radical to form a disulfide. The formation of five- or six-membered ring structures can account for the effectiveness of compounds in which the –SH and –NH$_2$ groups are separated by two (or three) carbon atoms. However, as has been pointed out by

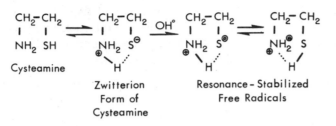

Fig. 11.16. Formation of free radical from cysteamine by interaction with OH°. This forms a resonance-stabilized free radical, according to the Radical Scavenger Hypothesis.

many authors, if the formation of these rings was the sole basis for the protective action of the substances, one would expect protection from other molecules with similar spatial distributions of functional groups.

(2) Repair by donation of hydrogen atoms. Another mechanism for protection which has been demonstrated with polymers is repair by donation of hydrogen atoms. If a molecule (RH) is converted by radiation (either by direct or indirect mechanisms) into a free radical R°

$$RH \rightarrow R° + H \qquad (11\text{-}1)$$

it may undergo a variety of reactions such as crosslinking

$$R° + R° \rightarrow R - R \qquad (11\text{-}2)$$

or peroxidation.

$$R° + O_2 \rightarrow RO_2° \qquad (11\text{-}3)$$

A protective agent may donate a hydrogen atom to the radical, restoring it to its orginal state.

$$R° + PH \rightarrow RH + P° \qquad (11\text{-}4)$$

This type of protection was demonstrated by Ormerod and Alexander (1963), using irradiated salmon sperm nuclei as a test system. The radicals were detected by means of electron spin resonance. In the presence of cysteamine the rate of disappearance of radicals associated with irradiated

DNA (from the sperm nuclei) was increased and there was a concomitant increase in radicals formed from cysteamine. Protection by hydrogen transfer is not limited to systems of nucleoproteins. Similar repair of molecules probably occurs in many complex biological systems.

(3) **Interaction with cellular components.** Several mechanisms for protection have been suggested which involve a bonding of the chemical protectant with intracellular components. This may be either a loose enzyme-substrate complex, a tight binding of some kind, or some form of a mixed disulfide. The latter hypothesis, suggested by Pihl and Eldjarn (1958), proposes that protective agents with free sulfhydryl groups form transient mixed disulfides with sulfhydryl groups of tissue proteins. When one of these

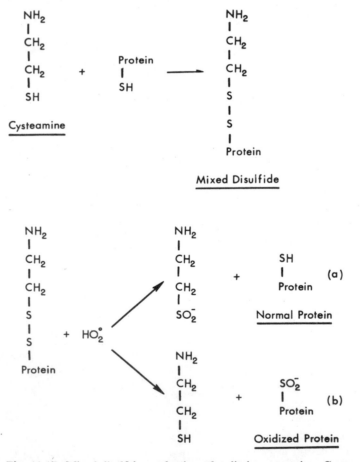

Fig. 11.17. Mixed-disulfide mechanism of radiation protection. Cysteamine reacts with a sulfydryl-containing protein to form a mixed disulfide. Oxidation of the mixed disulfide by a free radical results in either (a) an unchanged protein or (b) an oxidized protein.

mixed disulfides is attacked by a free radical, one of the sulfur atoms is oxidized and the other is reduced. If the sulfur from the protein is reduced and the sulfur from the protective agent is oxidized (Figure 11.17A), the protein is undamaged by the radiation. Alternatively (Figure 11.17B), the protein may be oxidized and is, therefore, altered. However, it is damaged in only 50% of such disulfide radical interactions. It is "protected" from 50% of the oxidations.

A striking correlation has been found between protector action *in vivo* and ability to form mixed disulfides with glutathione or cysteine, although there are exceptions. This theory requires that oxidation of sulfhydryl groups of proteins is a major factor in radiation injury. Evidence for this is conflicting. Furthermore, there are compounds that will form disulfides but that are not protective agents. Thompson (1962) suggests the additional possibility that disulfides protect by altering metabolic pathways since many sulfhydryl-containing enzymes become inactive when the sulfhydryl groups are blocked. Perhaps this promotes a shift from aerobic to anaerobic pathways.

(4) Production of tissue hypoxia. Thiols are readily oxidized. Their oxidation, according to another theory, consumes enough oxygen to reduce the tissue oxygen tension significantly. Since hypoxia has been demonstrated to protect, this is (or was) an attractive hypothesis. Recent experiments, however, have suggested that cysteine, AET, and other thiols exert a protective effect without altering the partial pressure of oxygen in tissues such as bone marrow and spleen.

Also at odds with the hypoxia theory are the findings that certain systems (T_2 phage) have no oxygen effect, but do show protection by cysteamine, and that cysteamine can protect a noncellular system by pathways unrelated to hypoxia. It appears likely that production of hypoxia may, under some conditions, be a contributing factor in radiation protection by the thiols, but that other mechanisms must also be involved.

(5) Other theories. Numerous other mechanisms have been proposed. For example, the protective thiols apparently interfere with mitosis of cells and with DNA synthesis. If cells are arrested in a radioresistant phase of cell division, they are protected. Also, if division is delayed, there may be a greater opportunity for repair of radiation-induced damage. Interference with carbohydrate and energy metabolism, possibly through mixed disulfide formation with enzyme –SH groups, may also be related to protection.

Sulfhydryl compounds are often good chelating agents. This has suggested to some workers that protection may be provided, for example, by chelating copper in copper enzymes and, thus, preventing oxidation of copper.

Present evidence indicates that there may be more than one mechanism by which the thiols protect against radiation. Several of the mechanisms which have been presented may be more or less important, depending on the system being irradiated and the specific conditions of irradiation.

OTHER PROTECTIVE AGENTS

Some degree of radiation protection has been demonstrated with a variety of substances which are familiar in clinical and experimental pharmacology. In general, they have DRF's of 1.3 to 1.5, and are, therefore, less effective than the thiol compounds that were discussed in the previous section. In most instances they apparently protect by virtue of their capacity, by one mechanism or another, to decrease the oxygen tension of radiosensitive tissues. Only a few will be mentioned in this section.

Respiratory depressants such as morphine, heroin, or ethyl alcohol have some protective action in mice, but only when given in very large doses. Indeed, the amount of ethanol required to protect, if translated from mouse to man, corresponds to about a quart of 100 proof whiskey. This quantity is usually lethal, if ingested at one time.

Epinephrine, a vasoconstrictor drug, produces a decreased blood flow through visceral organs. The decreased oxygen tension of the spleen is considered to be the basis for its slight protective action. Serotonin (5-hydroxytryptamine), a potent vasoconstrictor, is one of the few compounds that shows a protective action (DRF: 1.3 to 1.8) at doses considerably below the toxic level. Most studies with serotonin have been done on rats and mice.

Certain of the vasodilator drugs such as choline esters and histamine also have some protective action. These substances lower the blood pressure and thus produce tissue hypoxia.

Carbon monoxide protects due to its capacity to bind hemoglobin and thus impair oxygen transport. Good protection is obtained when an animal breathes carbon monoxide until two-thirds of its hemoglobin has been converted to carboxyhemoglobin, but at this level the animal is in a critical state.

A different type of protection has been demonstrated with certain of the central nervous system depressants such as pentobarbital. Administration of these drugs to guinea pigs which have received massive doses of radiation partially protects them from the central nervous system syndrome, and extends their survival from several hours to about 4 days. These drugs have no effect on survival time, however, with doses of less than 6000 rads.

Many other materials have been suggested or tested with varying, and often inconsistent, results. Until more is known about the initial steps in radiation damage and about the pharmacologic action of these substances, their possibilities as protective agents are difficult to assess.

TREATMENT OF RADIATION INJURY

So far, this chapter has dealt with procedures which when used prior to, or during, irradiation will modify the acute radiation syndrome. Most of them are ineffective when given after exposure. Certain postirradiation treatments, however, do increase the survival of irradiated animals and have shown promise in the treatment of humans who have been accidentally exposed to sources of radiation.

No treatment is successful in decreasing lethality following total-body exposures to very high doses (in excess of 2000 rads for humans) which give rise to the central nervous system syndrome. Total-body exposures between 500 and 2000 R (human) result in a severe gastrointestinal syndrome which can be treated with some degree of success. However, individuals receiving this much radiation usually will die of a hematopoietic syndrome, even when there has been satisfactory therapy for the GI effects. Below 500 rads, there may be some GI effects, but the hematopoietic syndrome will predominate. Therapy is possible for some of the symptoms in this dose range.

Cronkite (1964) emphasizes that "the cardinal rule in the management of radiation injury is to do nothing unless there are clear-cut clinical indications for a specific agent or maneuver." One treats the symptoms only as they appear. For example, antibiotics should be withheld unless signs of infection develop. If they do, large doses of broad spectrum antibiotics and sulfonamides can be used. Likewise, fluid and electrolyte imbalance should be corrected only as needed. The platelet count should be carefully followed and the individual observed for signs of bleeding. If bleeding occurs, there should be massive platelet transfusions, equivalent to that found in about one-third of the blood volume of the patient. Red cells should be given only if indicated by the hematocrit. Unneeded cells will overload the circulation.

Under certain conditions, in animals, bone marrow cell transfusions are effective in repopulating the marrow of irradiated recipients, especially if isologous (genetically identical) or autologous (same animal) cells are used. Heterologous (different species) or homologous (different strain) transplants may repopulate the marrow but immunological complications somewhat later result in a graft-host reaction which may be fatal. If an individual has received an exposure in excess of 600 R, homologous transplantation may be indicated. However, the radiation dose must be very high in order to obtain a "take" (that is, high enough to inactivate the host's immune system). With present techniques, transplants are not generally considered to be very successful in humans.

In many instances, therapy may be unnecessary, and possibly even detrimental. The treatments which are recommended at present should be used with great care in individuals as needed. They should not, indeed cannot, be applied on a broad population basis.

GENERAL REFERENCES

Bacq, Z. M., and Alexander, P. *Fundamentals of Radiobiology*, Chapters 19 & 20, 2d ed. Pergamon Press, New York (1961).

Cronkite, E. P. The Diagnosis, Treatment and Prognosis of Human Radiation Injury from Whole-Body Exposure. *Ann. New York Acad. Sci.*, 114: 341–349 (1964).

Errera, M., and Forssberg, A. *Mechanisms of Radiobiology*, Vol. II, Chapters 4 & 5. Academic Press, Inc., New York (1960).

Hollaender, A. (ed.). *Radiation Protection and Recovery*. Pergamon Press, New York (1960).

Thomson, J. F. *Radiation Protection in Mammals*. Reinhold Publishing Corp., New York (1962).

ADDITIONAL REFERENCES CITED

Casarett, A. P., and Brayer, F. T. Relation of Adrenal Cortical Volume to Survival after X-irradiation. *Radiation Res.*, 14: 748–759 (1961).

Gray, L. H., and Scholes, M. E. The Effect of Ionizing Radiations on the Broad Bean Root. VIII. Growth Rate Studies and Histological Analysis. *Brit. J. Radiol.*, n.s., 24: 285–291 (1957).

Jacobson, L. O., Marks, E. K., Robson, M. J., Gaston, E., and Zirkle, R. E. The Effect of Spleen Protection on Mortality Following X-irradiation. *J. Lab. Clin. Med.*, 34: 1538–1543 (1949).

Jamieson, D., and Van den Brenk, H. A. S. Effect of Progressive Changes in Body Temperature of Rats on Tissue Oxygen-Tensions in Relation to Radiosensitivity. *Intern. J. Radiation Biol.*, 6: 529–540 (1963).

Kohn, H. I., and Kallman, R. F. The Influence of Strains on Acute X-ray Lethality in the Mouse. I. LD_{50} and Death Rate Studies. *Radiation Res.*, 15: 309–317 (1956).

McLaughlin, M. M., Dacquesto, M. P., Jacobus, D. P., and Horowitz, R. E. Effects of the Germfree State on Responses of Mice to Whole-Body Irradiation. *Radiation Res.*, 23: 333–349 (1964).

Neal, F. E. Variation of Acute Mortality with Dose-Rate in Mice Exposed to Single Large Doses of Whole-Body X-irradiation. *Intern. J. Radiation Biol.*, 2: 295–300 (1960).

Omerod, M. G., and Alexander, P. On the Mechanism of Radiation Protection by Cysteamine: An Investigation by Means of Electron Spin Resonance. *Radiation Res.*, 18: 495–509 (1963).

Pihl, A., and Eldjarn, L. Pharmacological Aspects of Ionizing Radiation and of Chemical Protection in Mammals. *Pharmacol. Rev.*, 10: 437–474 (1958).

Rossi, H. H. Distribution of Radiation Energy in the Cell. *Radiology* 78: 530–535 (1962).

Sacher, G. A., and Grahn, D. Survival of Mice Under Duration of Life Exposure to Gamma Rays. I. The Dosage-Survival Relation and the Lethality Function. *J. Natl Cancer Inst.*, 32: 277–314 (1964).

Schambra, P. E., and Hutchinson, F. The Action of Fast Heavy Ions on Biological Material. II. Effects on T1 and ØX-174. Bacteriophage and Double-Strand and Single-Strand DNA. *Radiation Res.*, 23: 514–526 (1964).

Storer, J. B., Harris, P. S., Furchner, J. E., and Langham, W. H. The Relative Biological Effectiveness of Various Ionizing Radiations in Mammalian Systems. *Radiation Res.*, **6:** 188–288 (1957).

Storer, J. B., and Hempelmann, L. H. Hypothermia and Increased Survival Rate of Infant Mice Irradiated with X-rays. *Am. J. Physiol.*, **171:** 341–348 (1952).

Weiss, L. Some Effects of Hypothermia and Hypoxia on the Sensitivity of HeLa Cells to X-rays. *Intern. J. Radiation Biol.*, **2:** 20–27 (1960).

12 *LATE EFFECTS*
OF RADIATION

Adult animals which have received radiation doses which are not acutely lethal usually appear to recover from the initial radiation syndromes in a month or two after exposure. They look grossly normal and indistinguishable from animals which have not been irradiated. As the animals grow older, however, they may have a higher incidence of certain tumors than do control animals of the same age and they may develop age-specific diseases sooner than do nonirradiated animals. They appear to age sooner and die earlier than unexposed animals. Life-shortening, carcinogenesis, and aging are all long-term or late effects of ionizing radiation. These effects are not independent. Obviously, an increased incidence of cancer will contribute to a shorter overall life span in a population. Likewise, the acceleration of age-specific changes will contribute to earlier mortality. For convenience, however, these subjects will be treated separately in this chapter.

RADIOLOGIC AGING

Normal aging involves a progressive deterioration of tissues, accompanied by a decline of functional reserves and adaptive powers. This leads eventually to disease and inevitably to death. Irradiation of experimental animals increases the incidence and/or the severity of clinically recognized diseases at given ages. G. Casarett (in Harris, 1963) has interpreted this as an indication that radiation causes a nonspecific diffuse deterioration of the body tissues. Such deterioration acts to advance the time of onset of many diseases.

Histopathologically, the changes associated with normal aging are a decrease in the number of parenchymal cells, a decrease in the fine blood vessels, and an increase in the density and amount of connective tissue. As a result of the latter change, there is an increase in the amount of connective tissue between the blood and parenchymal cells (histohematic barrier) and, therefore, an impediment to the flow of gases and metabolites between blood and vital tissue cells. The three histopathologic aspects of aging — changes in parenchymal cells, connective tissue, and fine vasculature — are interrelated; it is not clear which are the primary and which the secondary effects. For example, due to the increased histohematic barrier and the resulting decrease in blood supply to tissues, the number of parenchymal cells is decreased. The reverse situation is also possible; that is, death of fixed postmitotic parenchymal cells may lead to replacement by connective tissue and, thus, an increase in the histohematic barrier.

On a long-term basis, animals which have been irradiated have an increase in the density and amount of connective tissue, a reduction of the fine vasculature, and a decrease in the number of parenchymal cells. These changes appear in the irradiated animals at an earlier age than in the unexposed control animals. Thus, these changes have apparently been accelerated by radiation.

Some idea of how this may occur can be realized by considering again the relative sensitivities of various types of mammalian cells (discussed in Chapter 8). In general, the endothelial cells of fine vasculature are less sensitive than dividing cells (types I and II)[1] but more sensitive than postmitotic cells (types III and IV). Cells of types I and II, however, are usually replaced or repaired. An animal which has received a sublethal radiation exposure large enough to produce the acute radiation syndrome, but which has apparently recovered, may be in the following situation.

Radiosensitive cells of types I and II will have been replaced (perhaps incompletely); radioresistant cells of types III and IV will not have been extensively damaged; the fine vasculature will have been damaged and not completely repaired. Secondary to the vascular endothelial damage will have been an increase in the interstitial connective tissue. Therefore, the histohematic barrier has been increased irreversibly, and will continue to increase with time. Eventually, there will be a loss of parenchymal cells due to the progressive change in vascularization. The animal will appear to "age" more rapidly than animals in the nonexposed control group.

Biochemical evidence is scant for the postulated increase in the histohematic barrier. Sobel[2] has shown a premature decrease in the hexosamine-collagen ratio following irradiation of skin. This indicates a relative decrease

[1]As indicated in Chapter 8, major categories of parenchymal cells are: Type I, vegetative intermitotic; type II, differentiating intermitotic; type III, reverting postmitotic; type IV, fixed postmitotic.
[2]Sobel in *The Biology of Aging*, quoted by G. Casarett (1962).

in ground substance and a relative increase in collagen. Whether this is a primary or secondary effect of radiation is not clear. The solubility of connective tissue has been reported to decrease following irradiation, and matting of connective tissue (which may be due to an increased crossbonding of the collagen molecules) has been demonstrated *in vivo* and *in vitro*.

LIFE-SHORTENING

When a population of animals is exposed to significant but sublethal radiation doses and kept until death, the animals usually die somewhat sooner than do comparable control animals. Moreover, the larger the radiation exposure, the greater is the total life-span shortening.

Figure 12.1 illustrates the results of some experiments which have measured life-shortening. Radiation dose is plotted as a percentage of the $LD_{50(30)}$ for each irradiated colony in order to compare results from different animal species and from colonies with different radiosensitivities. The life-shortening is plotted as a per cent of total life span of the animals.

The relation between total dose and life-span shortening depends upon the dosage pattern. Life-shortening following a single acute exposure may or may not be linear with increasing dose. As illustrated in Figure 12.1, the life-shortening effect appears to be greater with larger exposures. There is some question, however, as to whether the greater life-shortening in experiments using high doses may be due to factors other than the radiation. With low doses, the curve certainly appears linear and there is about a 1 to 1.5% shortening per 100 rads. With exposures close to the $LD_{50(30)}$ certain experiments have indicated a total shortening of between 20 and 50% of the remaining life span.

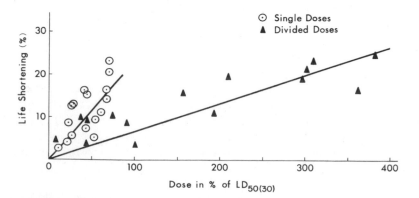

Fig. 12.1 Life-shortening in per cent of the normal life span from single doses of x- or gamma radiation and from divided doses administered in equal daily increments. (Courtesy H. Blair.)

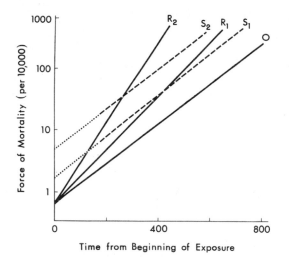

Fig. 12.2. Schematic representation of the long-term effect of single or repeated exposure to ionizing radiation on the Gompertz function (logarithm of the age-specific rate of mortality) for mammals. S_1 and S_2 represent populations given single exposures at time zero; R_1 and R_2 are populations given repeated or continuous exposure beginning at time zero. The early portion of lines S_1 and S_2 are dotted to indicate time needed for displacement to attain its steady value. (Courtesy G. Sacher [1959] and Little, Brown & Co.)

Chronic irradiation produces less life-shortening than acute irradiation for the same total dose. The data in Figure 12.1 are for chronic irradiation at constant weekly doses until death. High weekly dosages are not included since, with these, death is due in part to acute injury which has not had time to repair. Also, with the high dose rates, the irradiation may be continued for some time after the animal has sustained acute lethal damage and is virtually dead. Thus, the total dose may frequently be greater than that required for lethality in the period under study. Life-shortening from chronic irradiation is slightly less than 1% per 100 rads.

The effect of a given dose of radiation on life span in mammals is often expressed by the use of the Gompertz function. This provides an indication of the rate at which animals are dying in a population. Figure 12.2 shows a schematic representation of the effect of single or repeated radiation exposures on the Gompertz function. Time (age of animals) is plotted against the logarithm of the rate of mortality. A relatively straight line is obtained for most populations.[3]

---

[3]Whenever possible, it is convenient to omit the mortality due to factors which are not age-specific. For example, certain infectious diseases in young animals may not be included.

A single dose of x-rays given in early adult life is often followed (after a latent period of 100 to 200 days) by a simple displacement of the line to the left. However, the line will still be parallel to the control line. The rate of mortality has not changed, but the animals are dying sooner. This displacement of the Gompertz function is taken to be an indication of a certain residue of irreparable damage which resulted from the irradiation; the amount of displacement depends in part on the radiation dose. The term *precocious aging* has been applied in such situations where animals die sooner but where there is no change in the *rate* of death.

In contrast, chronic irradiation usually results in a change in the slope of the Gompertz function. This is consistent with the idea that each radiation exposure produces a certain amount of unrepaired damage and, therefore, a constant displacement of the Gompertz function. Added together, each of these displacements gives a constantly increasing divergence from the controls. Different daily doses, then, produce a fan of Gompertz curves. This increase in the *rate* of development of terminal diseases is usually called *accelerated aging*.

There are exceptions to these generalizations. For example, relatively high single doses of radiation often lead to a change in the slope of the Gompertz function. Also, certain experiments with chronic irradiation have produced a simple displacement in the Gompertz curve rather than a change in slope.

Some authors have attempted to estimate the life-shortening effect of various radiation doses without maintaining the animals for their entire life span. They assume that the amount of life-shortening is related to the amount of unrepaired damage which the animal has sustained. They further assume that the unrepaired damage is reflected in a decrease in the $LD_{50(30)}$ dose. Accordingly, animals which have received sublethal radiation are permitted

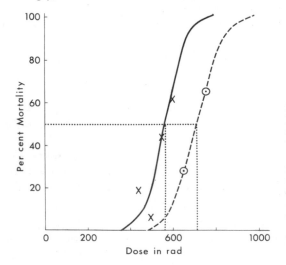

Fig. 12.3. Change in $LD_{50(30)}$ as a result of previous radiation exposure. This method is used as an estimate of life-shortening from irradiation. The right curve is for previously unexposed rats. The left curve is for rats which had received a single exposure to 600 R 60 days previously (see text). (Courtesy H. Blair.)

to recover from the acute syndrome. They are then reirradiated and the new $LD_{50(30)}$ is determined. The difference between this $LD_{50(30)}$ and one for animals which had not received the initial exposure is taken to be a measure of the irreversible injury from the first exposure. An example of such an experiment is shown in Figure 12.3. The dose-mortality curve on the right is for young adult rats which had received no previous exposure. The $LD_{50(30)}$ is about 700 rads. The left curve is for similar rats, of the same colony, who had been given a single dose of 600 rads 60 days previously. The $LD_{50(30)}$ for these animals is 550 rads. It is expected that all the reversible injury from the initial dose is repaired during the 60 day period. The difference between the 700 rads lethal dose and the 550 rads lethal dose (150 rads or about 21% of the 700 rads dose) is assumed to be a measure of the irreversible injury left by the 600 rads dose. In another experiment 600 rads was shown to reduce the life span by about 20%. It appears from these results that under at least some conditions the amount of life-shortening can be estimated by this method.

The biological significance of the technique is less clear. Measuring un-repaired damage (presumably at least part of which is vascular and early connective tissue change) by a difference in $LD_{50(30)}$ (largely determined by the response of GI and hematopoietic systems) implies a direct connection between damage to the vasculo-connective tissue and the acute response of the radiosensitive parenchymal cells of certain specific organ systems. While the condition of the vasculo-connective tissue undoubtedly has an influence on the recovery of parenchymal cells from radiation and, therefore, on the survival of the individual, many other factors must also be involved.

The reduction in life span by radiation has been attributed clinically to the induction of specific diseases, such as cancer, and also to the acceleration of normal aging processes which precipitate certain diseases and cause death. As has been indicated in the previous sections, the reduced life span has been expressed mathematically by postulating an accumulation of "irreversible injury." When an organism is exposed to radiation, certain deleterious changes occur. Recovery proceeds by processes which appear to be mathematically exponential with time. However, some of the injury is not repaired. This is the irreversible component of the radiation injury. Biologically, this may be the change in fine vasculature which was described in the previous section or somatic mutations which have been produced by the radiation. Figure 12.4 illustrates one way that the irreversible injury can be diagrammatically correlated with life-span shortening. Time (age of animal) is plotted on the abscissa; injury (in arbitrary units) on the ordinate. At some level of injury, terminal diseases result which lead to death in a short time. Normal accumulation of aging injury is expressed as a line increasing with time. A typical animal which had not been exposed to radiation would, under an ideal aging situation, accumulate injury along this line, finally develop a terminal disease, and die at point (A).

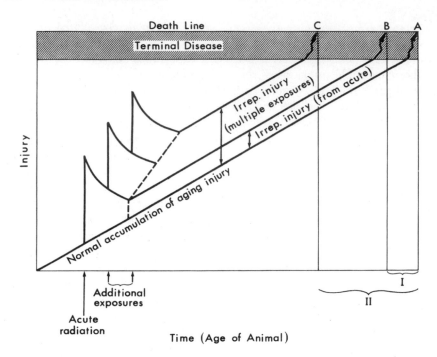

Fig. 12.4 Theoretical diagrammatic scheme for life-shortening caused by radiation injury. Irreparable injury from a single acute exposure appears as a line parallel to the "normal" aging line. Irreparable injury from additional exposures will add to that from the initial exposure. The total accumulated injury is represented during the exposures by a dotted line with a steeper slope than the "normal" line. When exposures are terminated and recovery of the reparable components has occurred, the irreparable injury which has been accumulated is represented by a line parallel to the control line. This line is displaced upward by an amount proportional to the total irreparable injury. I is life-shortening due to the single exposure. II is life-shortening due to the multiple exposures.

Acute, sublethal irradiation superimposes another injury on the "normal" line. A given dose, presented as a single exposure, results in a large amount of injury, some of which is reparable, some irreparable. The recoverable injury will be repaired and the total injury will decline exponentially, as shown, while the irreparable injury remains. This process displaces the normal aging line upward. A typical animal so treated may develop a terminal disease and die at (B), sooner than the average nonirradiated individual. The time between (A) and (B) is the life-shortening due to the radiation (interval I on Figure 12.4).

Another individual who receives more than one radiation exposure will recover from each exposure with exponential kinetics but will accumulate some irreparable injury from each treatment. The accumulated irreversible injury is expressed as a dotted line in Figure 12.4. Such an animal might die at (C), and the time between (A) and (C) would be the total life-shortening due to the multiple exposures (interval II on Figure 12.4).

If many small doses are given, extending over a large portion of the life span of the animal, the line for irreversible injury will have a slope which is steeper than the normal aging injury for the entire period of irradiation.[4] If radiation continues until death, the animal may die soon after the total injury reaches the terminal disease region. The nature of the terminal "disease" may depend on the relative contributions of the irreparable components and the acute, reparable injury from the final irradiation.

Any valid explanation of the radiation-induced life-shortening must ultimately be based on the physiological processes involved. Terminal diseases which require a certain induction period may appear sooner if the irreparable injury in some way shortens the induction period. Diseases which appear only when the reserve functional capacity of the body is insufficient to resist a specific stress may appear sooner if the irreparable injury component decreases this functional reserve capacity. Moreover, since radiation changes are nonspecific, they can add quantitatively to other forms of damage. Thus, when quantity of injury is important in the appearance of a disease, radiation may contribute to the appearance of that disease.

LIFE-LENGTHENING

In certain life-span experiments, when the radiation dose has been quite low, the irradiated animals live longer on the average than do the control animals. Some of the cases of apparent life-lengthening have been tentatively explained on the basis of nonradiation factors such as different environments for the groups of animals (for example, chronically irradiated animals were kept in a special radiation room) or on statistical probability when the population groups were small.

In some experiments, however, low radiation doses have apparently increased the life span slightly, but significantly. Two explanations have been suggested: (1) Radiation may have a therapeutic effect on already existing diseases, and (2) it may have a prophylactic effect in discouraging the appearance of new diseases of the infectious type (not age-dependent diseases). Small doses of radiation have been shown to activate the reticulo-endothelial system and cause a slight increase in the number of circulating lymphocytes or neutrophils. These stress-induced factors may help the ani-

---

[4] There is a marked superficial similarity between the change in Gompertz function (Figure 12.2) and the change in "injury level" (Figure 12.4) following single and repeated irradiations. However, the former applies to the rate of death within *populations* of control and exposed animals, while the "injury level" as drawn is an attempt to explain the accumulation of injury within irradiated or control *individuals*.

mal to resist disease processes. In fact, animals in some of the irradiated groups do not have as high an incidence of certain diseases as is found in control groups.

In certain specific experiments, the radiation exposure was sufficient to sterilize female mice in a colony in which repeated pregnancy decreased life span. The irradiated animals lived longer than the control animals in which pregnancies were permitted; nonirradiated virgin mice had a longer life span than did the irradiated mice.

RADIATION CARCINOGENESIS

Sufficient radiation to almost any portion of the body increases the incidence of cancer. The type of tumor formed depends on such factors as area irradiated, dose of radiation, age of animal, and species. Sometimes radiation does not increase the absolute incidence of cancer, but causes it to appear sooner in the population. This is called *temporal advancement*. An increase in absolute incidence is usually referred to as *true induction*. In some situations both temporal advancement and true induction occur.

There is often a longer lag between the application of a carcinogenic agent (such as radiation) and tumor formation than can be accounted for by the time required for the cell divisions. This is used as evidence that cancer induction is a multistage process. Some investigators have chosen to distinguish between initiating factors (changes which make the cells of origin of the cancer potentially cancerous) and promoting factors (changes which permit initiated cells to become cancerous). Thus, one agent initiates a precancerous change; there is a delay time, called the *latent period;* then another promoting agent produces a second change, and a tumor grows. Many carcinogenic agents (including radiation) can function as either initiating or promoting agents. Moreover, certain combinations of factors, called *co-carcinogens*, can be given simultaneously, and will result in tumor formation with a much shorter latent period. Presumably, both initiating and promoting factors are provided at the same time.

There is no general agreement on the mechanism by which exposure to radiation (or other carcinogenic agents) causes cancer. Certain theories for cancer production have been suggested which are more or less applicable for different carcinogenic agents. According to the somatic mutation theory, cancerous tissue is formed from a cytogenetically aberrant cell which is capable of reproducing itself.[5] The agent which produces such an aberrant

[5]Under a broad definition of mutation, all proposed mechanisms of carcinogenesis fall within this genetic mutation theory. It becomes merely a description of a malignancy rather than a mechanism. Usually, however, the theory refers to simple gene mutations or single-hit changes.

cell may be an initiating factor. The mutagen may also be a promoting factor if the mutation occurs in a site which is favorable for tumor development and if the uncontrolled cell reproduction proceeds immediately. Since radiation has been shown to cause both genetic and somatic mutations, this may be one mechanism by which radiation induces cancer.

It is also known that even low doses of radiation cause chromosome aberrations. A widely debated question is the role of chromosomal changes in carcinogenesis. A specific chromosomal abnormality has been demonstrated in many cases of myeloid leukemia. Also, many malignant tumors have aberrant chromosome numbers. There is no indication, however, that the chromosomal changes cause the tumors. They may, rather, be a by-product of the rapid cell division which characterizes tumors.

Viruses have been suggested as possible agents in radiation-induced leukemias. Depression of host immunity may permit infection by an exogenous carcinogenic virus. An alternative hypothesis suggests that radiation may actuate a latent carcinogenic virus infection.

Generalized tissue disorder, such as may follow radiation exposure, has also been considered to promote the formation of a cancerous growth. Such a "precancerous lesion" characteristically has a poor blood supply, and is, therefore, hypoxic. Persistent partial and abortive regenerations result in bizarre cell forms which may be cancerous. The appearance of certain skin cancers has been correlated with prior tissue disorganization and poor vascularity.

The mechanisms which have been mentioned require that the source of carcinogenic cells be exposed directly to the carcinogenic agent. In the case of radiation, this means that the tumor site must be within the radiation field. This is not always the case, however, especially in tumors of the endocrine system. For example, in some situations, irradiation of the thyroid gland increases the incidence of pituitary tumors. In certain instances, both direct and indirect mechanisms seem to be involved. The induction of ovarian tumors depends on destruction of oogonia and oocytes by direct irradiation and also on stimulation of the remaining ovarian stroma by pituitary hormones.

Perhaps one of the most pressing questions in radiation carcinogenesis concerns the presence or absence of a threshold dose; that is, is there a radiation dose below which cancer is not induced? To answer this question requires the exposure of a very large population to low doses and a comparison of the cancer incidence in that population with the incidence in a control population. Many factors such as diet, genetic constitution, and change in life span must be considered, making such a comparison extremely difficult. Some useful information can be obtained, however, by extrapolating dose-effect curves to very low doses.

It currently appears that some radiation-induced tumors probably have a threshold; others probably do not. Linearity in the low-dose regions of some dose-response curves has suggested to some authors that single event interactions have occurred between radiation and tissue with no threshold for the production of tumors. Other experiments on tumor production have indicated that the tumor incidence does depend on the exposure rate. The dose-response curves sometimes are not linear. Such results suggest that there may be a threshold dose in these situations. Obviously, the dose-response relationships pertain only to the particular experiment involved. Until more is known about the mechanisms of cancer production, generalizations from one experimental situation to another are most difficult.

RADIATION CARCINOGENESIS IN EXPERIMENTAL ANIMALS

Many different types of tumors have been produced in experimental animals by total-body irradiation or by local irradiation of specific tissues. Generally, radiation exposures of several hundred roentgens or more have been used. Most of the animal experiments have shown that very low doses of radiation produce no detectable increase in the incidence of certain tumors. Does this mean that there is a threshold dose below which certain neoplasms cannot be induced or caused to appear earlier? It may be that the tumor incidence at such low doses is too small to be significantly different from the control incidence, or the latent period for tumor induction may exceed the life span in the short-lived laboratory animals.

Based on a series of single dose studies on rats, Maisin and co-workers (1958) have commented on the probable existence of thresholds of carcinogenesis as follows:

> In effect, the threshold for carcinogenesis in various tissues appears to be specific for each one. At the level of the kidneys, cancers appeared only in those animals in which this organ had received a dose of 850 R or more. The latent period for appearance of cancer of the kidney is, moreover, particularly long (10 to 32 months). In the case of the digestive tract, although 600 R seems to be a dose sufficient to cause a cancer, the yield nevertheless becomes much more important after doses of 850 R and more. Finally, for the lung, only one cancer appeared in the more than 700 animals irradiated at this level with doses varying from 600 to 850 R. However, among the 10 animals irradiated over a single lung with a dosage of 2000 R, 3 bronchial epitheliomas were noted.
>
> The skin, mammary glands, salivary glands, seminal vesicles, suprarenals, smooth muscle tissue and even subcutaneous connective tissue may become cancerous after doses of 600 R. Thus, there appears to exist a gradation of thresholds for cancer production in the various tissues irradiated by a single dose of x-rays.

A tumor type which has been studied in some detail by Kaplan and co-workers (1952) is the thymic lymphoma. Moderate radiation doses will produce these neoplasms in considerable numbers in mice, especially in C-57 and RF strain animals. There appears to be a threshold for the induction of these tumors. The tumor incidence (13%) following a single exposure of

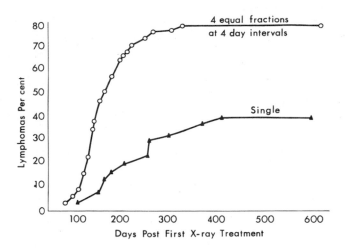

Fig. 12.5. Cumulative incidence of lymphomas in mice exposed to 475 R in a single dose or given as four equal fractions at 4 day intervals. (Modified from Kaplan & Brown [1952], courtesy National Cancer Institute.)

283 R was not significantly different from the spontaneous incidence (6 to 8%), although some of the tumors appeared earlier in the irradiated mice than in the control animals. Increasing the exposure to 475 R increased the incidence to 41%. Moreover, the dose-response curve did not appear to be linear. It is interesting to note that division of the total dose into four equal fractions at 4 day intervals almost doubled the lymphoma production following 475 R. This is illustrated in Figure 12.5.

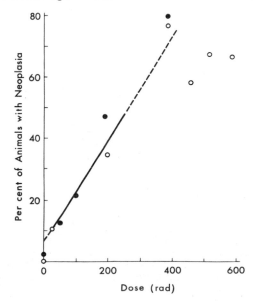

Fig. 12.6. Incidence of mammary gland neoplasia in irradiated rats suggesting a linear relationship between dose and cancer incidence. The open and closed circles represent results from different experiments. (From Bond *et al.* [1960], courtesy Academic Press, Inc.)

The absence of a threshold is suggested in a series of experiments by Shellabarger, Bond, and others on the induction of mammary tumors in female Sprague-Dawley rats.[6] This tumor is especially convenient for radiation studies since it has a low spontaneous incidence in young rats, a high incidence following irradiation, and a latent period following irradiation of only about 3 months. As indicated in Figure 12.6, there is approximately a linear increase in tumor incidence (scored at 10.5 to 11 months) between exposures of 25 and 400 R. Above 400 R the incidence either remains constant or decreases. The shape of the curve suggests that a nonthreshold mechanism may be responsible for tumor induction in this particular experiment.[7]

Direct radiation injury to the breast and the functioning of an intact ovarian system are both required for maximum incidence of the mammary tumors. It may be postulated that primary radiation damage to the mammary gland may be the initiating factor and that there may be no threshold for this effect. Intact ovarian function may provide the promoting factor. This second factor is present immediately after the initiating event occurs, so there is no long time lag (latent period) preceding tumor development. Bond and co-authors (1960) point out that lack of linearity in the dose-response curve of other experiments does not rule out a nonthreshold somatic mutation as the basis for a primary event. Nonlinearity might indicate that the necessary factor is not operative.

RADIATION CARCINOGENESIS IN MAN

The production of tumors in humans by radiation has been observed for many years. Skin cancers were common in early radiologists and dermatology patients. Radiation from radon and radon-daughters is considered one of the contributing factors in the high incidence of lung cancers among miners from the Jachymov and Schneeberg pitchblende mines. Bone tumors resulted from accidental ingestion of radium by the girls who painted luminous watch dials and from radiation given medically for various complaints. Thorotrast (thorium dioxide), used as a contrast medium for tissues in diagnostic x-irradiation, localizes in liver and spleen and its presence has been associated with numerous cases of carcinoma of the liver. More recently, Japanese atomic bomb survivors and patients given therapeutic radiation

---

[6]See, for example, Bond et al. (1960).

[7]It is also possible that the apparent absence of a threshold is the result of experimental factors other than radiation. For example, tumors in irradiated animals were scored at 10.5 to 11 months; most neoplasms in control animals occur after this time. Therefore, the tumors in irradiated animals may not all represent true induction. In addition, all types of mammary tumors were scored together. The linear response then does not represent the kinetics of formation of a single tumor type, but may be the composite of several different mechanisms. Linearity may be fortuitous. Nevertheless, without further information, the linear, nonthreshold relationship is a convenient curve to express the data which are available.

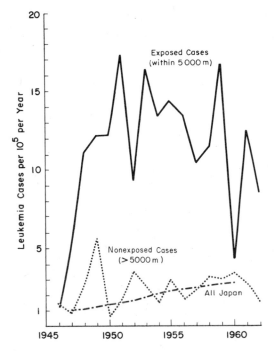

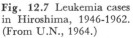

Fig. 12.7 Leukemia cases in Hiroshima, 1946-1962. (From U.N., 1964.)

for various illnesses have provided more information on radiation carcinogenesis in general, and especially on the induction of leukemia.

Tumors in Japanese survivors. It is clear that the incidence of various types of leukemia was increased among the Japanese bomb survivors. Figure 12.7 shows a comparison of the overall incidence in Japan, in the "nonexposed population" (beyond 5000 meters from hypocenter) and the "exposed" population (within 5000 meters) of the survivors of Hiroshima. Even with the best estimates, there is great uncertainty about the actual doses received, the size of the populations, and the incidence of leukemia. Furthermore, the studies are of necessity made on the survivors, which at high dose levels are presumably the more radioresistant members of the initial population.

From the data which are available, it appears that in the range between 100 and 900 rads the average rate of increase of the incidence with dose was approximately linear and was between 1 and 2 cases/10^6/year/rad in both Nagasaki and Hiroshima. There is also a moderate degree of correlation between the year of onset of leukemia and the distance from the hypocenter. It appears from this that the latent period is shorter with higher doses.

Surveys of the Japanese survivors suggest that the incidence of thyroid carcinoma has been increasing in the irradiated population of Hiroshima and Nagasaki. The incidence varies inversely with distance from the hypocenter.

Other forms of malignant diseases have also probably been increased in the Japanese survivors. The mean latency for radiation-induced leukemia is, however, apparently less than for other neoplasms and the induction of leukemia may occur more frequently than that of other tumors. One report has estimated that the "doubling dose" for total cancer incidence was received at about 1300 meters from hypocenter, where the dose was approximately 400 rads.

Leukemia in radiologists. Until rather recently, when the dangers of radiation were appreciated, radiologists did not utilize suitable shielding and consequently received considerable exposures during radiographic examinations. The average annual incidence of death from leukemia in radiologists between 1948 and 1961 was $253/10^6$/year as compared with the expected incidence (based on total population) of $85/10^6$/year. Although these figures represent a comparatively few actual number of cases (12 leukemias in radiologists), it appears possible that radiation has increased the leukemia incidence in this group.

Tumors in radiation therapy patients. During the past 40 years, irradiation of localized portions of the body has been employed for the treatment of certain rheumatic diseases (for example, ankylosing spondylitis) and for "enlarged thymus" in children. The incidence of various tumors in the irradiated individuals has been studied. In evaluating the results of such studies, one important factor must be considered; that is, the possibility that the disease which is being treated predisposes the individual to cancer.

In a survey by Court-Brown and Doll, it was found that a significant excess of deaths from all types of leukemia occurred in the irradiated spondylitics, as compared with the expected numbers in a nonspondylitic, non-irradiated population. There may, however, be a correlation between leukemia and ankylosing spondylitis; also, some of the other forms of therapy used on the spondylitics may have been leukemogenic. Radiation doses received by these patients varied considerably in size, fractionation pattern, and tissue volume exposed. Taking these factors into account as well as possible, a regression line fitted to the incidence observed at doses to the spine between approximately 300 and 1500 rads has a slope of $0.5/10^6$/year/rad. Considering the marked differences in dose rate, population, and tissue irradiated, this value agrees well with that for the Japanese survivors (1 to $2/10^6$/year/rad).

A series of studies have examined the incidence of leukemia in children who received therapeutic irradiation for enlarged thymuses. One study, a survey by Hempelmann and co-workers, showed an increased incidence of leukemia in the irradiated children. However, in another study, no increase in leukemia was reported in a group of children who were similarly irradiated but who did not appear to have the condition called thymic enlargement. It is possible, then, that the leukemia cases in the Hempelmann study can be attributed to the thymic condition rather than to the irradiation.

In children irradiated for thymic enlargement, the incidence of thyroid carcinoma is increased over the general population. Again, a wide variety of doses, dose rates, fractionation schedules, and reasons for treatment make a dose-effect analysis difficult; but within the exposure range of 100 to 300 R, for an average follow-up time of about 16 years, the estimate of risk is probably between 0.5 and 1.5 cases/10^6/year/rad. This range is based on the slope of a straight line fitted through the data.

Tumor incidence related to diagnostic radiation. Many studies have been made of children irradiated *in utero*. The results have varied considerably, but it seems apparent that irradiation of fetal tissues can result in leukemia and that the risk per unit dose is higher than for postnatal irradiation, possibly by a factor of 5. Malignancies other than leukemia may also occur (as a result of the irradiation) in children who were irradiated *in utero*. At present, however, the data on such cases are limited.

Tumors in individuals containing internal emitters. A summary has recently been made of the results of a survey of 361 people with skeletal burdens of ^{226}Ra and ^{228}Ra (mesothorium). Approximately two-thirds of these were radium dial painters; the remainder were contaminated as a result of laboratory work, from injection of radium solutions or from ingestion of radium or radium-mesothorium mixtures. The United Nations Report (1964), with regard to data of Maletskos and co-workers, has stated:

> At the present stage of the statistical study of these data ..., no clinically significant signs or symptoms are seen with residual or terminal body burdens of <0.5 μCi Ra .... With higher residual body burdens, beginning in the neighborhood of 1 μCi Ra and extending to about 25 μCi Ra, the fraction of the total number of cases which involve either of these types of malignancies (osteogenic sarcomas and carcinomas of the paranasal sinuses and mastoids) amounts to roughly $\frac{1}{4}$.

It is thus apparent that the presence of the radium isotope has increased the incidence of bone tumors. The latent period for the appearance of these tumors is of the order of 20 to 30 years.

Thorium dioxide in a colloidal suspension (thorotrast) was widely used in diagnostic radiology between 1928 and 1945 for the visualization of body cavities and certain organs. Numerous cases of malignant tumors have been reported which apparently relate to the thorotrast injection. The mean latent period for the appearance of these tumors is about 18 years. The alpha radiation dose to liver and spleen of these individuals has been estimated to be as high as hundreds of rads per year.

OTHER LATE EFFECTS OF
RADIATION

In addition to generalized aging changes and induction of cancer, some specific organ systems show characteristic long-term effects of radiation. Many of these contribute little, if at all, to the life-shortening or "aging" of the individual. Certain of these changes have already been discussed in Chapter 9 but are mentioned again in this chapter for completeness.

EFFECTS ON FERTILITY

The very marked effect on the developing germ cells of both testis and ovary have been described. If all of the stem cells are destroyed, complete sterility will result. Even if some stem cells survive irradiation, a partial or functional sterility will occur in a male if the number of functional sperm which are produced is decreased. This may involve an absolute decrease in numbers of ejaculated sperm or an increase in abnormal sperm. In addition, irradiation may result in a hastening of the normal involutional changes associated with advancing age.

EFFECTS ON BONE

Figures 9.7A and 9.7B illustrate the inhibition of bone growth which occurs following either local or total-body irradiation. Experimental studies in rats have also indicated that chronic total-body irradiation can decrease growth at radiation levels which are too low to decrease the hemoglobin or neutrophil levels. Radiation osteitis is a late degeneration effect of massive radiation (several thousand rads) of bone. After many years, the degenerative processes lead to necrosis, fractures, and osteogenic sarcomas.

RADIATION CATARACTS

Exposure of the optic lens to radiation results in the formation of lens opacities or cataracts. The latent period between irradiation and opacity may be months or years, and the opacities may vary greatly in extent.

Doses in excess of about 15 rads of x-ray increase the incidence of lens opacities in mice. The critical dose for cataract formation in humans has been estimated by some authors as between 20 and 45 rads. Neutrons are especially efficient in the production of cataracts. Fractionation of dose delays the time of onset of cataracts and decreases the incidence of severe opacities.

GENERAL REFERENCES

Bacq, Z. M., and Alexander, P. *Fundamentals of Radiobiology*, Chapter 17, 2d ed. Pergamon Press, New York (1961).

Casarett, G. W. Radiologic Aging, Generalized and Localized, pp. 251–268 in Dougherty, T. F., Jee, W. S., Mays, C. W., and Stover, B. J. (eds.), *Some Aspects of Internal Irradiation*. Pergamon Press, New York (1962).

Casarett, G. W. Experimental Radiation Carcinogenesis. *Progr. Exptl. Tumor Res.*, **7**: 49–82 (1965).

Harris, R. J. C. (ed.). *Cellular Basis and Aetiology of Late Somatic Effects of Ionizing Radiation*. Academic Press, Inc., New York (1963).

Neary, G. J. Aging and Radiation. *Nature*, **187**: 10–18 (1960).

Sacher, G. A. On the Statistical Nature of Mortality with Especial Reference to Chronic Radiation Mortality. *Radiology*, **67**: 250–257 (1956).

United Nations Scientific Committee on the Effects of Atomic Radiation. Nineteenth Session. Suppl. #14 (A/5814)(1964).

ADDITIONAL REFERENCES CITED

Bond, V. P., Cronkite, E. P., Lippincott, S. W., and Shellabarger, C. J. Studies on Radiation-Induced Mammary Gland Neoplasia in the Rat. III. Relation of the Neoplastic Response to Dose of Total-Body Radiation. *Radiation Res.*, **12**: 276–285 (1960).

Kaplan, H. S., and Brown, M. B. A Quantitative Dose-Response Study of Lymphoid-Tumor Development in Irradiated C57 Black Mice. *J. Natl. Cancer Inst.*, **13**: 185–208 (1952).

Maisin, J., Maldague, P., Dunjic, A., Pham-Hong-Que, and Maisin, H. Carcinogenic Effect of a Single Dose of X Rays in the Rat, pp. 134–144 in *U. N. Intern. Conf. Peaceful Uses At. Energy, 2nd, Geneva*, **22** (1958).

Sacher, G. A. Relation of Lifespan to Brain Weight and Body Weight in Mammals, pp. 115–141 in Ciba Foundation Colloquia on Aging, Vol. 5. *The Lifespan of Animals*. Little, Brown and Company, Boston (1959).

13 EFFECTS OF RADIATION ON HIGHER PLANTS AND PLANT COMMUNITIES

In many of the preceding chapters, the effects of radiation on animals have been emphasized. Ionizing radiation also produces changes in plants. The initial ion pair formation, the production of free radicals, the formation of biochemical lesions, and many of the cellular changes that are induced are similar in plants and animals. Plant tissues and structures are different in detail from animals, and, therefore, the gross, external manifestations of the biochemical and cellular changes are very different. At the cellular and subcellular levels the changes have much in common.

To understand what ionizing radiation can do to plants, it is desirable to have some familiarity with plants and plant structures. Figure 13.1 illustrates diagrammatically the regions of an idealized higher plant which are of greatest interest.

Mature higher plants possess a root system, leaves, and flowers or cones. Nutrients are absorbed from the soil by the root system and conducted up the stem to the leaves. Organic materials are synthesized in the leaves. The manufactured materials which are not used directly for growth may be stored in various organs (for example, starch in potato tubers). Cell division occurs mainly in terminal regions of stems and roots, called *meristems*. Also, certain growth stimulating hormones are produced in or near the meristematic regions. Cell elongation and differentiation occur in nearby areas.

In contrast to animal cells, most of the living cells of plants may divide and differentiate to form organized tissues if injury or some other abnormal condition arises. Thus, entire portions of plants may be replaced from cells of another portion, as, for example, in the formation of a new root system from the stem of a plant "cutting."

Within the floral buds of mature plants are germ cells which undergo meiotic division to form haploid microspores (male) and megaspores (female).

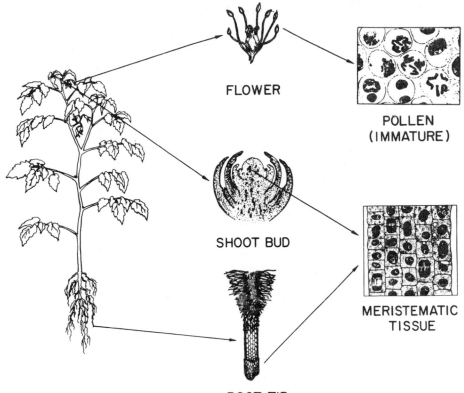

FLOWER

POLLEN
(IMMATURE)

SHOOT BUD

MERISTEMATIC
TISSUE

ROOT TIP

Fig. 13.1 A semidiagrammatic illustration showing various levels of organization in a plant from the mature plant (at left) through the major regions responsible for growth and reproduction (middle column) to the cellular level. (Modified from Sparrow [1962], courtesy Brookhaven National Laboratory.)

Subsequent mitotic division of the microspore nuclei produce binucleate (or trinucleate) haploid pollen grains. Similarly, mitotic division of the megaspores gives rise to the eight-nucleate haploid embryo sacs of the ovules. During pollen development or after germination, on the pistil of a flower, two sperm cells are formed, one of which fertilizes the egg nucleus in an ovule.

The seed which develops from the fertilized egg contains a plant embryo which developed from the diploid zygote, the endosperm (food storage tissue) and the surrounding seed coat. The embryo possesses an embryonic root (radicle), a stem (hypocotyl and epicotyl), and one or more leaflike cotyledons which may contain stored nutrient material. Germination occurs when the seed is exposed to water. The seed coat swells and splits open, and rapid cellular elongation of the radicle and cotyledon-bearing hypocotyl forces these structures beyond the seed coat. Subsequent cell division, growth, and differentiation result in the formation of the other plant structures.

RADIATION EFFECTS ON PLANTS

Within a year after the announcement of the discovery of x-rays by Röntgen in 1895, reports described injurious effects of x-irradiation on plants. Interest in radiobotany was sporadic for the next few decades, however, due perhaps to the lack of practical or useful applications at that time. Some interest resulted from the demonstration that radiation could dramatically increase mutation frequency (see Chapter 6), and this has now led to the use of induced mutations in plant breeding. Also, much of the early basic work on chromosome aberrations was done on plant material (see Chapter 5).

More recently, there has appeared a new field of investigation pertaining to the use of radiation in food processing and storage. The amount of fundamental research in radiobotany has greatly increased and important contributions to general radiobiological concepts have resulted. Much work remains to be done, however, before the mechanisms of radiation damage to plants are fully understood. The following sections describe briefly a few of the experiments which have been done and some of the information which has been accumulated on the subject.

RADIATION EFFECTS ON POLLEN

Pollen has been used for a variety of radiobiological experiments. The grains are relatively simple, numerous, haploid, adaptable to environmental insults such as anoxia and desiccation, and small enough to be used successfully with low penetrating radiations, including ultraviolet.

When fresh pollen is irradiated, massive doses are required to inhibit subsequent germination. Values of LD_{50} (dose to inhibit germination in 50%) have been reported between 35 kR and 550 kR depending on the species of plant (see Figure 13.2). It is interesting that there is an apparent correlation of radiosensitivity with pollen size. The large grains are generally more sensitive than the small (see Chapter 8 for a discussion of relative radiosensitivity). Doses which do not prevent germination may, however, greatly depress pollen tube growth, often resulting in short tubes with burst tips. Genetic effects may also be produced.

Since germination involves the elongation of cells and little mitotic activity, it is not surprising that high radiation doses are required to inhibit germination. In their extensive article on the effects of radiation on pollen, Brewbaker and Emery (1961) indicate that the sigmoid curve and high lethal dose for pollen suggest a cumulative type of physiological action, probably affecting cell membranes and the cellular machinery involved in the synthesis of cellulose, pectin, and callose. However, very much lower doses to pollen will reduce seed set to 50%. This indicates that germination and

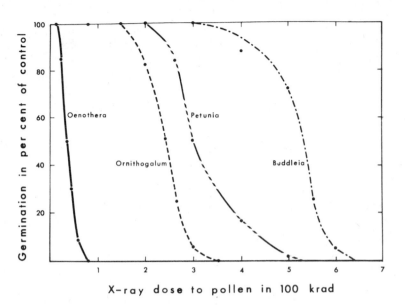

Fig. 13.2. Representative dose-response curves for pollen germination. Pollen diameters are estimated to be: *Oenothera* 140 μ, *Ornithogalum* 72 μ, *Petunia* 35 μ, *Buddleia* 18 μ. (From Brewbaker & Emery [1961], courtesy Pergamon Press.)

ultimate survival of pollen are very different with respect to radiobiologic injury.

Environmental conditions may markedly influence the radiosensitivity of pollen. Desiccated pollen is about twice as sensitive, as measured by germination and tube growth, as pollen maintained at ambient humidity if the irradiation is carried out in the presence of oxygen. Under conditions of anoxia, the water content does not greatly influence radiosensitivity. Storage will also increase radiation damage in pollen which is irradiated in the presence of oxygen. Freezing, however, does not alter sensitivity.

Inhibition of seed formation and impairment of seedling development in a number of species has resulted from pollen irradiation with doses in the range of 2 to 20 krads depending on the species used. These effects have been attributed, in part, to the production of dominant mutations.

Irradiation of pollen is also a useful method of inducing mutations. Since only one haploid nucleus in a pollen grain will contribute to the formation of a gamete, all of the cells in an F_1 plant contain the genetic material from that nucleus. Any mutation which has occurred will be present in all cells of the embryo.[1] Dominant mutations will appear in the first generation. Recessive mutations may appear in a subsequent generation. Aberrant leaf

[1]In contrast, the F_1 plants derived from irradiated seeds may contain a mixture of mutated and nonmutated cells since the embryo in a seed contains many cells.

shapes and sizes, reduced flower size, and abnormal chlorophyll content have been reported in F_1 plants derived from irradiated pollen.

Chromosome and chromatid aberrations can also be induced and studied in irradiated pollen since in many species there is division of the haploid nucleus during pollen germination. Figures 5.7 and 5.10 illustrate some chromosomal aberrations which have been produced in pollen of *Tradescantia*.

RADIATION EFFECTS ON DEVELOPING PLANT EMBRYOS

After a haploid egg cell is fertilized by the sperm from a pollen grain, cellular division and differentiation occur which result in the formation of a structurally complex seed. This development may be considered to be roughly analogous to animal embryogenesis.

As is the case with the mammalian embryo, the developing plant embryo is radiosensitive. The degree of radiosensitivity depends greatly on the species, the stage of the embryo at irradiation, and the criteria used to measure the effect. For example, the following criteria have been used by Mericle and Mericle (1961) to measure radiosensitivity in barley embryos: abortion (failure of an ovary containing an irradiated embryo to continue development and form a complete seed), lethality at germination (failure of embryo root and shoot emergence, following water inbibition), seedling lethality (after germination, death occurring during subsequent seedling development), survival to maturity (ability of the plant material to survive from the embryonic stage through the production of seeds), and mutation rates for gametically-induced abnormalities.

In barley,[2] moderate radiation exposure (500 to 1000 R) of zygotes or early "proembryos" (only a few cells) markedly reduces the number which will develop into mature plants. Many fail to develop into seeds or, if seeds are formed, they do not germinate (see Figure 13.3). The incidence of developmental abnormalities is small.

Mid- or late proembryos are somewhat less radiosensitive than the earlier stages and show a different pattern of response. Seeds are usually formed, but the incidence of nongermination is high. The plants which do form may have specific morphologic anomalies.

Irradiation at a still later stage, when the embryo is differentiating, often results in the formation of abnormalities as the plant develops. The anomalies are stage specific; that is, the effective period for induction of a particular abnormality is often restricted to one or a few embryonic stages. The reader is reminded that such critical periods during organogenesis have been described in the section on animal embryos (see Chapter 10).

---

[2]The radiosensitivity of developing embryos may vary by as much as a factor of 100 in different species. For example, Devreux and Scarascia Mugnozza (IAEA, 1965) have reported a decrease in the size of zygotes of *Nicotiana* after 4 and 8 kR.

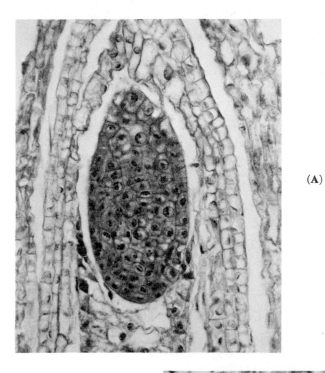

(A)

(B)

Fig. 13.3 (A) Nonirradiated late proembryo of bar-
ley; (B) embryo 4 days
after x-irradiation (700 R)
at early proembryo stage.
Should appear as in (A).
(Courtesy Mericle & Meri-
cle [1961] and Brookhaven
National Laboratory.)

Embryos in the final stages of maturation are somewhat less radiosensitive, although irradiation in these stages may result in growth inhibition in roots and shoots during subsequent development.

RADIATION EFFECTS ON SEEDS

Dormant seeds are less radiosensitive than seeds with developing embryos. The decreased sensitivity of dormant seeds is probably related to their quiescent state (little division or differentiation) and to their much lower water content. One of the most commonly used criterion of seed radiosensitivity is the seedling height at specific times after germination. Results from a typical experiment are illustrated in Figure 13.4. Radiation of seeds dramatically decreases the height of the corn seedlings. The highest exposures result in the least growth. Notice in Figure 13.4 that the root development is inhibited by radiation, especially at the highest exposure level.

There have been reports of a reversal of the seedling height curve at high exposures; that is, for low doses there is a progressive decrease in seedling heights with increasing exposure. With high doses, however, there is a slight reversal or apparent increase in leaf height relative to that at a slightly lower exposure. This reversal has been explained as the result of interaction between chromosome damage and mitotic inhibition. Extensive chromosome damage in irradiated dividing cells will result in a high degree of lethality of

Fig. 13.4. Seedling height 13 days after planting corn seeds which had been exposed to 10,000, 20,000, or 40,000 R.

daughter cells. At higher doses, the chromosome damage is greater, but mitosis has been so completely inhibited that cells do not die in a mitotic-linked death (see Chapter 5). They survive and are apparently still capable of elongation to produce leaves. The radiation exposure at which the reversal occurs differs with species. For example, it occurs between 5000 and 15,000 R in corn, but between 50,000 and 150,000 R in wheat and barley.

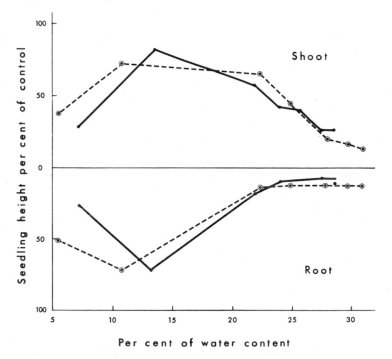

Fig. 13.5. Effect of water content (% wet weight) on radiosensitivity of wheat and barley seeds. Seeds received 20 kR gamma rays (●——● Wheat, ○--○ barley). (From Biebl & Mostafa [1965], courtesy Pergamon Press.)

The water content of seeds has been shown to markedly influence their radiosensitivity, as illustrated in Figure 13.5. Radiation has a minimum effect on both shoot and root growth when the exposed seeds are hydrated at about air-dry conditions (water concentration between 10 and 15% of wet weight). Above and below this water content, growth is much more severely inhibited by the radiation. Similarly, germination of seeds, survival of seedlings, and seed production of the plant originating from irradiated seed are all inhibited more when irradiated at low or high water contents than when air dried.

It appears that the variation in radiosensitivity with water content is related to the free radical concentrations in the seeds at various times after irradiation. Based on electron spin resonance measurements, Ehrenberg

(1961) has concluded that the number of free radicals produced in seeds is independent of water content, but the survival time of the free radicals depends on the degree of hydration. At medium degrees of moisture content (air dry) where the radiation sensitivity is minimum, free radicals probably persist for a number of hours or less. Then they apparently recombine to form relatively harmless products. It has been suggested that the hydration water of air-dry seeds might act by increasing mobility of radicals and facilitating recombination.

Very dry seeds have a high radiosensitivity, especially in the presence of air or oxygen. It appears that free radicals are quite stable for several days at a low water content. This may be the result of restricted mobility. When exposed to water, radical recombination occurs during the first hours of germination. If oxygen is present prior to or during germination, there will be an opportunity for the formation of biologically damaging peroxyl radicals $(R° + O_2 \rightarrow RO_2°)$ before radical recombination occurs. The molecules formed from peroxyl radicals may be harmful and the effects on the resulting plant will be great. In contrast, if the dried seeds are irradiated and kept in nitrogen and are brought into contact with air-free water to produce germination, many of the radicals will apparently recombine at germination to form harmless products. Damage is therefore less.

Seeds are also radiosensitive at high water content. It has been suggested that free radicals react very rapidly (seconds or less) at high water content to form a variety of deleterious products. A hydrated embryo is also in a more active physiological state and is, therefore, probably more radiosensitive.

Seeds which are presoaked and then dried back to 16% water for irradiation are just as sensitive as if they are irradiated wet. Since soaking permits a leaching of materials into the water, some authors[3] have inferred that a permanent change in radiosensitivity occurs during soaking and that this change is associated with the loss of the leached amino acids, sugars, and inorganic substances. Such a leaching effect may contribute to the high radiosensitivity of wet seeds.

Under certain conditions the degree of damage which results from seed irradiation is increased as the time is increased between irradiation and planting. In one experiment by Curtis et al. (1958) damage increased by as much as a factor of 20 after storage for a few weeks. In the same paper, it was demonstrated that there are two components to the curve of storage vs. damage (see Figure 13.6). The first component is rapid and nearly complete in 4 hours; the second component is slow and lasts several weeks. The storage effect is especially pronounced wih desiccated seeds which show a large first component to the curve. If desiccated seeds are planted immediately after

[3]Ehrenberg, A. et al. Acta Chem. Scand., **11**: 199–201 (1957), quoted by Curtis, et al. (1958).

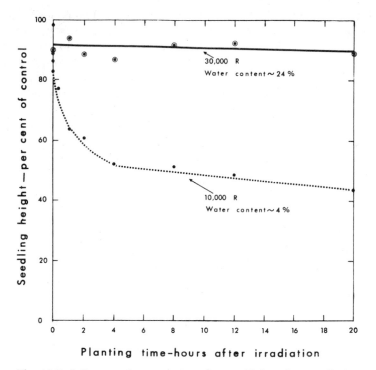

Fig. 13.6. Influence of storage time (up to 20 hours) on radiation damage in barley seeds irradiated with gamma rays. (From Curtis *et al.* [1958], courtesy Academic Press, Inc.)

irradiation, the inhibition of seedling growth is very similar to that found with air-dry seeds. If seeds are planted 1 week after irradiation, however, plants from desiccated seeds are considerably more inhibited than those from air-dry seeds. These results are presumably correlated with the lifetime of the free radicals in the seeds. Immediate planting (and hydration) of desiccated seeds may result in recombination of the radicals before peroxydation can occur.

Under certain experimental conditions, a stimulation of seedling growth has been observed following seed irradiation. This phenomenon will be discussed in a later section.

RADIATION EFFECTS ON GROWING PLANTS

Radiation of growing plants may result in death, growth inhibition, altered metabolism, morphological abnormalities, or mutations. The severity of the effect depends on the radiation factors, the species studied, and the age of the plant.

Extensive studies of radiation effects on growing plants have been done by a number of investigators at the Brookhaven National Laboratory on Long Island, N. Y. These investigators use gamma radiation from cobalt-60 or cesium-137 in the gamma radiation fields, in shielded growth rooms, or in greenhouses. For chronic exposures the plants are irradiated for 20 hours each day for several weeks or more. (The remaining 4 hours are used for necessary cultivation, planting, observation, and so forth.) Daily exposure rates range from about 1 R/day to about 10,000 R/day. Acute exposures may last from a few minutes to a few hours.

The usual procedure is to transplant young seedlings into the field during the spring or early summer. Several rows (or arcs) of each species are planted at various distances from the source. It is then possible to follow the growth (if any) as well as morphologic and other changes at the various exposure levels.

Table 13.1. Percentage of the Daily Dose Causing 100% Mortality (LD_{100}) Required to Produce Various Responses in Plants Chronically Exposed to Cobalt-60 Gamma Radiation[a]

Responses	No. of species observed	Daily dose as percentage of LD_{100}[b]
Normal appearance	14	<11
10% growth reduction	23	26 ± 2.5
Failure to set seed	8	31 ± 5.5
50% growth reduction	12	34 ± 3.5
Pollen sterility (100%)	4	41 ± 4.5
Floral inhibition or abortion	21	44 ± 3.5
Growth inhibition (severe)	41	58 ± 3.0
LD_{50}	17	75 ± 2.5
LD_{100}	41	100

[a]From Sparrow and Woodwell (1962).

[b]Average (± one standard error of the mean) based on data available for various numbers of species indicated. Most species were herbaceous annuals exposed for 8 to 12 weeks in the Brookhaven gamma field.

Sparrow and Woodwell have summarized some of the effects produced in plants by radiation (Table 13.1). They express the dose required for each effect as a percentage of the LD_{100}, the dose required to kill all of the plants of the species studied. Many of these effects will be considered in greater detail in the following sections.

Fig. 13.7. View of a portion of the Brookhaven gamma field. The location of the ^{60}Co gamma ray source is indicated by the shadow of the post in the lower right corner. Dose rates are shown. (Courtesy A.H. Sparrow and Brookhaven National Laboratory.)

GROWTH INHIBITION

A commonly used end point for comparing radiosensitivity is the exposure which causes "severe growth inhibition." This is the exposure which reduces the growth of the plants to about 15% of that of the control plants. Presumably, the reproductive capacity of most of the cells in the meristems has been lost.

Figure 13.7 shows a portion of a Brookhaven gamma field. The shadow of the post which contains the source can be seen at the lower right corner of

Fig. 13.8. Composite photograph of 4 to 5 year old seedlings of *Pinus strobus* grown for 17 months at the daily exposure rates indicated. (Courtesy A.H.Sparrow and Brookhaven National Laboratory.)

the picture. The exposure rates are indicated. Gladioli in the foreground show considerable growth inhibition after several weeks at exposure levels of about 5000 R/day. Other gladiolus plants receiving about 1000 R/day are essentially normal. *Gladiolus* is considered to be a very radioresistant species.

As a contrast to the radioresistance of the *Gladiolus*, Figure 13.8 shows the marked radiosensitivity of seedlings of white pine. Plants received exposures ranging from 2.5 R/day to 20 R/day for 17 months. Some normal growth is apparent at the 10 R/day level. Severe growth inhibition has occurred at a dose level somewhere between 7.5 and 10 R/day.

The wide variability in sensitivity to chronic radiation is illustrated by the examples in Table 13.2. The most sensitive plant listed shows severe growth inhibition from exposure to about 40 R/day over an 8 week period. Growth of the most resistant plant is severely inhibited by an exposure of about 6000 R/day for 12 weeks.

Table 13.2. RADIOSENSITIVITY OF VARIOUS PLANTS TO CHRONIC GAMMA RADIATION[a]

Plant	Length of exposure(weeks)	Exposure to produce effect (R/day)		
		none[b]	mild	severe
Lilium (hybrid) h.v. Tangelo (lily)	8	10	20	40
Tradescantia paludosa (spiderwort)	12	15	20	40
Vicia faba (broad bean)	15	30	60	90
Nicotiana rustica (tobacco)	15	50	100	400
Sedum spurium (stonecrop)	14	255	528	760
Sedum sieboldi (stonecrop)	13	1000	2500	4100
Gladiolus (hybrid) h.v. Friendship	12	800	1500	5000
Luzula acuminata (wood rush)	12	1720	2600	6000

[a]Modified from Sparrow (1962).
[b]No gross external evidence of radiation damage.

As a further illustration of the difference in radiosensitivities of two plants, Figures 13.9 and 13.10 show the survival curves for two species of woody plants, *Acer saccharum* (sugar maple) and *Sambucus canadensis* (elderberry). In Figure 13.9 radiation dose vs. survival (log) is plotted. Clearly, *A. saccharum* is the more radioresistant of the two species. The same data are plotted on a probit scale (see Chapter 11 for a description of probit plots) in Figure 13.10 to facilitate a quantitative comparison (LD_{50} for *S. canadensis* = 1.0 kR; LD_{50} for *A. saccharum* = 4.9 kR).

As indicated in Chapter 8, among the many factors which probably contribute to the wide range of radiosensitivities exhibited by various or-

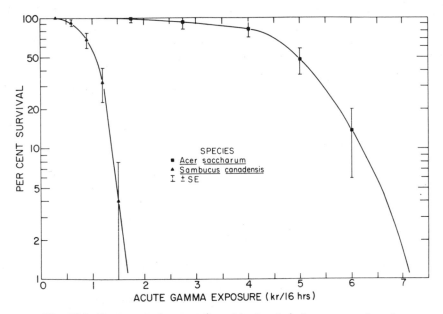

Fig. 13.9. Dose-survival curve (logarithmic plot) for two species of woody plants exposed to a single dose of gamma radiation. (Courtesy A. H. Sparrow [1965] and Genetics Society of Japan.)

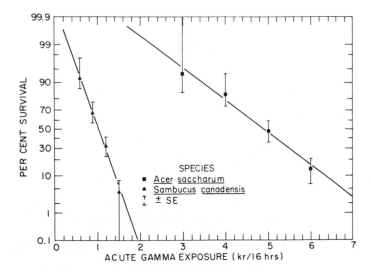

Fig. 13.10 Dose-survival curve (probit plot) for two species of woody plants exposed to a single dose of gamma radiation. (Courtesy A. H. Sparrow [1965] and Genetics Society of Japan.)

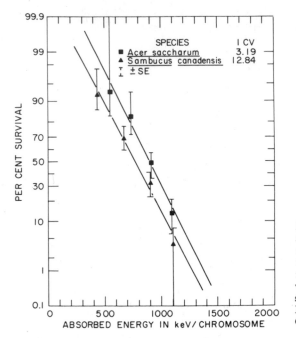

Fig. 13.11. Survival vs. absorbed energy per chromosome for two species of woody plants showing correlation of radiosensitivity with interphase chromosome volume. (Courtesy A. H. Sparrow [1965] and Genetics Society of Japan.)

ganisms, the interphase chromosome volume seems to correlate most directly. The relationship can be demonstrated using the data from Figure 13.9 for the two species of woody plants. The per cent survival (probit) vs. absorbed energy per chromosome is plotted in Figure 13.11, using interphase chromosome volumes of 3.19 and 12.84 for *A. saccharum* and *S. canadensis*, respectively. Since the curves are so similar, it appears that the high radiosensitivity of *S. canadensis* is probably related to its larger interphase chromosome volume.

Growth inhibition of plants can be shown to result from a direct effect on the meristematic regions. In an experiment reported by Read (1959) the growth rate of *Vicia faba* roots was reduced by a single small radiation exposure. In order to decide whether this growth reduction was due to general damage to the plant or to local damage, specific parts of the plant were irradiated or shielded as indicated in Table 13.3. The results indicate that irradiation of the root tips (meristematic region) caused the growth inhibition. In another experiment cited by Read, an exposure of about 10,000 R was given to the elongation zone of the root. The growth inhibition which occurred was attributed to radiation scattered from the elongation zone into the meristem. Thus, it was concluded that root growth inhibition produced by radiation arises primarily from effects on the meristem. It is likely that both inhibition of synthesis of the growth-stimulating auxins (see later section) and inhibition of cell division in the meristem are involved in this effect. Radiation of the root meristem will also reduce the production of lateral rootlets on an irradiated root or on a new root formed from an irradiated meristem.

Table 13.3. Mean Root Growth of Vicia Faba During 10 Days After a Single Small Exposure to x-irradiation[a]

Treatment	Root growth (cm)
Whole plant irradiated	5.22
Only 5 mm of root tip irradiated	6.10
All except 5 mm of root tip irradiated	20.16
Total plant shielded	21.17
Untreated control	21.12

[a]From Read (1959).

GROWTH STIMULATION

Although high doses of ionizing radiation inevitably inhibit growth in plants, there are many reports of an apparent growth stimulation by small exposures (usually less than 5000 R). This may take the form of taller plants, longer roots, faster germination, increased leaf thickness, or earlier flowering (see Figure 13.12). The stimulatory effect has been noted in many species of plants. In some species, the effect cannot be repeated regularly, and in other species it is never seen.

Growth enhancement usually involves the early period of stem or root elongation and does not generally produce a significant change in the final

Fig. 13.12. *Nicotiana glauca x langsdorffii* exposed to gamma radiation at rate indicated (R/day) in gamma greenhouse at Brookhaven National Laboratory. Exposure started at first true leaf stage. Photograph taken after 147 days of exposure. Note the earlier flowering of 15 R/day plants. (Courtesy A. H. Sparrow and Brookhaven National Laboratory.)

fresh or dry weight yields.[4] It has been suggested that low doses of radiation may disturb the activity of certain enzymes involved in the synthesis of growth hormones (auxins). Considerably more work must be done in the area, however, before the mechanisms involved are clear.

Occasionally, there has been interest in the use of radioactive fertilizers in order to stimulate greater crop production. Although a few reports (primarily from the Soviet Union) have suggested that such a procedure is feasible, extensive field tests by the United States Department of Agriculture have failed to show a stimulatory effect with radioactive fertilizers. The use of radioactive fertilizers to stimulate plant growth is, therefore, not only prohibited for reasons of public health but is also highly questionable as to its supposed stimulatory effects.

MORPHOLOGIC EFFECTS

Irradiated plants may show morphologic and histologic changes in roots, stems, buds, leaves, or flowers. The type of response depends primarily on the level and duration of exposure, on the species, age, and physiological condition of the plants, and on the environment during and after exposure.

Root growth and the formation of new roots can be inhibited by suitable doses of radiation. Moreover, in some species, acute irradiation of local stem areas can cause the formation of new adventitious roots above the irradiated area. This abnormal root formation has been used occasionally to propagate plants in which root cuttings are difficult to obtain. It may be due to an imbalance of plant growth hormones produced by the radiation. Stem swelling, dwarfing, fasciation, and excessive branching of stems characteristically result from irradiation. Much of the branching is apparently due to growth of extra axillary buds which result from destruction of auxin in the plant (see p. 303) rather than from direct bud stimulation by x-rays. The decrease in auxin level may either initiate new meristem formation or permit growth in inhibited meristems.

The leaves which are already present at the beginning of a sublethal chronic exposure show relatively little change other than thickening, coarseness, or drying. Young leaves, however, may show dwarfing, asymmetrical development or thickening of the leaf blade, distorted venation, a change in texture, or mosaic-like color changes. Leaf thickening in the leaves of *Antirrhinum* (snapdragon) was shown to increase progressively with increasing doses up to more than twice the control thickness at 600 R/day. This thickening was associated with increased cell division, cell enlargement, and a pulling apart of the cells. Similar changes result from acute exposure.

Flowering is generally retarded by radiation, although the flowering of some plants is stimulated by low doses (see previous section). Fasciation

[4]Several attempts to use this commercially have been abandoned or were inconclusive.

Fig. 13.13. Effects of exposure of dormant tulip bulbs to acute x-irradiation. Note abnormal petals at the higher dosage. (Courtesy A. H.
Sparrow and Brookhaven National Laboratory.)

(branching) in the flower head of sunflowers, snapdragons, and other species
has been described. Frequently, there is a modification in the form and number of flower parts, particularly of petals and stamens. Abnormalities commonly include modified petal lobes (Figure 13.13), increase or decrease in the
number of petals or stamens or portions of these, double and open ovaries,
multiple or fused styles, color changes, and dwarfing or enlargement of
flower parts. Inflorescences often show the formation of sterile flowers or
modified leaf clusters in the floral position.

TUMOR FORMATION

Spontaneous tumor formation in stems and floral stalks of plants has been
repeatedly described. Ionizing radiation has been shown to markedly increase the tumor incidence in certain species. Sparrow and co-workers (1956)
at the Brookhaven National Laboratory studied the tumors produced in a
hybrid of *Nicotiana* exposed to chronic gamma radiation. They noted that
control plants branched regularly at each node and that each auxiliary shoot
terminated in a flowering stalk (Figure 13.14A). Irradiated plants, in contrast, branched and flowered irregularly and a significantly larger percentage
showed more tumors than the controls as early as 37 days after the start of
irradiation (Figure 13.14B). The increase in number of plants with tumors
was accompanied by an increase in the amount of tumor per plant. For

Fig. 13.14. Hybrid tobacco plants showing response to chronic gamma irradiation. (A). Control. Note regular branching and terminal flower stalks. Tumors few and not apparent at this magnification. (B). Plant exposed to 210 R/day for 11 weeks. Short plants like this did not flower. Branching was infrequent and obscured by large, leafy tumor masses. The control plant (A) is much larger than the irradiated plant (B). (Courtesy A. H. Sparrow and Brookhaven National Laboratory.)

example, after 64 days exposure to 385 R/day, an average of 53% of the top wet weight was tumorous as compared to 0.1% in the controls. Slightly fasciated leafy tumors, compact smooth tumors, and tumors of the floral stalks were seen as well as other types. An increase in root tumors was also noted.

The specific changes which lead to the initiation of such radiation-induced plant tumors are still unknown. It is probable, however, that physiological disturbances such as alteration in growth hormones or nutritional factors may play a major role. Somatic mutation cannot be ruled out as a possible cause.

RADIATION EFFECT ON PLANT HORMONES

Reference is made throughout this chapter to the possible role of growth hormones in mediating the radiation response of plants. However, the nature and action of these compounds in normal plants are still not entirely clear.

Growth is controlled in plants by small quantities of specific organic compounds called *growth regulators, plant hormones,* or *phytohormones.* Some of the most commonly discussed phytohormones are the naturally occurring auxins, gibberellins, and kinins. Auxins are generally acids with an unsaturated cyclic nucleus. Indoleacetic acid (IAA) (Figure 13.15) is now accepted as the most common naturally occurring auxin in plants. Its primary effect appears to be on cell elongation, but it has also been shown to affect other processes in plants. It will, for example, at certain concentrations inhibit growth of lateral buds, especially those that form branches.

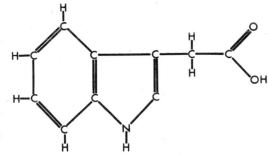

Fig. 13.15. Structural formula for indoleacetic acid, a naturally occurring plant auxin.

In 1935, Skoog showed that x-rays reduced the content of IAA *in vivo* and *in vitro.* Twenty years later, Gordon (1957) re-examined this phenomenon and found that the overall decrease in IAA was too great to be accounted for by radiation-induced oxidation of the auxin molecule alone. The effect seemed to be a disruption of the auxin-synthesizing enzymes. Gordon further showed that the enzyme which converts indoleacetaldehyde to indoleacetic acid was a critical part of the radiosensitive process.

Radiation sensitivity of the auxin synthesizing system may account for several of the typical responses of plants to radiation. The inhibitory effect of small doses on overall plant growth and the branching or fasciation which is reported after somewhat higher doses may be related to a decrease in auxin.

Some papers, primarily from workers in Russia, suggest that radiation may, under certain conditions, result in an increase in physiologically active substances, possibly by destruction of inhibitory substances. The controversial growth stimulation effect (see previous section) might be attributed to such a change.

Phytohormones and other growth regulators can also influence the response of irradiated tissues. For example, large doses (300 to 600 krads) of gamma radiation will cause a delay in the initiation of germination of lettuce seeds. This effect can be reversed by treatment of the irradiated seeds with gibberellic acid, another phytohormone. Radiation-inhibited geotropic response in corn and pea seedlings is reversed by auxin; and radiation-induced decay susceptibility of potato tubers is reversed by the methyl ester of indonyl 3-acetic acid. Much work remains to be done in this area before the interactions are clear.

RADIATION EFFECTS
ON PLANT COMMUNITIES

The surface of our earth may be characterized by natural groupings of vegetation and animals. Each grouping or community is, at its mature state, a well-structured, self-sustaining unit in which a variety of biological interactions contribute to a stable pattern of behavior. If the composition of an established community is substantially altered, there will be a succession of new developments toward the establishment of a new stable structure which may or may not resemble the original system. The character of a community is largely determined by the type of plants which are present.

One would expect two major types of radiation effects at the community level. There should be an increase in the frequency of deleterious mutations and a decrease in the survival and vigor of the irradiated organisms. However, despite the increase in mutations following acute or short-term irradiation, the overall genetic consequences may be of lesser importance than the acute effects on the organisms. Most of the mutants which would be formed would be those that occur spontaneously and would not be new to the population. They would be present in increased numbers. If the radiation exposure is of limited duration such that it does not produce a long-term change in the mutation rate (and if breeding is at random, and if selective forces within the ecosystem are not changed), then the incidence of a given mutant gene should become stabilized at the level determined by the pressures of natural selection, like that of any spontaneous mutation. There should be no long-term major increase in mutations. However, with continuing or chronic irradiation, the increased incidence of mutant genes would be sustained in the population.

It is likely that the major effect on a community is related to the survival and vigor of the irradiated organisms. One can predict grossly the radiosensitivities of the various populations in a community on the basis of laboratory studies. For example, the lethal exposure for most mammals lies between 200 and 1000 R; for insects between 1000 and 100,000 R; for flowering plants, between 1000 and 150,000 R; for microorganisms, between a few thousand

and a million R. One might expect, then, that the effect on the radiosensitive mammals and to some extent on the insects would determine the overall effect on the community. However, the animals and motile insects are dependent on the plants, the primary food producers of the community. They will move into or out of the area as the population of plants changes. Therefore, unless the mobility or availability of the animals is restricted naturally or artificially, most changes in communities are associated with alterations in the plant population.

There is known to be wide variation in the response of plants to radiation, and many species have not yet been studied. However, it is possible to estimate the probable radiosensitivities of the various members of a plant community. For example, it has been determined that herbaceous species are more radioresistant than woody species and that dormant plants are appreciably more resistant than the same plants when actively growing. In addition, the radiosensitivity of a plant species has been shown to be related to the interphase chromosome volume of the meristematic cells in the manner described by Sparrow (see Chapter 8). Using the established relationship with chromosome volume, Woodwell and Sparrow (1965) have estimated the exposures required to affect the major components of a variety of vegetations. For example, Table 13.4 lists the estimates for the major trees in one type of North American Eastern Deciduous system, the hemlock-hardwood forests. Average interphase chromosome volumes range from about 50 μ^3 for pines to about 2 μ^3 for some of the hardwoods.

Table 13.4. ESTIMATED ACUTE EXPOSURES REQUIRED TO AFFECT DOMINANTS IN THE NORTH AMERICAN, EASTERN DECIDUOUS HEMLOCK-HARDWOOD FOREST[a]

Species	Somatic chromosome number	Interphase chromosome volume ($\mu^3 \pm$ S.E.)	Sensitivity range: slight inhibition of growth to mortality (R)
Tsuga canadensis (hemlock)	24	21.3 ± 0.8	420–1100
Betula lutea (yellow birch)	84	2.2 ± 0.1	3860–10,120
Pinus strobus (white pine)	24	46.5 ± 2.8	190–500
Pinus resinosa (red pine)	24	43.2 ± 3.5	210–540
Acer saccharum (sugar maple)	26	3.2 ± 0.2	2800–7360

[a]Modified from Woodwell and Sparrow (1965). Estimates are based on correlation between radiosensitivity and interphase chromosome volume.

In general, the gymnosperms have much higher chromosome volumes and, therefore, presumably are more radiosensitive than the angiospermous trees. The estimated LD_{100} values for acute exposures range from about 500 R for pines through 10,000 to 13,000 R for birch, beech, and hickories. The forests that contain gymnosperms as major dominants would thus be more sensitive than deciduous forests; a few hundred to 2000 R should kill most trees in the coniferous forests, while about five times that exposure range would be required to kill most trees in hardwood forests. Agricultural areas and grasslands could probably be severely damaged by exposures in the range of 5000 to 10,000 R. The nature of the community, then, is a major determining factor in its response to radiation.

The amount of the radiation exposure will determine the overall effect on a given community. Low exposures may inhibit growth and reproductive capacity of sensitive species temporarily. Recovery should be rapid and there should be no change in the composition of the plants. Higher doses would cause a selective elimination of the more radiosensitive populations and reduce the diversity of species. There might then follow a temporary increase in the more resistant elements such as insects or bacteria, which are capable of rapid propagation. These might produce a secondary effect such as damage to plants from herbivorous insects. This effect should be short-lived and the damage minor. Within 2 or 3 years there should be established an orderly succession leading to an ecosystem basically similar to the system destroyed. Severe effects would be produced by massive exposures which are sufficient to reduce the capacity of the site for supporting life or greatly alter the types of organisms present.

Woodwell and Sparrow (1965) have estimated the radiation exposures required to produce these categories of effects on major natural communities (Table 13.5). Their estimates are based on the correlations between radiosensitivity and interphase chromosome volume and on limited experience with irradiated natural communities.

One may also consider the radiation sensitivity of man-dominated communities such as farming areas. Table 13.6 indicates the acute exposures

Table 13.5. ESTIMATED RADIATION EXPOSURES (R) REQUIRED TO DAMAGE MAJOR NATURAL COMMUNITIES[a]

	Level of damage		
Communities	Minor	Intermediate	Severe
Coniferous forest	200	200–2000	> 2000
Deciduous forest	200	200–10,000	>10,000
Grassland	2000	2000–20,000	>20,000
Herbaceous annuals	4000	4000–70,000	>70,000

[a]Modified from Woodwell and Sparrow (1965).

Table 13.6. PREDICTED LD$_{100}$ VALUES AND ACUTE EXPOSURES REQUIRED TO PRODUCE SLIGHT EFFECTS ON VEGETATIVE GROWTH OF VEGETABLE AND FIELD CROPS[a]

Species	Estimated interphase chromosome volume ($\mu^3 \pm$ S.E.)	Predicted exposures required to produce	
		Slight effects (R)	LD$_{100}$ (R)
Allium cepa (onion)	39.3 ± 2.3	377	1491
Triticum aestivum (wheat)	14.6 ± 1.1	1017	4022
Zea mays (corn)	14.0 ± 0.6	1061	4197
Solanum tuberosum (potato)	4.6 ± 0.3	3187	12,608
Oryza sativa (rice)	3.0 ± 0.1	4974	19,677
Phaseolus vulgaris (kidney bean)	1.6 ± 0.1	9137	36,149

[a]Modified from Sparrow *et al.* (1965).

which are predicted to produce slight growth inhibition and lethality for a number of vegetable and field crops. It is obvious that the range of relative sensitivities is very broad, and one might expect a selective removal of the more sensitive plants at exposures in excess of a few thousand R. Woodwell and Sparrow (1965) point out, however, that in cities and most agricultural communities the most sensitive dominant organism is man. The severity of the radiation effects on the system would be determined by his ability to survive and function normally. Exposures greater than 200 R would seriously affect men, and few, if any, would survive whole-body exposures in excess of 1000 R. Therefore, one would expect that the character of the community might be altered indirectly by radiation exposures greater than a few hundred roentgens.

The knowledge of the radiation effects on plant communities is not based entirely on theoretical calculations or laboratory observations. A number of natural plant communities have been exposed to radiation and the effects have been observed. Among these are an oak-pine forest, an old-field at the Brookhaven National Laboratory, and a forest stand around the Lockheed reactor in northern Georgia.

At the Brookhaven National Laboratory on Long Island, under the direction of Dr. Woodwell, a 9500 Ci cesium-137 source provides chronic gamma radiation, 20 hours a day, to a pine-oak forest. Rates of exposure vary from several thousand R per day within a few meters to about 1 R per day at 130 meters from the source.

In the spring of 1962, approximately 6 months after the start of exposure, effects were obvious as much as 40 meters from the source (40 R/ day). Differences in sensitivity among plant species produced five clearly defined zones of vegetation (Figure 13.16): A zone of total kill of all higher plants (>350 R/day; >63,000 R accumulated total); a sedge zone (150 to 350 R/day; 27,000 to 63,000 R total); a heath shrub zone (60 to 150 R/day; 11,000 to 27,000 R total); an oak zone (20 to 60 R/day; 3600 to 11,000 R total); and the oak-pine forest (background). At exposures less than 10 R/ day, there was reduced shoot growth of all tree species, but the trees did not die. Principal changes in insect populations followed changes in the abundance of food.

Fig. 13.16. Oak-pine forest at Brookhaven National Laboratory after exposure to chronic gamma irradiation for 6 months. All vegetation near the source has been destroyed; progressively radiosensitive species are in zones at increasing distances from the source. (Courtesy G. Woodwell and Brookhaven National Laboratory.)

An established field containing herbaceous annuals was also exposed to daily gamma irradiation at the Brookhaven National Laboratory. The exposure required to reduce diversity to 50% was about 1000 R/day, or a total of about 100,000 R at the time of study. This represents more than five times the exposure required to produce a similar effect in the pine-oak forest and indicates that the herbaceous annuals in the field were considerably more resistant than the forest vegetation.

The effect of a series of intermittent acute exposures was studied around an unshielded (air shielded) reactor in Northern Georgia. In such a facility, neutrons and gamma radiation are released from the reactor without heat or blast. Within 1 week after the first large burst of radiation was released into the environment in 1959, pine trees which had received doses of 2000 rads or more began to turn brown and in a few weeks were completely dead. As other pines at greater distances accumulated doses of about 7000 rads, they died in 90 to 120 days. No other vegetation showed immediate changes, but the following spring it was apparent that almost all terminal buds of the hardwood trees had been killed. As in the Brookhaven Forest, pines were considerably more radiosensitive than other vegetation.

These experimentally irradiated communities are neither sufficiently large nor sufficiently isolated to restrict the movement of animals into or out of the experimental area. Accordingly, the distribution of animals can be assumed to have been regulated by the relative availability of food within or outside the area.

CYCLING OF RADIONUCLIDES
WITHIN A COMMUNITY

The preceding discussion has considered the effects of external radiation sources on the plants in a community. Additional factors must be considered if radionuclides are introduced into a system either as fallout from a nuclear detonation, effluent from a reactor, or accidental or deliberate contamination of an area. In such a situation, the animals of the community must be considered as well as the plants.

A radionuclide will behave in the same manner as its stable isotope; for example, iodine-131 will follow the same path as stable iodine; strontium-90 will behave in the system as would stable strontium. Therefore, any element which is taken up or metabolized by organisms will be similarly taken up in its radioactive form. The radiation from such internally contained radionuclides will then contribute to the radiation exposure of the organisms. It is necessary to know the extent of uptake in order to evaluate the radiation exposure. Perhaps this can be best explained by a few examples.

Any nuclide (radioactive or not) in a community will be distributed according to its metabolic behavior in the various components of the food chain through which it passes. Some organisms may concentrate it; some may discriminate against it. This is illustrated in Table 13.7 which gives the approximate concentration factors of several elements in various marine organisms. Certain elements (such as phosphorus) are present in the organisms in greater concentrations than they are in the seawater. The marine organisms are able to concentrate these elements to the extent necessary to

Table 13.7. APPROXIMATE CONCENTRATION FACTORS OF DIFFERENT ELEMENTS IN MEMBERS OF THE MARINE BIOSPHERE[a]

Element	Concentration in seawater $\mu g/l$	Concentration Factors[b]				
		Algae (noncalcareous)	Invertebrates		Vertebrates	
			Soft	Skeletal	Soft	Skeletal
Na	10^7	1	0.5	0	0.07	1
Sr	7000	20	10	1000	1	200
P	70	10,000	10,000	10,000	40,000	2,000,000

[a]Modified from Eisenbud (1963).
[b]Concentration factor is the ratio of the concentration of an element in an organism to the concentration in seawater, (based on liveweight measurement.)

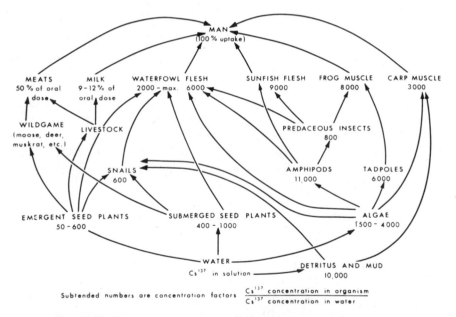

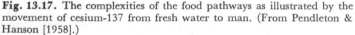

Fig. 13.17. The complexities of the food pathways as illustrated by the movement of cesium-137 from fresh water to man. (From Pendleton & Hanson [1958].)

meet their metabolic requirements. Other elements (such as sodium) are required in the same or even lower concentrations than in seawater and are not concentrated.

A very complex pathway may be followed by a trace element in passing from the water in an aquatic system to man. This is shown in Figure 13.17, which illustrates the movement of cesium-137 from water to man through a

Pathways from Fallout Through Food Chain
STRONTIUM 90

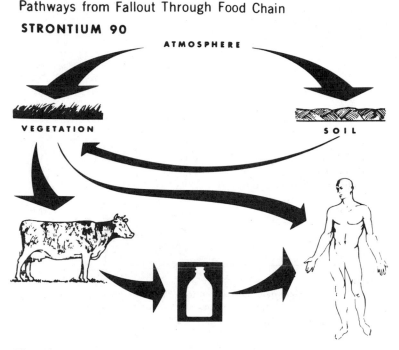

Fig. 13.18. Pathways of strontium-90 through the food chain. (Courtesy C. L. Comar.)

variety of plants and animals, each of which concentrates it to a different extent. Each radionuclide has its own path through the food chains. Accurate evaluation of the radiation dosage received by each organism in the chain requires a knowledge of the paths and of the concentration factors at each step.

Some radionuclides are incorporated into an organism in a way that is governed by another element, usually one that is essential and similar chemically. For example, the movement of strontium is related to the simultaneous movement of calcium. As illustrated in Figure 13.18, strontium-90 reaches man primarily through his consumption of dairy products and foods of plant origin. In practically every step of the food chain, calcium is preferentially utilized relative to strontium. The degree of this discrimination is important. Table 13.8 gives some typical data on the stable strontium and strontium-90 levels relative to calcium in several constituents of the food chain. Consider first the ratio of stable strontium to calcium (note that all values are normalized to a value of 1 for total diet). The total diet actually contained 1059 mg/day calcium and 1.61 mg/day stable strontium. Cattle feed had twice as much strontium relative to calcium as did the diet of man. But the milk produced by the cattle was at a relative level of 0.2. This, there-

Table 13.8. STABLE STRONTIUM TO CALCIUM, AND STRONTIUM-90 TO CALCIUM RATIOS OF FOOD CHAIN ELEMENTS AS MEASURED IN NEW YORK CITY IN 1965[a]

	Relative contribution to dietary calcium	Ratio stable Sr/Ca[b]	Ratio Sr-90/Ca[b]
Cattle feed		2.0	(7.5)
Foodstuffs			
milk	62%	0.2	0.75
grain	17%	1.4	1.6
root vegetables	3%	4.5	1.9
leafy vegetables and fruits	9%	3.7	2.0
other	9%	1.7	0.5
Total diet	100%	1.0	1.0
Man		0.25	(0.25)
Mother's milk		0.1	(0.1)
Newborn		0.1	(0.1)

[a]Data courtesy of C. L. Comar.

[b]Each column separately normalized to a value of 1 for total diet. Values in parentheses represent estimates based on discrimination factors.

fore, represents a discrimination factor of 10 from the diet of the cow to the milk produced. Likewise, in man there is a discrimination factor of 10 for milk production (1.0 total diet to 0.1 mother's milk) and for transfer from the mother's diet to the fetus (1.0 to 0.1). Between diet and body of man there is a factor of 4 (1.0 to 0.25). Whereas milk contributes 62% of the total calcium of the diet, it contributes only a small proportion (12%) of the stable strontium because of the low Sr/Ca ratio (due to discrimination in the cow). Because of the relatively high strontium content in other dietary constituents, they contribute a large proportion of the total stable strontium.

One would expect that the same Sr/Ca ratios would be found with strontium-90 as with stable strontium. However, they are not always the same as demonstrated in the last column in Table 13.8. This represents relatively short times after the occurrence of contamination before equilibrium of the strontium isotopes is reached. The proportion of strontium-90 (relative to calcium) in superficial vegetation, such as cattle feed, is higher than the proportion of stable strontium. Because of this variation, and because the same discrimination factors apply for the various strontium isotopes, the proportion of strontium-90 which is ingested in milk is high. Therefore, paradoxically, even though milk is the greatest overall contributor of strontium-90, reducing the intake of milk tends to increase the ^{90}Sr/Ca of the intake and, therefore, increases the body burden of strontium-90 to be attained. In time, one would expect to approach the situation where strontium-90 is distributed in the same way as stable strontium. Then, plant sources would become increasingly important as the dietary source of strontium-90.

These examples illustrate the complexity of most food chains and the vast amount of information which must be considered before one can estimate the quantity of a radionuclide which is deposited in an organism as a result of environmental contamination. If the amount of radionuclide within the organism is known, then the radiation dose and the biological effect on the organism can be estimated on the basis of factors such as the distribution of the nuclide within the organism, the type of radiation which is emitted, the time that the nuclide remains in the organism, and the relative sensitivity of various tissues.

GENERAL REFERENCES

Gunckel, J. E., and Sparrow, A. H. Aberrant Growth in Plants Induced by Ionizing Radiation, pp. 252–279 in *Abnormal and Pathological Plant Growth*, Brookhaven Symposia in Biology No. 6. Brookhaven National Laboratory, Upton, New York (1954).

Gunckel, J. E., and Sparrow, A. H. Ionizing Radiations: Biochemical, Physiological and Morphological Aspects of Their Effects on Plants, pp. 555–611 in Ruhland, W. (ed.), *Encyclopedia of Plant Physiology*, Vol. XVI. Springer-Verlag, Berlin (1961).

IAEA. Symposium on the Effects of Ionizing Radiation on Seeds and Its Significance for Crop Improvement, Proc. Karlsruhe, Germany (1960).

IAEA. The Use of Induced Mutations in Plant Breeding. *Suppl. Radiation Botany*, **5** (1965).

Platt, R. B. Ecological Effects of Ionizing Radiation on Organisms, Communities and Ecosystems, pp. 243–255 in Schultz, V. S., and Klement, A. W., Jr. (eds), *Radioecology*. Reinhold Publishing Corp., New York, and AIBS, Washington, D. C. (1963).

Romani, R. J. Biochemical Responses of Plant Systems to Large Doses of Ionizing Radiation. *Radiation Botany*, **6:** 87–104 (1966).

Sparrow, A. H., Binnington, J. P., and Bond, V. *Bibliography on the Effects of Ionizing Radiations on Plants, 1896–1955*, BNL (504). Brookhaven National Laboratory, Upton, New York (1958).

Sparrow, A. H., and Konzak, C. F. The Use of Ionizing Radiation in Plant Breeding: Accomplishments and Prospects, pp. 425–452 in Tourje, E. C. (ed.), *Camellia Culture*. The Macmillan Co., New York (1958)

Woodwell, G. M. Effects of Ionizing Radiation on Terrestrial Ecosystems. *Science*, **138:** 572–577 (1962).

ADDITIONAL REFERENCES CITED

Biebl, R., and Mostafa, I. Y. Water Content of Wheat and Barley Seeds and Their Radiosensitivity. *Radiation Botany*, **5:** 1–6 (1965).

Brewbaker, J. L., and Emery, G. C. Pollen Radiobotany. *Radiation Botany*, **1:** 101–154 (1961).

Curtis, H. J., Delihas, N., Caldecott, R. S., and Konzak, C. F. Modification of Radiation Damage in Dormant Seeds by Storage. *Radiation Res.*, **8:** 526–534 (1958).

Ehrenberg, A. Research on Free Radicals in Enzyme Chemistry and in Radiation Biology, pp. 337–350 in Blois, M. S. *et al.* (eds.), *Free Radicals in Biological Systems.* Academic Press, Inc., New York (1961).

Eisenbud, M. *Environmental Radioactivity.* McGraw-Hill Book Co., New York (1963).

Gordon, S. A. The Effects of Ionizing Radiation on Plants: Biochemical and Physiological Aspects. *Quart. Rev. Biol.*, **32:** 3–14 (1957).

Mericle, L. W., and Mericle, R. P. Radiosensitivity of the Developing Plant Embryo, pp. 262–286 in *Fundamental Aspects of Radiosensitivity*, Brookhaven Symposia in Biology No. 14, Brookhaven National Laboratory, Upton, New York (1961).

Pendleton, R. C., and Hanson, W. C. Absorption of Cesium-137 by Components of an Aquatic Community, pp. 419–422 in *U. N. Intern. Cong. Peaceful Uses At. Energy 2nd, Geneva*, **18** (1958).

Read, J. *Radiation Biology of* Vicia faba *in Relation to the General Problem.* Charles C. Thomas, Springfield, Illinois (1959).

Skoog, F. The Effect of X-radiation on Auxin and Plant Growth. *J. Cellular Comp. Physiol.*, **7:** 227–270 (1935).

Sparrow, A. H. *The Role of the Cell Nucleus in Determining Radiosensitivity.* Brookhaven Lecture Series No. 17, BNL-766, Brookhaven National Laboratory, Upton, New York (1962).

Sparrow, A. H. Comparisons of the Tolerances of Higher Plant Species to Acute and Chronic Exposures of Ionizing Radiation, pp. 12-37 in Mechanisms of the Dose Rate Effect of Radiation at the Genetic and Cellular Levels. *Suppl. Japan J. Genet.*, **40** (1965).

Sparrow, A. H., Gunckel, J. E., Schairer, L. A., and Hagen, G. L. Tumor Formation and Other Morphogenetic Responses in an Amphidiploid Tobacco Hybrid Exposed to Chronic Gamma Irradiation. *Am. J. Botany*, **43:** 377–388 (1956).

Sparrow, A. H., Sparrow, R. C., Thompson, K. H., and Schairer, L. A. The Use of Nuclear and Chromosomal Variables in Determining and Predicting Radiosensitivities, pp. 101–132 in The Use of Induced Mutations in Plant Breeding, *Radiation Botany*, Suppl. 5 (1965).

Sparrow, A. H., and Woodwell, G. M. Prediction of the Sensitivity of Plants to Chronic Gamma Irradiation. *Radiation Botany*, **2:** 9–26 (1962).

Woodwell, G. M., and Sparrow, A. H. Effects of Ionizing Radiation on Ecological Systems, pp. 20–38 in *Ecological Effects of Nuclear War*, BNL-917, Brookhaven National Laboratory, Upton, New York (1965).

APPLIED
RADIATION
BIOLOGY

14

Preceding chapters have dealt with the interaction of ionizing radiations with biological systems — with the immediate and long-term effects of small and large doses on many different levels of biological organization. In this chapter another phase of radiation biology will be discussed briefly; namely, the role of radiation in the world today: Where and how is radiation being used; how is it contributing to our industrial, medical, and scientific progress; of what magnitude are the resulting radiation exposures; and what effects on man can be expected to result from them?

USES OF RADIATION

In the past few decades it has become evident that ionizing radiation has many practical applications. Radionuclides and x-rays are used in medicine, industry, and agriculture. Many recent advances in biological, medical, and environmental research would not have been possible without the use of radionuclides. Basic physical research also utilizes sources of ionizing radiation. A few of the applications of ionizing radiation are described in the following sections.

MEDICAL USES OF RADIATION

Radiation is a valuable tool in medicine. It is used both for the diagnosis of disease or improper body function and for therapy of existing malignant or

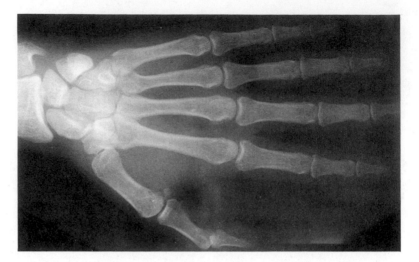

Fig. 14.1 Radiograph of human hand showing bone as white image and soft tissues in various shades of gray. (Courtesy E. C. Showacre.)

nonmalignant conditions. Until recently, x-rays have been the chief source of radiation for medical application, but the use of radionuclides is now extensive and is increasing.

Diagnosis. Radiation can be used to visualize certain body structures by passing a beam of x-rays through the body tissue to a photographic plate. The degree of blackening of the developed plate is related to the amount of radiation which impinges on it. Dark areas on the film correspond to "soft tissue" which has not scattered or absorbed the radiation; light areas on the film correspond to dense or radio-opaque tissue which absorbs or scatters the radiation. Organs and structures with different radio-opacities can thus be visualized against each other as shadow pictures. For example, bone, a strong absorber, is seen as a bright image on a radiograph (see Figure 14.1). Soft tissues, which are much less absorbent, appear as a variety of gray shades. Changes in size, shape, or density of either bone or soft tissue can often be detected and related to a known disease process.

Additional techniques can sometimes improve the diagnostic process. For example, a highly radio-opaque substance such as barium sulfate can be introduced into the gastrointestinal tract before an x-ray is taken. The digestive tract then shows up clearly on the film as a well-defined light region, and obstructions may appear as a narrowing of a segment of the tract. The passage of the barium through the gastrointestinal tract can also be followed by repeated photographs or by a series of viewings on a fluoroscope screen (direct visualization of the x-ray image instead of photographs). The rate of movement of the barium can then be determined and intestinal obstructions can be detected.

Radionuclides also can be used in medical diagnosis. When certain elements are injected into the body they localize in specific tissues. If a radioisotope of the element is used, its presence can be detected by external counting of the emitted radiation. The pattern of localization of the element can be determined and correlated with possible disease processes.

An example of this technique is the use of iodine-131 in the visualization of the size, shape, and iodine-concentrating ability of thyroid tissue. Patients are given a small amount of radioactive sodium-iodide solution to drink (this is the so-called "atomic cocktail"). Several hours later the gamma radiation from thyroid-contained radioiodine is measured. Total uptake can be measured by a count of the entire thyroid or the distribution of the radioiodine can be determined with a linear photoscanner. This scanner is collimated to record impulses from only a very small area. It is moved back and forth over the organ to give a visual map of the relative activities at intervals over the field. The size and densities of the active areas can then be correlated with organ size and with iodine-concentrating ability. As illustrated in Figure 14.2, the pattern of iodine uptake is obviously very different in a normal thyroid (A), an enlarged thyroid (B), and a cancerous thyroid (C).

A similar technique is used to locate fragments of cancerous tissue which migrate to other parts of the body (metastatic cancers). If thyroid metastases are capable of concentrating iodine (less than 10% of them are), they can be located by scanning the whole body in the manner that was just described.

Iodine-131 has been incorporated into a dye, rose bengal, which is used for liver function tests. When this dye is injected into a vein, it is removed from the blood by the liver. The rate of disappearance of the dye from the blood is determined by the decrease in count rate and is used as a measure of liver activity. Radioiodine is also used to locate brain tumors or to label blood serum albumin for measurement of the volume of circulating plasma.

The body localization of many other radionuclides has been utilized for a variety of diagnostic tests. Arsenic-74 or phosphorus-32 will concentrate in certain brain tumors, and hence can be used in scanning the head for the presence of brain tumors. Chromium-51 attaches to red blood cells and is used to measure total red blood cell volume, red cell life time, or cardiac output. Vitamin B_{12} contains cobalt; hence cobalt-60 incorporated into Vitamin B_{12} can be used to study pernicious anemia. Iron-59 is used to measure red blood cell production or gastric absorption of iron in studying the cause of low body iron, which is sometimes referred to as *tired blood* in drug advertisements.

Many other radionuclides are also used as diagnostic tools and the number of applications is continually expanding. In each case their usefulness depends on the fact that the activity in an organ or tissue can be measured by external counting of the radiation emitted from the site of localization. In all cases advantage is taken of the exquisite sensitivity of radiation measuring devices so that the radiation dose to the tissue is negligible.

Therapy. Radiation has another important role in medicine in the treatment of disease, particularly cancer. The doses used in radiation therapy are much larger than in diagnostic applications. Although any cell can be killed by a sufficient amount of radiation, advantage is taken in radiation therapy of techniques for selective irradiation of tumor tissues and of the increased radiosensitivity of some rapidly dividing tumor cells.

Whenever a radionuclide can be found that will selectively localize in a tumor, it is potentially possible to use the nuclide to irradiate (and possibly destroy) the cancerous growth. Large amounts of iodine-131 (in the millicurie range) have successfully destroyed certain types of thyroid tumors. This treatment is especially valuable when there are metastases of the type which will concentrate iodine. The radioiodine will "seek out" the cancerous fragments wherever they are located and, as a result of the preferential uptake of the element, will selectively irradiate the thyroid tissue. Of course, normal thyroid tissue is also destroyed as a consequence of this technique.

A new technique which may be very useful in the future is the incorporation of a radionuclide into a molecule which is selectively taken up by a tissue. For example, antibodies against certain tumor tissues can be made and then a radionuclide can be incorporated into the antibody molecules. When injected into an individual who has the tumor, the molecules will localize in the cancerous tissue and the attached radionuclide will provide local irradiation.

As another application of radionuclides in radiation therapy, boron has been used experimentally in the treatment of inoperable brain tumors. This element, when injected into the body, will localize in brain tumors but not in normal brain cells. When the head of a boron-injected individual is irradiated with a beam of slow neutrons (from a nuclear reactor), the boron-10 absorbs the neutrons, becoming boron-11 and almost immediately gives off an alpha particle [$^{10}B(n,\alpha)^7Li$]. The tumor thus receives local alpha irradiation.

Beads and needles containing radionuclides such as radium-226, gold-198, or yttrium-90 have been implanted into tumors to provide internal localized irradiation. They are usually left in place until the desired dose is delivered.

Applicators containing a concentrated radionuclide (such as strontium-90) or high-intensity sources of cobalt-60 or cesium-137 are used as external radiation sources. X-rays are used in a similar way. Beta-emitting radionuclides or low-energy x-rays are suitable for local exposure of a portion of the body surface and have been used successfully in the treatment of certain skin tumors.

The treatment of internal tissues by external radiation sources is complicated by the presence of overlying skin and tissue. If a beam of radiation is directed at a tumor, the skin and superficial tissues receive a high dose relative to the tumor. This problem has been partially solved by the use of rotational therapy. High-energy radiation is used that will give a reasonable dose

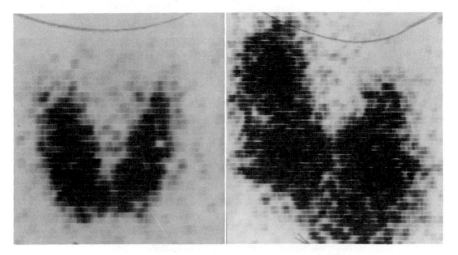

(A)   (B)

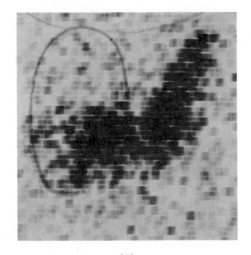

(C)

Fig. 14.2. Thyroid images produced by linear photoscan in individuals given iodine-131 for diagnostic tests. These represent: (A) a normal thyroid, (B) an enlarged thyoid, and (C) a cancerous thyroid. (Courtesy T. C. Evans.)

rate at the site of the internal tumor. The patient is rotated in the radiation beam so that the tumor is always being irradiated, but any particular portion of the skin is exposed for only a fraction of the treatment time. The tumor dose is thus maximized relative to the skin dose. Conversely, the radiation source may be rotated around the patient with the same effect. Instead of complete rotation, the exposure may be delivered from two, three, or more directions. The nature and location of the tumor determine the type and length of exposure.

Radiation doses in the range of hundreds or thousands of rads are used for most treatments. Often this much radiation would cause severe injury to the individual if delivered in a single exposure. Therefore, most radiation treatments are delivered as a series of doses over a total period of days or weeks. The fractionation schedule takes into account the recovery ability of normal and tumor tissues.

INDUSTRIAL USES OF RADIATION

A multitude of industrial applications of radionuclides have been developed. A few of these will be mentioned to demonstrate the wide variety of potential uses of radiation and to illustrate some of the sources of potential hazard to living things should these uses not be kept well under control.

Power. The development of power from radiation is perhaps one of the most important industrial applications. During the past 10 years, a number of nuclear reactors have been built to convert the energy released in the fission of uranium-235 or plutonium into usable electric power. As fission products and neutrons released in the fission process collide with surrounding molecules, their kinetic energy is partially converted into heat. This heat, which is removed from the reactor core by liquid or gaseous coolants, is used to generate steam, turn a turbine, and produce electricity. The reactor thus substitutes for the traditional boiler in conventional steam-electric power plants. Atomic power plants promise to be of particular benefit in sections of the world in which the cost of fossil fuels (coal, oil, or gas) is high. They will also help to curb the rate of depletion of the fossil fuel reserves which are now being exhausted at a rapid rate. Atomic power plants produce none of the common air pollutants which are released by large coal- or oil-burning installations. With atmospheric pollution being a major problem as power demands rise, the substitution of nuclear units is being given serious thought. Current designs make release of radioactivity from these plants a most unlikely event.

Reactors are also used for propulsion. In this country they have provided power for many types of ships such as the submarines *Nautilus*, *Triton*, USS *Skipjack*, and USS *George Washington;* the merchant ship *Savannah;* the destroyer USS *Bainbridge;* the cruiser USS *Longbeach;* and the aircraft carrier USS *Enterprise*. Increased range and cruising speed, as well as greater capacity for sustained submersion for submarines, have been demonstrated. A very practical use of nuclear power was shown by the Soviet Union's nuclear-powered icebreaker, *Lenin*. Ships of this type should help to overcome the difficulties of all-year navigation in the polar regions. There is one major disadvantage of the use of reactors in ships — at present the capital costs of nuclear propulsion equipment are substantially higher than those of conventional diesel equipment. Eventually, it is expected that the balance will shift in favor of nuclear propulsion.

Reactors have not as yet found extensive applicability for manned aircraft. Their use in vehicles for space travel, however, promises to be of great importance. Under the SNAP Program (Systems for Nuclear Auxiliary Power), the U.S. Atomic Energy Commission has developed a series of devices to supply power for a variety of space (and terrestrial) uses. There are two different types of devices, both of which convert heat into electricity. In one system, the heat is obtained from small nuclear reactors; in the other, it comes from the decay of certain radionuclides such as plutonium-238, polonium-210, strontium-90, or cesium-137. Since it has been possible to make such devices which are relatively compact, light weight, durable, reliable (functioning without human attention), and long-lived, their potential use is great in such varied power-requiring situations as space travel, orbiting satellites, weather stations in out-of-the-way places, and navigational buoys. It is interesting to note, parenthetically, that such systems have a major advantage in space travel over another recently developed source of power — the solar cell, which operates on the energy from sunlight. Solar cells are often damaged by radiation during passage through bands of high radioactivity surrounding the earth (Van Allen belts). In addition, as satellites carrying solar cells move toward the sun, the extra heat absorbed reduces the cell's efficiency. Furthermore, as a space craft moves away from the sun, the intensity of solar energy drops. Also, of course, during lunar and planetary nights and under opaque atmospheres, as on Venus, there is no sunlight and solar-powered fuel cells would be useless.

Nondestructive testing. Many industries have found that radiation is a valuable tool for various methods of nondestructive testing; that is, the examination of the structure or composition of an object without impairing its future usefulness. It might be helpful to think of diagnostic x-irradiation as a medical application of nondestructive testing. Dental x-ray pictures are taken to show structural changes (caries) in the teeth. The teeth are not harmed in the process.

Objects of many sizes and shapes can be radiographed to show structural flaws. Cracks or holes will not absorb or scatter as much of the radiation as, for example, the solid portion of a pipe or motor. The cracks will appear on a film as dark spots against the light "solid" areas. Thick or very dense objects require the use of high-energy radiation; thin or less dense objects can be tested with low-energy sources.

Radiation can be used to gauge the thickness of sheets of paper or metal (see Figure 14.3). A radiation source is placed on one side of the sheet and a radiation detector on the other side. A particular count rate

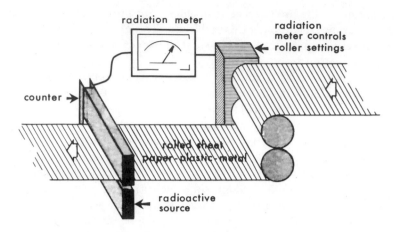

radiation meter

radiation
meter controls
← roller settings

counter →

rolled sheet
paper-plastic-metal

← radioactive
source

Fig. 14.3. Method of gaging the thickness of a sheet of metal, paper, etc., using a radioactive source. (Courtesy C. Shilling.)

will be associated with a desired thickness of the tested material. If the material is thinner, the count rate will increase since less of the radiation will be absorbed or scattered; if the material is thicker, the count rate will go down. This technique has been used on assembly lines to measure thickness of material as it is formed. In a refinement of the technique, an electronic signal from the detector may be fed back by a servomechanism to control the thickness of the sheet by changing a roller setting. Thus the device not only can measure and monitor the thickness of material continuously but can also control it.

Irradiation of food. In the food industry, radiation may be used in three ways: (1) To prevent sprouting of root crops, such as potatoes (see Figure 14.4); (2) to eliminate insects from grain before storage; and (3) to preserve food by inhibiting or destroying bacteria and other microorganisms. The amount of radiation required for each of these effects differs greatly. A dose of 1000 to 4000 rads is highly effective as a sprout inhibitor when applied to onions or potatoes. Grains and cereal can be disinfested of insects at 20,000 to 50,000 rads, and at 50,000 rads it is possible to sterilize the larvae of insects that lodge inside fruits. Pasteurization doses, generally in the range from 200,000 to 500,000 rads, will prolong shelf life or storage time. For example, the refrigerated storage life of fresh fish can be extended up to 30 days by such doses. Much higher doses, between 2 and 4.5 million rads are required for "sterilization" of foods for long-time storage without refrigeration. Bacon and other pork products, chicken, and beef can be prepackaged and irradiated at 4.5 million rads. These meats are reported to be well preserved (but not particularly palatable) after a year's storage at room temperature following the treatment.

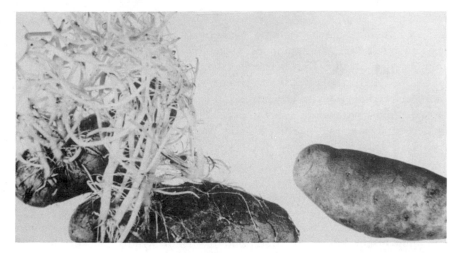

Fig. 14.4 Demonstration of the effectiveness of radiation in preserving foods. Both potatoes are the same age. The potato on the right was irradiated. (Courtesy USAEC.)

There are, however, certain problems associated with food sterilization that have not yet been solved. Chemical changes may be produced in foods which can result in unpleasant flavors, loss of color, or change in texture. For example, lemons become spongy and dairy products develop an odor. The latter is probably from peroxides formed in the milk and butterfat. Cheese is reported to become rancid faster after treatment. The vitamin content of some foods may also be decreased. (The vitamin C content declines by 50% in lemons subjected to 200,000 rads or more.) Many of these changes are apparently associated with oxidation since, if oxygen is removed from the system, the flavor and odor are much less affected. Furthermore, many of these changes will disappear during postirradiation storage; the unpleasant flavor may disappear and color return. It is possible that improved techniques of irradiation will minimize these problems.

There is also a question as to whether toxic products are produced in certain foods during irradiation. Preliminary studies have shown that irradiated juices contain a material which can produce chromosomal aberrations in certain plants. However, exhaustive studies have demonstrated no toxic effect in mammals which were fed irradiated foods.

Insect eradication. Radiation has been used successfully to control populations of certain insects. Males are raised in the laboratory and irradiated with a dose which renders them sterile but not impotent. They are then released into the environment in large numbers. A normal female mating with one of these males produces infertile eggs. Continued introduction of sterile males into the population for a few generations will drastically reduce the total number of this species of insect.

The eradication of the screwworm by this method was studied on an isolated island (Curaçao in the Netherland Antilles) which had a high natural infestation (see Baumhover *et al.*, 1955). The species was completely eliminated from the 170 square mile island by systematic weekly releases of males rendered sterile but not impotent by gamma rays. A similar success was achieved against the screwworm in Florida. The success with the screwworm stimulated work on other insects. The Mediterranean fruit fly, white-pine weevil, and the Oriental fruit fly have all been studied with varying results.

USE OF RADIATION IN RESEARCH

There are numerous examples of experimental situations in which an understanding of the effects that radiation has an a system can contribute to knowledge of fundamental mechanisms. For example, without the use of radiation to increase the spontaneous mutation rate in *Drosophila*, many productive genetic studies would not have been possible (see Chapter 6). Studies of the types of abnormalities produced by irradiation of developing embryos have demonstrated the developmental time sequence which takes place during normal organogenesis (see Chapter 10). In chemistry, physics, and engineering research, radiation has also been widely used. Radiation effects on the rate of chemical reactions can contribute to an understanding of the processes. The effect of strain produced in metals can give basic information on structure.

Perhaps the most widespread and unique research application of radiation is the use of radionuclides as tracers. In this application, use is made of the ability of instruments to detect and identify very small quantities of radionuclides. A molecule, cell, or even animal can be "tagged" or "labeled" with a suitable radionuclide, and the movement of the label can be followed within an animal, plant, or entire ecosystem. Several examples of such techniques illustrate this application.

Tritium, carbon-14, and phosphorus-32 have been most extensively used as tracers in biological research since the corresponding stable isotopes are present in practically all important cellular components. It is, therefore, possible to label molecules that participate in many important cellular processes. Organic chemists have been able to incorporate radionuclides into many molecules during synthesis. Other molecules will incorporate a label by exchange reactions; that is, nonlabeled molecules will exchange a stable isotope for a radioactive isotope (stable hydrogen for tritium) under proper conditions.

Other molecules which cannot be manufactured can be synthesized and labeled biologically. Plants are particularly useful for synthesis of labeled compounds because of their ability to incorporate carbon-14 from $^{14}CO_2$ into

Fig. 14.5. Plant growth chamber at Argonne National Laboratory used for the production of "labeled" plant products. Radioactive carbon-14 in the form of CO_2 is supplied to the plants in the sealed chamber. By photosynthetic processes, the plants produce many ^{14} C-containing products. (Courtesy Argonne National Laboratory.)

complex molecules through the process of photosynthesis. A growth chamber which is used for these preparations is shown in Figure 14.5.

Studies of the replication of DNA. Studies of the synthesis of DNA have made use of thymidine which has some of its hydrogen (^{1}H) replaced by tritium (^{3}H) (usually called *tritiated thymidine*). The radioactive thymidine is incorporated into nuclear DNA only during replication of the DNA prior to mitosis. Therefore, when autoradiograms are prepared of cell populations which have been exposed to tritiated thymidine, it is possible to identify the cells in which DNA replication occurred after the introduction of the radionuclide. Furthermore, the ^{3}H can be seen to be associated with chromosomes in those cells which are in metaphase or anaphase at the time of examination (see Figure 3.9). There are many extensions of this basic technique. For example, cells can be "pulse labeled" by exposing them to ^{3}H for only a short period of time. Then the cells which synthesized DNA during the pulse time can be followed and identified by the label.

Similar methods can be used to identify RNA in cells and to measure its rate of formation. The nucleoside uridine is a precursor of RNA that is predominantly, although not exclusively, incorporated into RNA. It can be labeled with either ^{3}H or ^{14}C and its incorporation measured. Labeled amino acids such as ^{14}C-leucine are used to measure protein formation in similar experiments. These examples illustrate the extremely valuable contribution that radionuclides can make to our understanding of cellular processes. The possibilities for use are endless.

Agricultural research. Radionuclides have played a major role in recent studies in the subject area of soil-plant relationships. For example, the uptake of different radionuclides from soil by growing plants has been measured using different fertilizers, soil compositions, or moisture contents. Other experiments use fertilizers which are labeled with radionuclides. Their pattern and rate of movement into plants are followed. Topical application of labeled nutrients to plant foliage has been studied to determine the feasibility of this as a more efficient route than soil fertilization.

Metabolic pathways in animals have also been extensively studied with the use of radionuclides. For example, the movement of calcium from diet into milk has been determined in cows. The incorporation of dietary phosphorus into bone and the extent of placental transfer of iodine have been measured.

Radiation has been used very successfully in plant-breeding studies. While most radiation-produced mutations are deleterious to the organism, a few are beneficial either to the organism or to man's use of the organism. If large plant populations are irradiated, it is often possible to isolate a few mutants in the next generation which have an increased resistance to disease, mature earlier, or yield larger crops. Strains of oats with increased winter-hardiness have been isolated from irradiated populations; dominant mutations have been found in tomatoes for early ripening and increased weight; and early-maturing mutants have been reported in many cereal grains. This technique has tremendous potential for increasing the agricultural quality and yield of many plants.

Radiation in ecological research. An interesting application of radionuclides is shown in ecological studies where the movement of animals is followed. Insects or small rodents are labeled either by mechanically attaching a radioactive source, by dipping in a radioactive solution, or by incorporating a radionuclide metabolically. The animals are then released into the environment and their movement followed.

Jenkins (1963) reported a related and interesting application of this technique which involved the use of labeled mosquitoes. Three million mosquitoes were metabolically labeled with phosphorus-32 and released. Over a period of one and a half months, one and a half million mosquitoes were caught in a 4 square mile region around the release site. These were "counted," and 141 were found to be labeled. It was determined from the dilution in labeled mosquitoes that the area must have contained a population of about 30 billion mosquitoes. The author points out that the value may be in errror as a result of factors such as "dilution" by other species immigrating into the area.

RADIATION EXPOSURE

Radiation has been shown to be very beneficial in many ways; however, it can also produce deleterious biological effects. In order to estimate the extent of the damage which is being produced in the population in return for the benefits which are being received, it is necessary to have an idea of the radiation levels to which the population is exposed.

The human population has always been exposed to natural radiation at relatively constant low levels. A consideration of these levels gives a "background" against which to consider the additional radiation exposures from man-made sources. The latter have resulted primarily as a result of the medical use of radiations, particularly in highly-developed countries, and to a much lesser extent from industrial and research uses and from fallout from nuclear explosions. The exposures from each of these sources will be considered in the following section.

RADIATION FROM NATURAL SOURCES

Natural radiation arises from two sources: Cosmic rays which enter the atmosphere from outer space, and radioactive materials in the earth's crust.

Cosmic rays. Cosmic radiation consists of primary radiation entering the atmosphere from outer space and of secondary radiation produced by interactions of the primary radiation with nuclei in the atmosphere. The primary component consists largely of highly-energetic (2000 Mev average energy), positively-charged nuclei. Protons are the major component (about 85%); alpha particles are of next greatest frequency (10 to 15%); nuclei of higher atomic numbers and electrons make up the balance. Most of the primary radiation is of galactic origin. During large sunflares there is an additional component, mostly protons of lower energy (10 Mev range), from the sun. These are sometimes called *solar* cosmic rays. Secondary cosmic radiations, such as muons (previously called *μ-mesons*), neutrons, protons, electrons, positrons, and photons, result from interactions of the primary particles with nitrogen, oxygen, or argon nuclei in the upper atmosphere.

The average dose rate to soft tissue at sea level from cosmic radiation has been estimated as between 0.3 and 1.1 mrads/yr from neutrons and about 28 mrads/yr from all components other than neutrons. This dose rate approximately doubles for every 1500-meter increase in altitude for the first few kilometers from sea level.

Terrestrial radioactivity. Terrestrial radioactivity comes from radioactive materials in the earth's crust. These materials are considered as having been present since the earth was formed, or, in the case of shorter-lived radionuclides, as continually being produced by radioactive decay or nuclear reactions. Terrestrial sources contribute both external radiation, in the form of penetrating gamma rays, and internal radiation (alpha, beta, and gamma) from the decay of radionuclides which have been incorporated into the body.

A large proportion of the natural radiation of terrestrial origin comes from radionuclides of the uranium, thorium, and actinium series (see Chapter 2). Because of their relative abundances, decay patterns, and other factors, uranium-238, thorium-232, and radium-226 (plus their shorter-lived daughters) are the most important radionuclides of the series.

There are also a few naturally occurring radionuclides that are not members of the series. Some of these have half-lives sufficiently long to have enabled them to exist from the time of formation of the earth's crust. Potassium-40 (half-life 1.31×10^9 years) is the most important of these.

Other natural radionuclides are produced continually by nuclear reactions between components of cosmic radiation and stable nuclei. Carbon-14 is formed by the action of neutrons upon the nitrogen of the upper atmosphere; tritium is formed by several different reactions.

External radiation is produced mainly by the gamma emitters in soil, rock, and construction materials (uranium-238, thorium-232, radium-226, and potassium-40). The average dose rate from these external sources is about 50 mrads/yr. There is considerable variation in the exposure levels at different geographical locations, depending on the mineral composition of the soil and rock. Granites have a higher level than basalts; limestones and sandstones are generally low, but certain shales may be high, especially those containing potassium salts in the fossil portions. In some areas there are large quantities of uranium ore or large deposits of monazite (the principal thorium-bearing mineral).

Radionuclides in soil may be either absorbed by plants or leached from soil into ground water and so may enter the food chain and eventually be ingested by man. Potassium-40 is relatively abundant in nature and is the largest contributor to internal radiation. Another source of internal radiation, the radioactive gas radon (from the disintegration of radium), does not enter the food chain but escapes from soils and rock into the atmosphere and may be inhaled and thus be incorporated into the body. Carbon-14 is produced in the atmosphere and is incorporated into all living material. Radium, rubidium, and polonium also contribute to the natural internal radiation. The internal dose to humans from these terrestrial radionuclides amounts to approximately 21 mrads per year.

High natural radiation and special food chain concentrating mechanisms in certain parts of the world have produced internal radiation concentrations which are much greater than the average values. Examples of these processes are evident in the midwestern United States where concentrations of radium-226 in teeth are 2 to 3 times greater than normal, due mainly to the higher ^{226}Ra levels in water in these areas. In the Arctic regions, high concentrations of lead-210 and polonium-210 in the lichens and in caribou and reindeer meat result in a high body content of these radionuclides in the humans who subsist primarily on these animals. A study by Hill (United Nations Report, 1966) shows lead and polonium concentrations in the placenta of Eskimo women eating large amounts of caribou, reindeer, or moose meat to be more than ten times the placental concentration in women in England and Canada who ate their normal local diet.

An additional exposure to natural radiation may result from other factors such as cigarette smoking. Although this is not a large contribution (the average alpha radiation dose to the respiratory tract of smokers from excess ^{210}Po deposition is not thought to exceed about 1 mrad/yr), it is possible that a local concentration of the polonium on the bronchial epithelium might result in higher dose rates at some points. The biological significance of this

Table 14.1. DOSE RATES DUE TO EXTERNAL AND INTERNAL RADIATION FROM NATURAL SOURCES IN "NORMAL" AREAS[a]

Source of radiation	Dose rate (mrad/year)		
	Gonads	Bone	Bone marrow
External			
Cosmic, ionizing component	28	28	28
neutrons	0.7	0.7	0.7
Terrestrial	50	50	50
Internal			
^{40}K	20	15	15
^{87}Rb	0.3	<0.3	<0.3
^{14}C	0.7	1.6	1.6
^{226}Ra	—	0.6	0.03
^{228}Ra	—	0.7	0.03
^{210}Po	0.3	2.1	0.3
^{222}Rn	0.3	0.3	0.3
	100	99	96

[a]From United Nations Report (1966).

is not known as yet. However, the exposure from this source is not likely to have a significant biological effect in view of the low total doses involved.

The dose-rate contributions from the various sources of natural radiation are summarized in Table 14.1. Gonads, bone, and bone marrow doses are presented separately to show the relative contributions of the different sources to these three tissues. Gonads are selected because of possible genetic effects; bone and bone marrow are given because of the possibility of bone tumors and leukemia or other blood disorders. The dose rates given for the gonads may be taken as the average dose rates in the whole body.

RADIATION FROM MEDICAL DIAGNOSIS AND THERAPY

The radiation exposures (both in and out of the exposure field) which are delivered when a diagnostic x-ray is taken depend on such varied factors as the field size, tissues irradiated, energy, dose rate, and total exposure time. Since the potential genetic changes are considered to be of particular importance when considering the population risk from small exposures, it has become customary to evaluate the exposure in terms of the gonadal doses received. A comprehensive compilation of gonadal exposures from various types of diagnostic radiation in a dozen countries is given in the United Nations Report (1962). A few selected values are presented in Table 14.2. It is apparent that there is a wide range of exposures which may be incurred in a given type of diagnostic examination.

Taking into account many factors such as the frequency of each type of examination and the "average" gonadal exposure from each, one may estimate an average yearly gonadal dose received by the population as a whole from diagnostic x-rays. This amounts to about 30 mrads/yr. This figure is a

Table 14.2. AVERAGE GONAD DOSE PER RADIO-GRAPH FOR SEVERAL TYPES OF DIAGNOSTIC EXAMINATIONS.[a]

	Gonad dose (mrads)	
Area irradiated	Male	Female
Chest (mass survey)	0.1–6	0.1–15
Stomach (barium meal)	3–123	6–1108
Retrograde pyelography	423–3700	403–1940
Colon (barium enema)	25–1310	20–2530
Lumbar spine	16–767	47–700

[a]From United Nations Report (1962). The range represents the extremes of the values reported by 12 countries.

sweeping estimate and is only given to provide an idea of the order of magnitude of the exposure, relative to natural background, fallout, etc.

Sometimes the radiation dose is considerable to tissues other than gonads. For example, during a routine full-mouth series of dental x-rays the lens of the eye may receive as much as 5 to 25 rads. Every attempt should be made to decrease exposures of this magnitude by the use of "faster" film, better beam direction, etc.

The use of radionuclides in medical diagnosis also results in a small additional population exposure. At the present level of usage this can be viewed as negligible relative to the beneficial effects of the improved medical diagnosis.

Most radiation-therapy treatments require doses of hundreds or thousands of rads to the treated tissue. It is not really meaningful to discuss an "average" dose to an irradiated individual, since there are so many different factors involved in each type of treatment. Likewise, a consideration of average dose to the population has even less meaning. For readers who are interested in an estimate of the relative contribution of therapeutic irradiation to the total radiation received by the population, a figure of 5 mrads per year per person has been suggested. The significance of the value is questionable, since the individuals treated by radiation therapy represent a highly selected population segment. It is doubtful that even the small value of 5 mrads/yr is pertinent to include as a contribution to an overall estimate of radiation exposure.

RADIATION FROM NUCLEAR EXPLOSIONS

The release of energy in a nuclear detonation is accompanied by the production of fission products and by the induction of radioactivity in the environment caused by the neutrons involved in the explosion. The radionuclides which are produced will have various half-lives ranging from a few seconds to several thousand years. Thus, the composition of radioactive debris will vary depending on the time which has elapsed since an explosion.

When an atomic device is detonated in the atmosphere,[1] the large amount of heat produced makes the resulting fireball rise (Figure 14.6). Coarse particles caught in the explosion will fall out on the ground in the vicinity of the explosion site. These are heavily contaminated with fission products and constitute the "local" fallout. Small particles and condensed vapors rise with the cloud. They may remain in the lower layer of the atmosphere (troposphere) or, in the case of powerful explosions, a large proportion of them may be injected into the stratosphere. Material in the troposphere will undergo mixing and movement but as a result of rainfall, gravitation,

[1]The following discussion relates to atmospheric detonations only. Underground explosions, when contained, do not produce fallout.

Fig. 14.6. Aerial photograph of cloud from nuclear detonation. This photograph was taken at a height of approximately 12,000 feet— 50 miles from the detonation site. Two minutes after zero hour, the cloud rose to 40,000 feet. Ten minutes later, as it neared its maximum, the cloud stem had pushed upward about 25 miles, deep into the stratosphere. The mushroom portion went up to 10 miles and spread for 100 miles. (Courtesy U.S. Air Force)

settling, etc., a large proportion will be deposited on the earth in a month or two. Debris injected into the stratosphere remains, creating a reservoir of radioactive debris. This is returned to earth over a period of years and thus continues to provide fallout for a long time.

External irradiation (primarily from gamma emitters) results from accumulated fallout deposits on the earth's surface. Internal irradiation (alpha, beta, or gamma) results from ingestion of radionuclides via the food chains or from inhalation. The radionuclides which are of greatest concern from fallout are strontium-90, cesium-137, carbon-14, strontium-89, and iodine-131.

Iodine-131 is the radionuclide that produces the greatest internal exposures in the first few weeks after a nuclear detonation. It is produced in relatively large quantities and is deposited on the surface of vegetation. The vegetation is eaten by dairy animals and the iodine is then secreted into milk. In this way it is transmitted very efficiently through the food chain to man. When ingested by man, it concentrates in the thyroid gland. Since the half-life of this radionuclide is short (about 8 days), it is significant primarily in local or tropospheric fallout. Almost all ^{131}I injected into the stratosphere disappears by radioactive decay before it is returned to earth.

Two radioisotopes of strontium which are produced with a high yield by nuclear detonations are strontium-90 with a half-life of 28 years and strontium-89 with a half-life of 51 days. Both are incorporated into bone, but because of its longer half-life and the presence of a short-lived, high-energy daughter (^{90}Y), ^{90}Sr is considered to represent the major hazard. Strontium reaches man primarily through his consumption of dairy products and foods of plant origin. Surface contamination of plants and contamination of the soil are both involved.

Cesium-137 (half-life 30 years) enters the food chain if it is deposited on plant surfaces. If it enters the soil, however, it tends to be retained there rather than be incorporated into plants through the root system. Cesium localizes in soft tissue and can represent a genetic hazard as a result of gonadal exposure. Because it is removed from the body rather rapidly (several months), cesium is not considered to represent as great a hazard as strontium in terms of life-time exposure.

Radionuclides can be produced by neutron activation of stable atoms following a nuclear explosion. Carbon-14 which is formed in this way represents a potential radiation hazard because of its long half-life (about 5800 years) and its ready incorporation into plants and animals.

The great diversity in types, times, and locations of nuclear detonations has resulted in a very varied pattern of fallout throughout the world. The estimate of an "average" exposure is, at best, a calculated guess. Furthermore, exposure from completed explosions will continue until the stratospheric reservoir is gone. For this reason, population exposures are usually presented as "dose commitments" — the *total* dose which will be accumulated over the past and future years as a result of the tests which have been carried out. The figures presented in Table 14.3 are estimated dose commitments from the nuclear explosions which occurred between 1954 and 1965. Dose commitments to gonads, bone (or cells lining bone surfaces), and bone marrow are given separately to illustrate the contributions of the various radionuclides to each tissue. Gonadal dose is probably fairly close to total-body dose.

Notice that the values given are for *total* dose commitment and cannot be compared with the figures in Table 14.1 which are dose rates *per year* from natural sources. The United Nations Report (1962) contains some estimates for dose commitment per year of future testing. If testing were to be continued at a constant rate, an equilibrium would be reached such that the dose commitment per year could be used to estimate the annual exposure from all previous detonations. Assuming that the yearly rate involves the injection of 10^6 Ci of ^{90}Sr and 10^{28} atoms of ^{14}C into the atmosphere, the total dose commitment from all sources other than ^{14}C is 7 mrads/yr of testing; because of its long half-life, the total dose commitment from ^{14}C (22 mrads/yr of testing) is distributed over many generations. If the ^{14}C dose commitment is only

Table 14.3. Dose Commitments (world population average) to Gonads, Bone, and Bone Marrow Due to Total Decay of Radionuclides from Nuclear Explosions for Period of Testing 1954-1965[a]

Tissue	Source of radiation	Dose commitment (mrads)
Gonads	External, short-lived	23
	^{137}Cs	25
	Internal, ^{137}Cs	15
	^{14}C	13[b]
	Total	76
Cells lining bone surfaces	External, short-lived	23
	^{137}Cs	25
	Internal, ^{90}Sr	156
	^{137}Cs	15
	^{14}C	20[b]
	^{89}Sr	0.3
	Total	239
Bone marrow	External, short-lived	23
	^{137}Cs	25
	Internal, ^{90}Sr	78
	^{137}Cs	15
	^{14}C	13[b]
	^{89}Sr	0.15
	Total	154

[a]From United Nations Report (1966).

[b]Only doses accumulated up to the year 2000 are given for Carbon-14. The *total* dose commitment to the gonads due to Carbon-14 from tests up to the end of 1965 is about 180 mrads.

considered up to the year 2000, the contribution from ^{14}C is only about 5% of the dose from fallout. In the immediate future, then, the figure of 7 mrads/yr from fallout might be compared with about 100 mrads/yr from natural background radiation.

Populations in certain parts of the world receive much higher exposure levels than the average values. These are due both to variation in fallout distribution and to peculiarities in food chains. The food chain in the Arctic illustrates the latter factor. Caribou and reindeer meat and milk products form a large portion of the human diet in the Arctic. The animals furnishing these foods subsist primarily on lichens during the winter months. These are slow-growing plants which derive their nutrients largely from atmospheric dusts. As a result, reindeer meat and milk and human diet contain high concentrations of fallout products, particularly cesium-137 and strontium-90.

RADIATION FROM INDUSTRIAL AND RESEARCH USES

The question inevitably arises as to how much radiation is received by individuals in the vicinity of power reactors. Normally, reactors are so well shielded that the radiation levels are negligible. Furthermore, reactors are so designed that they cannot explode like an atomic bomb. Should the reaction get out of control, the intense heat would cause the fuel to melt and disperse. The reaction would automatically stop. The most serious hazard in the event of any conceivable reactor accident is the possible escape of radioactivity. All possible precautions are taken to avoid this possibility including special gas-tight enclosures (sometimes enclosing the entire reactor installation). As an additional precaution, most large reactor sites are located in regions of low population density, with favorable terrain, meteorological conditions, etc. An impressive safety record has been achieved. The contribution to population exposure is negligible.

A potential radiation exposure problem exists with small reactors and radioisotope power generators. These devices are built to withstand very large stresses. In addition, proper shielding and care in handling of the materials should reduce the potential danger to negligible levels.

Industrial, medical, or research personnel who are involved with the use of radionuclides or x-ray facilities may be exposed to radiation. Most of the recorded cases have been the result of carelessness or bravado. Careful handling of equipment and proper shielding should minimize the exposures. State and federal regulations have been set up to insure that individuals in these occupations do not receive more than the "maximum permissible dose" of radiation. In most installations, if that level is reached, the individual is transferred to a job unrelated to the use of radiation. The estimated annual genetically significant dose from occupational exposure in several countries is given (U.N. Report, 1962) as between 0.2 and 0.4 mrad.

RADIATION FROM OTHER SOURCES

Radiation may come from other sources in addition to natural background, environmental contamination, and medical or occupational exposures. A few examples of these will illustrate typical exposure levels. Radium was used for many years for the luminous marking on watch and clock dials. Reports from a number of countries have indicated that as much as 0.5 microgram of radium was contained on a single clock face. Radiation from the radium resulted in exposure to individuals wearing or handling the watches. Estimates of the annual genetically significant doses from this source range from 0.5 mrad to 8 mrads. Moreover, a report from Germany estimated that the annual dose to members of a sales staff who handled luminous dial watches may have been as high as 90 mrads. In recent years, radionuclides such as tritium, in which the energy of the radiation is insufficient to penetrate the watch casing, have been used instead of radium on watch dials. Many radium-containing watches and clocks, however, are still in use and are contributing slightly to the population exposure.

X-ray fluoroscopy for shoe fitting has been prohibited in several countries and states since it is regarded as causing unnecessary radiation exposure. In 1959, on the basis of a survey in Germany, the average annual genetic dose to the population was estimated as only 4 to 7 microrads (10^{-6} rad). In this country, however, a number of girls who modeled shoes suffered such severe damage (presumably associated with repeated x-ray fluoroscopy) that amputation of their feet was required.

Television sets produce small quantities of radiation. At the normal operating voltage of black and white home television sets (15 kv) the dose rate at the surface of the screen is 1 mrad/hr. Viewers are not exposed to this much radiation, however, since most sets contain a plastic or glass covering which reduces the dose rate considerably. If TV sets are operated at a higher voltage (as is required in color television tubes), the dose rate is greatly increased. Additional shielding is then required. The average gonad dose from home television is small. In 1959 it was estimated to be much less than 1 mrad/year.

All of these and other miscellaneous sources of radiation probably contribute about 2 mrads/year. With increased use of various sources of radiation, none of which individually contribute an appreciable dose, the total genetically significant dose may be expected to increase slightly.

TOTAL RADIATION EXPOSURE

When one considers the average yearly gonadal exposure from each of the sources that have been described (see Table 14.4), it is apparent that exposures from the total of all the man-made sources of radiation are considerably lower than the natural background. On a total population basis, then, the

Table 14.4. Average Yearly Gonadal Exposure from Various Sources of Radiation[a]

Source	Dose (mrads/yr)
Background	100
Fallout	7
Medical — diagnosis	30
therapy	5
Occupational	0.3
Other sources	2

[a]See text for explanation of each item.

exposure level is low. If one considers specific geographic or occupational populations, however, the values may be considerably higher than the average figure. It is important to recognize the possible exposures both for the world population as a whole and for specific segments of the population.

MAXIMUM PERMISSIBLE EXPOSURE

The National Council on Radiation Protection and Measurements (NCRP) and the International Commission on Radiological Protection (ICRP) have developed recommendations for permissible dosage levels for the population. The levels are set to keep radiation exposure of the individual well below a level at which adverse effects are likely to be observed during his life time. Another objective is to minimize the incidence of genetic effects.

Under ordinary circumstances, persons in the general population should not receive a total-body dose greater than 0.5 rem per year. This is exclusive of natural background or deliberate medical exposure. The cumulative dose over a 30 year period should not exceed 5 rem. The point must be made that these are maximum values and that every attempt should be made, especially when children are involved, to minimize exposures.

Somewhat higher levels are set for radiation workers. For example, the accumulated dose for whole body, head and trunk, active blood-forming organs, gonads, or lens of eye should not exceed a value of five times the number of years beyond age 18; that is, a 30 year old individual should not receive an accumulated dose greater than 60 rem. A person under 18 should not be permitted to receive any industrial exposure.

The hands and forearms or feet and ankles of industrial workers should not receive more than 75 rem in a year or 25 rem in a given 13 week period. Values for other specific exposures can be found in many sources such as the Federal Radiation Council Reports.

ESTIMATE OF RISKS
FROM RADIATION

It is apparent from the preceding sections that most of the population will be concerned only with very low radiation levels, except in the instance of widespread nuclear disaster. Even small radiation exposures, however, may result in some risk to the individuals. It is of importance to consider what the extent of this risk will be.

The International Commission on Radiological Protection (ICRP, 1966) has suggested several ways to express the population risk from radiation. The absolute risk can be expressed as the number of disabilities expected per unit dose of radiation in the life times of a million members of a population, or as the number of disabilities per year in such a population (e.g., X cases per year in a million exposed adults for each rad received). Such a precise number implies that the value is known with great accuracy. That is not the case. Actually, any numbers which are given are only estimates based on the best guesses that can be made with our present knowledge. For this reason, the "order of risk" may be a more realistic expression. One may define a fifth-order risk, for example, as a risk of injury or death in the range of 1×10^{-5} to 10×10^{-5}; that is, 10 to 100 injuries would be expected per million individuals.

Risk can also be expressed in relative terms. For example, the risk of injury from radiation can be compared with the risk of the same injury from some other cause or from unidentified natural causes. This method of expression may give a clearer picture of the relative importance of the radiation.

The assessment of radiation risk may be separated into somatic risk to the exposed generation and genetic risk to either the first generation offspring of exposed individuals or to later generations. Obviously, the somatic risk and the first-generation genetic risk are of particular and immediate concern to those individuals who may receive exposures in excess of the general level. The population as a whole is primarily concerned with the overall long-term genetic effects.

SOMATIC RISKS

At the low levels of exposure which the population can normally expect to receive, the somatic effects which appear to be of interest are (1) cancer induction, (2) production of developmental abnormalities in the fetus, and (3) nonspecific reduction in life span. Other effects may be produced, but are of considerably less importance.

It is known that high doses of radiation will produce cancer, but there are no human data which provide direct evidence of an effect from total-body exposures of less than 100 rads. Since the dose-response curves of some neoplasms are at least approximately linear at higher doses, however, it has seemed expedient to extrapolate these curves to get some idea of the probable incidence of cancer production at low doses. Such extrapolation probably results in a considerable overestimate of the risk, especially if the dose rate is low.

Estimates determined in this manner of the risk of induction of cancer by radiation are summarized in Table 14.5. The figures refer to the annual risk under continuous exposure to 1 rad/yr or to the lifetime risk from a sin-

Table 14.5. Estimated Risk of Cancer From Exposure to 1 Rad[a]

Type of tumor		Estimated cases per 10^6 exposed persons[b]	Order of risk to the individual
Fatal neoplasms:	Leukemia[c]	20	5th
	Others	20	5th
Thyroid carcinoma[d]		10–20	5th

[a]Modified from ICRP (1966). A linear dose-response relationship below 100 rads is assumed.
[b]Effects would be experienced over 10 to 20 years.
[c]Risk may be enhanced by factor of 2 to 10 if the fetus is exposed.
[d]Estimate refers to exposure in childhood. Unlike other classes of cancer for which estimates are given, incidence is not equivalent to mortality.

gle exposure to 1 rad. It must be emphasized, however, that these are calculated extrapolations of high-dose effects, and there are no experimental or observed data to support such inferences for low-dose exposures.

Thyroid carcinomas are considered separately since most of the information on them comes from exposure of children and the clinical mortality may be relatively low. The calculated yield of these neoplasms is about the same as for leukemia.

If radiation exposure is received *in utero* the incidence of leukemia per rad is about five times higher than when adults are exposed. Other types of childhood malignancies appear to be increased by *in utero* irradiation in about the same proportion as leukemia.

Developmental abnormalities in animals which were irradiated *in utero* have demonstrated that the stage of development at the time of irradiation largely determines the effect produced. Irradiation before implantation tends to cause embryonic death; irradiation during the period of major organogenesis (weeks 2 to 6 in humans) results in malformations and neonatal death; irradiation of developing fetuses indicates that they are less radiosensitive than earlier stages, both with respect to death and the production of abnormalities.

In man, abnormalities of central nervous system, eye, and skeleton have been attributed to radiation (see Chapter 10). The evidence, however, is mostly inconclusive. So far, attempts to correlate either human congenital malformations or somatic mutations with low levels of radiation (background or diagnostic exposures) have not been conclusive. Because of the suggestive results from animal experimentation, however, ICRP (1966) has recommended that the pelvic irradiation of women of child-bearing age should be restricted as far as possible to the 10 days following the onset of menstruation in order to reduce the danger of unwittingly irradiating women in the early stages of pregnancy.

Large doses of irradiation clearly reduce the life span of irradiated individuals. The data from low-dose exposures is very uncertain. The conclusion of the ICRP (1966) with regard to this effect is as follows: "The possibility that small doses of radiation have a nonspecific and deleterious effect on life expectancy is not excluded; but the weight of evidence in favor of such an effect is not sufficient to justify making any quantitative estimate of the risk."

GENETIC RISKS

The genetic risks of radiation, especially from small doses, are much more difficult to assess than somatic risks. Some mutations result in a detrimental condition which is evident in the first generation, others require many generations for expression. Furthermore, there are limited quantitative data on genetic abnormalities in man. Table 14.6 contains some of the estimates of several types of potential damage (referred to as *genetic detriment*) for first-generation offspring of exposed individuals (ICRP, 1966).

Category A is for the birth frequency of dominant autosomal (not sex-linked) genetic traits. These are specific recognizable abnormalities such as cleft palate, hare lip, etc. It has been estimated that the natural frequency of mutations of this type which are of social importance is about 0.8%. In a population of one million births, this represents 8000 individuals. The number of mutations of this type in a population is ordinarily in equilibrium — some are continually being removed from the population by selection against the individuals who contain the mutation, while new mutations are continually appearing. The proportion of new mutants among all mutant individuals of this type arising in each generation is estimated to be about 4%. Therefore, the number of new mutants in a population of 10^6 is 4% of 8000 or about 300 individuals. These result from causes unassociated with man-made radiation.

Mutations resulting from radiation exposure must be added to the natural incidence. The ICRP has chosen to use a figure of 20 rads as the doubling dose to calculate the mutations arising from radiation.[2] This means that 20 rads will double the natural incidence or add an additional 300 mutants. An exposure to the population of 1 rad would result in $\frac{1}{20} \times 300 = 15$ additional dominant mutations. This can be expressed as a fifth-order risk. This figure represents only about 0.2% of the total natural incidence of the mutations and could be detected only in large populations by means of rigorous statistical analysis.

[2]This value is somewhat lower than the doubling dose suggested by Muller (see Chapter 6), but it appears to be a logical estimate when all known variables are considered.

Table 14.6. Estimated Genetic Detriment to First Generation of a Million Offspring in a Population Where the Whole Parental Generation Had Received 1 rad in Addition to Natural Background; a Linear Dose-Response Relationship is Assumed[a]

	Type of detriment	Number expected without parental man-made irradiation		Estimated additional number resulting from 1 rad	
		Total natural incidence	No. arising *de novo* in preceding generation	No. of cases	Order of risk[b]
(A)	Autosomal dominant gene traits (births)	8000	300	15	5th
(B)	Sex-linked gene traits (births)	250	<80	<5	6th
(C)	Chromosomal aberrations (births)	7000	7000	unknown[c]	
(D)	Abortions associated with chromosomal aberrations (zygotes)	35,000	35,000	unknown[c]	
(E)	"Genetic deaths" (zygotes)	240,000	6000	200	4th

[a]Modified from ICRP (1966).
[b]See previous section for definition of order of risk.
[c]No predictions can be made for these values, but current frequencies are shown to draw attention to their magnitude.

Category B, sex-linked gene traits (recessive mutations carried on the X chromosome) can be analyzed in a similar way. The total natural incidence is about 250 cases in a population of a million. A maximum of one-third of these, or somewhat less than 80, can be assumed to have arisen in the preceding generation without parental irradiation. Again, assuming the "doubling dose" to be 20 rads, the number of new mutants resulting from 1 rad would be less than 5, a sixth-order risk.

Two other categories of mutations (C and D) are related to chromosomal aberrations. In the limited studies that have been made, it is apparent that the natural incidence of chromosomal defects is very high. These defects are primarily changes in chromosome number (aneuploidy) although some are the result of structural changes in chromosomes. A high percentage of defects of this type result in abortion or still births (D). A smaller percentage are compatible with live birth (C). As a result of early death, sterility, or social restraint, those individuals who are born with chromosomal aberrations do not usually reproduce. Therefore, most chromosomal aberrations can be regarded as having arisen in the parental generation and causing their detrimental effect in the first-generation offspring only.

The natural risk of acquiring an altered chromosomal number or an aberration which results in prenatal death is of the second order (35,000 in a population of about a million). The further risk of an aberration causing less severe damage and being compatible with live birth is of the third order (about 7000 in a population of a million). It is against this high background incidence that the radiation-induced chromosomal aberrations must be evaluated. At present, it is not possible to estimate their frequency with any degree of confidence since a linear relationship between dose and chromosome aberrations cannot be assumed, and the nature and frequency of the aberrations produced in man by radiation are not fully understood. No values are given in Table 14.3 for radiation-induced chromosomal aberrations, but the naturally occurring frequencies are included, as a reminder of their high incidence.

A final estimate of genetic detriment is given as category E, "genetic deaths." This category includes many types of mutations, all of which lead eventually to the death of a zygote. The term *genetic death*, as introduced by Muller, means the extinction of a gene lineage through the premature death or reduced fertility of some individual carrying the gene. Each harmful mutation leads, on the average, to one genetic death. The persistence of the mutation in the population (in generations) will be determined by the extent of selection against it. Some are very detrimental, have a very high selection coefficient, and the mutation persists only for a few generations. Others have a very low selection coefficient and persist for many generations.

In a natural population not subjected to man-made radiation, it has been estimated that each individual contains an average of eight deleterious, partially-dominant mutations, and each of these mutations has a 2.5% chance of elimination in any one generation. The total risk of elimination of the individual is then $0.025 \times 8 = 0.2$. In a population of 10^6 births,[3] this would mean about 240,000 eliminations or genetic deaths per generation. If the gene frequencies are at equilibrium, the same number of new mutations should arise in a generation as are eliminated. Of the 240,000 new mutations, 2.5% or about 6000 are eliminated in the first generation. This means that there are 6000 genetic deaths in one generation as a result of the 240,000 new mutations arising in the parental generation.

It has been estimated by extrapolation from high-dose experiments, that 1 rad induces 0.0036 mutation per gamete of the type described in category

[3]A population of "10^6 births" is equivalent to 1,175,000 pregnancies, because of the high spontaneous abortion rate.

E. In the population of 10^6 births, 1 rad to each of the parents would result in about 8000 new mutations. Of these, 2.5% or about 200 would be eliminated in the first generation as genetic deaths. This represents a fourth-order risk for the first generation. Notice that although this number seems large, the actual number of genetic deaths in the first generation is a very low percentage (0.1%) of the natural incidence.

GENETIC EFFECTS ON LATER GENERATIONS

As indicated in the previous section, a genetic death will result from each mutation in category E. The death may involve failure of implantation, death *in utero*, shortening of life span or a failure to leave offspring, depending on the type and severity of the detriment. The elimination of an individual may occur within the first generation if the expression of the detriment in heterozygous individuals is severe, or it may have its effect many generations later if the detriment is a slight reduction in viability or a minor reduction in fertility. At some time, all the 8000 mutations (estimated to be produced by 1 rad to 10^6 people) will result in a genetic death. By assuming that 2.5% of the total effect is expressed in each generation, it can be estimated that, in this situation, 1900 genetic deaths would occur over the first 10 generations and the remainder during subsequent generations. The estimate for the first 10 generations may be compared with 2.4×10^6 genetic deaths which are expected to result from spontaneously arising mutations.

It is difficult to draw any general conclusions about the genetic risks from low-radiation levels. It is clear that radiation will produce mutations, both of the visible variety and of the less easily characterized genetic death type. These cannot be ignored. The detriment, in terms of both genetic death and visible mutations, will be greatest in the first-generation offspring of the irradiated individuals. However, the *total* number of deleterious conditions suffered by all subsequent generations will greatly exceed that experienced by the first generation.

It is important to recognize that the numbers of mutations induced by small doses of radiation are very low relative to the total natural incidence of similar mutations. Mutagenic drugs, industrial compounds, and other chemicals to which the population is exposed could well be inflicting more mutation damage in some populations than man-made radiations.

THE FUTURE: EXPOSURE VS. RISK

Man is continually discovering new and exciting uses for radionuclides and other sources of ionizing radiation. It is evident that these uses will increase in years to come. Concomitant with expanded use, however, will be an increase in the radiation exposure to the population. Despite careful control of radioactive sources, there will be associated with this added exposure the risk of additional somatic and genetic detrimental effects not only to the present generation but also to future generations. The benefits to be gained from each additional exposure must be weighed against the possible risks involved. Careful adherence to safety policies and to the rules which have been developed by groups such as NCRP, ICRP, and the Federal Radiation Council will make these benefits attainable without undue cost or risk.

GENERAL REFERENCES

United States AEC, *Understanding the Atom* series. USAEC Division of Technical Information Extension, Oak Ridge, Tennessee (1963–1966). Especially:

> Baker, P. S., Fuccillo, D. A., Jr., Gerrard, M. W., and Lafferty, R. H., Jr. *Radioisotopes in Industry*
> Berger, H. *Nondestructive Testing*
> Comar, C. L. *Fallout from Nuclear Tests*
> Corliss, W. R. *SNAP, Nuclear Space Reactors*
> Donnelly, W. H. *Nuclear Power and Merchant Shipping*
> Hogerton, J. F. *Nuclear Reactors*
> Kisieleski, W. E., and Baserga, R. *Radioisotopes and Life Processes*
> Martens, F. H., and Jacobson, W. H. *Research Reactors*
> Meed, R. L., and Corliss, W. R. *Power from Radioisotopes*
> Osborne, T. S. *Atoms in Agriculture*
> Phelan, E. W. *Radioisotopes in Medicine*
> Urrows, G. M. *Food Preservation by Irradiation*

Baumhover, A. H., Graham, A. J., Bitter, B. A., Hopkins, D. E., New, W. D., Dudley, F. H., and Bushland, R. C. Screw-worm Control Through Release of Sterilized Flies. *J. Econ. Entomol.*, **48:** 462–466 (1955).

Claus, W. D. (ed.). *Radiation Biology and Medicine.* Addison-Wesley Publishing Co., Inc., Reading, Massachusetts (1958).

Comar, C. L. *Radioisotopes in Biology and Agriculture.* McGraw-Hill Book Co., Inc., New York (1955).

Federal Radiation Council. *Estimates and Evaluation of Fallout in the United States from Nuclear Weapons Testing Conducted Through 1962.* Rept. No. 4, U. S. Govt. Printing Office, Washington, D. C. (1963).

International Commission on Radiological Protection. *The Evaluation of Risks from Radiation.* ICRP Publication 8. Pergamon Press, New York (1966).

Jenkens, D. W. Use of Radionuclides in Ecological Studies of Insects, pp. 431–440 in Schultz, V., and Klement, A. W., Jr. (eds.). *Radioecology.* Reinhold Publishing Corp., New York, and AIBS, Washington, D. C. (1963).

United Nations Scientific Committee on the Effects of Atomic Radiation. Seventeenth Session, Suppl. #16 (A/5216) (1962).

United Nations Scientific Committee on the Effects of Atomic Radiation. Nineteenth Session, Suppl. #14 (A/5814) (1964).

United Nations Scientific Committee on the Effects of Atomic Radiation. Twenty-first Session, Suppl. #14 (A/6314) (1966).

PUBLICATIONS OF
PERTINENT SYMPOSIA

Use of Isotopes in Plant and Animal Research, conference sponsored by Kansas State College, June 12-14, 1952. U. S. Atomic Energy Commission Publ. TID-5098 (1953).

Atomic Energy and Agriculture, Symposium of Atlanta Meeting of the American Assn. for the Advancement of Science, Dec. 27–29, 1955; Am. Assn. for the Advancement of Science, Publ. 49 (1957).

Radioactive Isotopes in Agriculture, conference sponsored by Argonne National Laboratory and U. S. Atomic Energy Comm. at Michigan State Univ., January 12–14, 1956. U. S. Atomic Energy Commission Publ. TID-7512 (1956).

Uses of Radioisotopes in Animal Biology and the Medical Sciences, sponsored by International Atomic Energy Agency, conference at Mexico City, Nov. 21–Dec. 1, 1961. Academic Press, Inc., New York (1962).

Radioisotopes in Animal Nutrition and Physiology, sponsored by International Atomic Energy Agency and the Food and Agriculture Organization, Symposium in Prague, Nov. 23–27, 1964. IAEA, Vienna (1965).

GENERAL

TEXT

REFERENCES

Alexander, P. *Atomic Radiation and Life*, 2d ed. Penguin Books, Baltimore (1965).

Bacq, Z. M., and Alexander, P. *Fundamentals of Radiobiology*, 2d ed. Pergamon Press, New York (1961).

Claus, W. D. (ed.). *Radiation Biology and Medicine*. Addison-Wesley Publishing Co., Inc., Reading, Massachusetts (1958).

Errera, M., and Forssberg, A. *Mechanisms in Radiobiology*, Vol. I & II. Academic Press, Inc., New York (1960).

Glasstone, S. *Sourcebook on Atomic Energy*, 2d ed. D. Van Nostrand Co., Inc., Princeton, New Jersey (1958).

Grosch, D. S. *Biological Effects of Radiations*. Blaisdell Publishing Co., New York (1965).

Hollaender, A. *Radiation Biology*, Vol. I (Pts. 1 & 2). McGraw-Hill Book Co., Inc., New York (1954).

Lea, D. E. *Actions of Radiations on Living Cells*, 2d ed. Cambridge University Press, Cambridge, England (1955).

Shilling, C. W. (ed.). *Atomic Energy Encyclopedia in the Life Sciences*. W. B. Saunders Co., Philadelphia (1964).

United Nations Scientific Committee on the Effects of Atomic Radiation Seventeenth Session Suppl. #16 (A/5216), United Nations, New York (1962).

USEFUL JOURNALS

Brit. J. Radiol., British Institute of Radiology, London.

Health Physics, Pergamon Press, New York.

Intern. J. Radiation Biol., Taylor & Frances Ltd., London.

J. Natl Cancer Inst., Nat'l Institute of Health, Bethesda, Maryland.

Nucl. Sci. Abst., USAEC Division of Technical Information Extension, Oak Ridge, Tennessee.

Radiation Botany, Pergamon Press, New York.

Radiation Res., Academic Press, Inc., New York.

AUTHOR INDEX

SUBJECT INDEX

A

Abnormalities, in embryos from radiation:
 human 231–231
 mammalian 226–231
 nonmammalian 231
 plants 288–290
 risk of 339
Abscopal effects of radiation:
 general 173
 on lymphocytes 179
Absorbed dose (*see* Dose)
Absorption coefficient 26–27
Absorption of nutrients in GI tract:
 in acute radiation syndrome 222–223, 226
 radiation effects on 188
Absorption of radiation (*see* types of radiation)
Abundance ratio 10
Accidental massive irradiation (man) 224–226
Acetic acid, effects of radiation on 75
Acrylamide, effect on peroxide formation 66
Action potential, radiation effect on 201
Active transport, radiation effect on 150, 188
Acute radiation (*see* specific effects)
Acute radiation syndrome (ARS): 220–226
 characteristics of 221–224
 from embryo irradiation 230
 in man 224–226
 organ interrelationships in 222–224
 treatment of 263
Adaptation syndrome, in irradiated mammals 211–212
Adrenal cortex:
 in acute radiation syndrome 223–224
 effect on survival 253
 radiation effects on 211
 response to radiation 211–212
 secretion after radiation 211
 in stress reaction 211–212
 threshold for cancer in 276
Adrenal medulla, radiation effects on 211
Adrenal volume, effect on survival after radiation 253

Adrenalectomy, effect on survival after radiation 253
AET (S,2-aminoethylisothiourea dihydrobromide):
 as chemical protector 257–261
 DRF for 257–258
 mechanisms of action 258–261
Age:
 influence on $LD_{50(30)}$ 249–250
 influence on radiosensitivity 249–250
Aging:
 accelerated by radiation 266–268, 270
 biochemical changes in 267–268
 histologic changes in 267
 normal 266–267
 precocious 270
Air dose 48
Albumin, radiation effect on 79–81
Alcohol, as radioprotectant 262
Algae, concentration of elements in 310
Alloxazin-adenine dinucleotide 71
Alpha particle(s) 11, 19–20
 absorption 19–20
 Bragg curve for ionization by 20
 detection:
  by autoradiography 43–46
  by ionization 37–38
 distribution of ionization 29
 dose from thorotrast 281
 dosimetry for 48
 emission 11
 energy 11, 20
 interaction with matter 19, 20
 as internal emitter:
  dose from 35
  dose rate from 35
 LET 28
 peroxide formation from 65
 quality factor (QF) for 33
 radical production in water, 65, 70
 radiolysis of water 62
 range 20, 27, 29
 RBE 243

351

growth stimulation by radiation 299–300
lethal exposure range 304
morphologic effects of radiation 300–302
radiation effects on 286–304
somatic mutations in 132–133
structure 284–285
summary of radiation effects on 294
in synthesis of labelled compounds 324–325
Plasma cell, in immune response 214
Plasma proteins, radiation effect on 183
Platelets:
  in acute radiation syndrome 224
  normal appearance of 174
  radiation effect on:
    in blood 182
    production 180–181
Ploidy, effect on radiosensitivity 163–164
Plutonium:
  autoradiography with Pu-239 44
  as power source (Pu-238) 320–321
  series 15–16
Poisson distribution, application to target theory
    143
Pollen:
  chromosome aberrations from radiation of 102–
    103, 106–107
  radiation effect on 286–288
  size, related to radiosensitivity 286
  sterility by radiation 294
Polonium-210:
  alpha emission from 11
  autoradiography with 45, 46
  background dose rate from 329
  in cigarettes 329
  discovery of 2–3
  LET of alpha from 28
  as power source 321
  source of background radiation 329
  vascular damage from 191
Positron(s) 17, 25
Potassium:
  dose from potassium-40 background 329
  loss from yeast 150–151
  potassium-40 as source of background radia-
    tion 328
Potato:
  chromosome volume of 307
  radiosensitivity of 307
Potsaid-Irie dosimeter 42
Precancerous lesion 275
Pregnancy (see also Embryos):
  after irradiation 210
  effects of radiation during 231–232, 339
Prenatal death:
  from embryo irradiation 226–228, 232
  risk of 342
Primary cellular site of radiation damage 115–116
Primary ionization 58

Probit plot:
  mammalian lethality 219
  plant lethality 297
Properdin, radiation effect on 212
Protection (see also Modification of radiation in-
    jury):
  molecular 71
  by nitrogen (microorganisms) 151
  radioprotective compounds 256–262
Proteins, radiation effects on:
  in vitro 79–82
  in vivo 82
Proteinuria following radiation 198
Proton(s):
  in atomic nucleus 8, 10
  in cosmic radiation 327
  description of 8
  effect on brain 200
  in formation of artificial radionuclides 14–15
  LET 28
  QF 33
  recoil from neutron bombardment 20, 21
Protozoa, radiation effect on regeneration 233
Pseudomonas, effect on survival of mice 251–252
Pulsed radiation:
  definition 244
  effect on lethality 245

Q

Quality factor (QF) 33

R

Rabbit, $LD_{50(30)}$ 220
Rad 32
Radiation (see also specific topics):
  detection 36–47
  dose (see Dose of radiation)
  exposure (see Exposure)
  interaction with matter 16–26
  measurement 36–47
  risks 337–343
  sickness 224–226 (see also Acute radiation syn-
    drome)
  units 31–33
  uses of:
    industrial 320–324
    medical 315–320
    research 324–327
Radiation chemical yield:
  G value 58
  M/N value 59
Radiation therapy:
  cause of acute radiation syndrome 224
  cause of cancer 278, 280
  as partial-body radiation 238
  technique 318–320